Multi-Objective Combinatorial Optimization Problems and Solution Methods

Multi-Objective Combinatorial Optimization Problems and Solution Methods

Edited by

Mehdi Toloo
Department of Business Transformation, Surrey Business School, University of Surrey, Guildford, United Kingdom, Department of Systems Engineering, Faculty of Economics, Technical University of Ostrava, Ostrava, Czech Republic, Department of Operations Management & Business Statistics, College of Economics, and Political Science, Sultan Qaboos University, Muscat, Oman

Siamak Talatahari
Department of Civil Engineering, University of Tabriz, Tabriz, Iran, School of Civil and Environment Engineering, University of New South Wales, Sydney, Australia

Iman Rahimi
University of Technology Sydney, Sydney, Australia

ACADEMIC PRESS
An imprint of Elsevier
elsevier.com/books-and-journals

Academic Press is an imprint of Elsevier
125 London Wall, London EC2Y 5AS, United Kingdom
525 B Street, Suite 1650, San Diego, CA 92101, United States
50 Hampshire Street, 5th Floor, Cambridge, MA 02139, United States
The Boulevard, Langford Lane, Kidlington, Oxford OX5 1GB, United Kingdom

Copyright © 2022 Elsevier Inc. All rights reserved.

No part of this publication may be reproduced or transmitted in any form or by any means, electronic or mechanical, including photocopying, recording, or any information storage and retrieval system, without permission in writing from the publisher. Details on how to seek permission, further information about the Publisher's permissions policies and our arrangements with organizations such as the Copyright Clearance Center and the Copyright Licensing Agency, can be found at our website: www.elsevier.com/permissions.

This book and the individual contributions contained in it are protected under copyright by the Publisher (other than as may be noted herein).

Notices
Knowledge and best practice in this field are constantly changing. As new research and experience broaden our understanding, changes in research methods, professional practices, or medical treatment may become necessary.

Practitioners and researchers must always rely on their own experience and knowledge in evaluating and using any information, methods, compounds, or experiments described herein. In using such information or methods they should be mindful of their own safety and the safety of others, including parties for whom they have a professional responsibility.

To the fullest extent of the law, neither the Publisher nor the authors, contributors, or editors, assume any liability for any injury and/or damage to persons or property as a matter of products liability, negligence or otherwise, or from any use or operation of any methods, products, instructions, or ideas contained in the material herein.

British Library Cataloguing-in-Publication Data
A catalogue record for this book is available from the British Library

Library of Congress Cataloging-in-Publication Data
A catalog record for this book is available from the Library of Congress

ISBN: 978-0-12-823799-1

For Information on all Academic Press publications visit our website at https://www.elsevier.com/books-and-journals

Publisher: Mara Conner
Acquisitions Editor: Chris Katsaropoulos
Editorial Project Manager: Mariana L. Kuhl
Production Project Manager: Sojan P. Pazhayattil
Cover Designer: Greg Harris

Typeset by Aptara, New Delhi, India

Dedication

This book is affectionately dedicated to *People who want to develop and lead others and make the world a better place* and also to *Our Family*.

Prof. Toloo
Dr. Talatahari
Dr. Rahimi

Mehdi Toloo is grateful for the partial support that he received for this research from the Czech Science Foundation (GACR: 19–13946S).

Contents

Contributors xv
Editors Biography xix
Preface xxi
Acknowledgments xxiii

CHAPTER 1 Multiobjective combinatorial optimization problems: social, keywords, and journal maps   **1**
Mehdi Toloo, Siamak Talatahari, Amir H. Gandomi and Iman Rahimi

    1.1 Introduction 1
    1.2 Methodology 1
    1.3 Data and basic statistics 2
    1.4 Results and discussion 3
        1.4.1   Mapping the cognitive space 3
        1.4.2   Mapping the social space 4
    1.5 Conclusions and direction for future research 7
        References 8

CHAPTER 2 The fundamentals and potential of heuristics and metaheuristics for multiobjective combinatorial optimization problems and solution methods   **9**
Ana Carolina Borges Monteiro, Reinaldo Padilha França, Rangel Arthur and Yuzo Iano

    2.1 Introduction 9
    2.2 Multiobjective combinatorial optimization 10
    2.3 Heuristics concepts 12
    2.4 Metaheuristics concepts 14
    2.5 Heuristics and metaheuristics examples 16
        2.5.1   Tabu search 16
    2.6 Evolutionary algorithms (EA) 16
    2.7 Genetic algorithms (GA) 17
    2.8 Simulated annealing 18
    2.9 Particle swarm optimization (PSO) 18
    2.10 Scatter search (SS) 19
    2.11 Greedy randomized adaptive search procedures (GRASP) 19
    2.12 Ant-colony optimization 19
    2.13 Clustering search 20
    2.14 Hybrid metaheuristics 20

2.15	Differential evolution (DE)	21
2.16	Teaching learning–based optimization (TLBO)	21
2.17	Discussion	21
2.18	Conclusions	23
2.19	Future trends	24
	References	24

CHAPTER 3 A survey on links between multiple objective decision making and data envelopment analysis 29

Amineh Ghazi and Farhad Hosseinzadeh Lotfi

3.1	Introduction	29
3.2	Preliminary discussion	31
	3.2.1 Multiple objective decision making	31
	3.2.2 Data envelopment analysis	32
3.3	Application of MODM concepts in the DEA methodology	35
	3.3.1 Classical DEA models	35
	3.3.2 Target setting	37
	3.3.3 Value efficiency	41
	3.3.4 Secondary goal models	43
	3.3.5 Common set of weights	46
	3.3.6 DEA-discriminant analysis	51
	3.3.7 Efficient units and efficient hyperplanes	54
3.4	Classification of usage of DEA in MODM	56
	3.4.1 Efficient points	56
3.5	Discussion and conclusion	59
	References	60

CHAPTER 4 Improved crow search algorithm based on arithmetic crossover—a novel metaheuristic technique for solving engineering optimization problems 71

S.N. Kumar, A. Lenin Fred, L. R. Jonisha Miriam, Parasuraman Padmanabhan, Balázs Gulyás, Ajay Kumar H and Nisha Dayana

4.1	Introduction	71
4.2	Materials and methods	73
	4.2.1 Crow search optimization	73
	4.2.2 Arithmetic crossover based on genetic algorithm	74
	4.2.3 Hybrid CO algorithm	74

4.3	Results and discussion	75
4.4	Conclusion	87
	Acknowledgments	88
	References	88

CHAPTER 5 MOGROM: Multiobjective Golden Ratio Optimization Algorithm 91
A.F. Nematollahi, A. Rahiminejad and B. Vahidi

5.1	Introduction	91
	5.1.1 Definition of multiobjective problems (MOPs)	92
	5.1.2 Literature review	93
	5.1.3 Background and related work	93
5.2	GROM and MOGROM	94
	5.2.1 MOGROM	95
5.3	Simulation results, investigation, and analysis	97
	5.3.1 First class	99
	5.3.2 Second class	101
	5.3.3 Third class	103
	5.3.4 Fourth class	111
	5.3.5 Fifth class	113
5.4	Conclusion	116
	References	116

CHAPTER 6 Multiobjective charged system search for optimum location of bank branch 119
Siamak Talatahari, Abolfazl Ranjbar, Mohammad Tolouei and Iman Rahimi

6.1	Introduction	119
6.2	Multiobjective backgrounds	120
	6.2.1 Dominance and Pareto Front	120
	6.2.2 Performance metrics	121
6.3	Utilized methods	122
	6.3.1 NSGA-II algorithm	122
	6.3.2 MOPSO algorithm	122
	6.3.3 MOCSS algorithm	122
6.4	Analytic Hierarchy Process	124
6.5	Model formulation	125
6.6	Implementation and results	127
6.7	Conclusions	131
	References	131

CHAPTER 7 Application of multiobjective Gray Wolf Optimization in gasification-based problems 133
Babak Talatahari, Siamak Talatahari and Ali Habibollahzade

7.1 Introduction 133
7.2 Systems description 134
    7.2.1 Downdraft gasifier 134
    7.2.2 Waste-to-energy plant 135
7.3 Modeling 136
7.4 Multicriteria Gray Wolf Optimization 138
7.5 Results and discussion 143
    7.5.1 Optimization at the gasifier level 143
    7.5.2 Optimization at the WtEP Level 150
    References 154

CHAPTER 8 A VDS-NSGA-II algorithm for multiyear multiobjective dynamic generation and transmission expansion planning 157
Ali Esmaeel Nezhad, Mohammad Sadegh Javadi, Alberto Borghetti, Morteza Taherkhani, Alireza Heidari and João P.S. Catalão

8.1 Introduction 157
8.2 Problem formulation 159
    8.2.1 Master problem 160
    8.2.2 Slave problem 161
    8.2.3 TC assessment objective of the MMDGTEP problem 161
    8.2.4 $EENS_{HL-II}$ evaluation procedure of the MMDGTEP problem 162
8.3 Multiobjective optimization principle 163
8.4 Nondominated sorting genetic algorithm-II 164
    8.4.1 Computational flow of NSGA-II 164
    8.4.2 VDS-NSGA-II 165
    8.4.3 Methodology 165
    8.4.4 VIKOR decision making 169
8.5 Simulation results 170
8.6 Conclusion 173
    Acknowledgment 174
    References 174

CHAPTER 9 A multiobjective Cuckoo Search Algorithm for community detection in social networks 177
Shafieh Ghafori and Farhad Soleimanian Gharehchopogh

9.1 Introduction 177
9.2 Related works 178

9.3	Proposed model	180
	9.3.1	Community diagnosis	180
	9.3.2	Multiobjective optimization	180
	9.3.3	CD based on MOCSA	181
	9.3.4	Fitness function	184
9.4	Evaluation and results	185
9.5	Conclusion and future works	190
	References	190

CHAPTER 10 Finding efficient solutions of the multicriteria assignment problem 193
Emmanuel Kwasi Mensah, Esmaeil Keshavarz and Mehdi Toloo

- **10.1** Introduction — 193
- **10.2** The basic AP — 194
- **10.3** Restated MCAP and DEA: models and relationship — 195
  - 10.3.1 The multicriteria assignment problem (MCAP) — 196
  - 10.3.2 Data envelopment analysis — 198
  - 10.3.3 An integrated DEA and MCAP — 202
- **10.4** Finding efficient solutions using DEA — 202
  - 10.4.1 The two-phase algorithm — 203
  - 10.4.2 The proposed algorithm — 205
- **10.5** Numerical examples — 206
- **10.6** Conclusion — 209
- Acknowledgments — 209
- References — 209

CHAPTER 11 Application of multiobjective optimization in thermal design and analysis of complex energy systems 211
A. Baghernejad and E. Aslanzadeh

- **11.1** Introduction — 211
  - 11.1.1 System boundaries — 211
  - 11.1.2 Optimization criteria — 211
  - 11.1.3 Variables — 212
  - 11.1.4 The mathematical model — 212
  - 11.1.5 Suboptimization — 212
- **11.2** Types of optimization problems — 213
  - 11.2.1 Single-objective optimization — 213
  - 11.2.2 Multiobjective optimization — 213
- **11.3** Optimization of energy systems — 215
  - 11.3.1 Thermodynamic optimization and economic optimization — 215

	11.3.2 Thermoeconomic optimization	215
11.4	Literature survey on the optimization of complex energy systems	217
11.5	Thermodynamic modeling of energy systems	217
	11.5.1 Mass balance	217
	11.5.2 Energy balance	217
	11.5.3 Entropy balance	218
	11.5.4 Exergy balance	218
	11.5.5 Energy efficiency	218
	11.5.6 Exergy efficiency	219
11.6	Thermoeconomics methodology for optimization of energy systems	220
	11.6.1 The SPECO method	221
	11.6.2 The F (fuel) and P (product) rules	222
11.7	Sensitivity analysis of energy systems	222
11.8	Example of application (case study)	222
	11.8.1 Integrated biomass trigeneration system	222
	11.8.2 Results and discussion	226
	11.8.3 Sensitivity analysis	232
11.9	Conclusions	234
	References	235

CHAPTER 12 A multiobjective nonlinear combinatorial model for improved planning of tour visits using a novel binary gaining-sharing knowledge-based optimization algorithm 237
Said Ali Hassan, Prachi Agrawal, Talari Ganesh and Ali Wagdy Mohamed

12.1	Introduction	237
12.2	Tourism in Egypt: an overview	238
	12.2.1 Tourism in Egypt	238
	12.2.2 Tourism in Cairo	238
	12.2.3 Planning of tour visits	239
12.3	PTP versus both the TSP and KP	240
	12.3.1 The Traveling Salesman Problem and its variations	240
	12.3.2 Multiobjective 0–1 KP	240
	12.3.3 Basic differences between PTP and both the TSP and KP	243
12.4	Mathematical model for planning of tour visits	244
12.5	A real application case study	247
	12.5.1 Ramses Hilton Hotel	248
12.6	Proposed methodology	250
	12.6.1 Gaining Sharing Knowledge-based optimization algorithm (GSK)	251
	12.6.2 Binary Gaining Sharing Knowledge-based optimization algorithm (BGSK)	254

12.7	Experimental results	257
12.8	Conclusions and points for future studies	259
	References	261

CHAPTER 13 Variables clustering method to enable planning of large supply chains — 265
Emilio Bertolotti

13.1	Introduction	265
13.2	SCP at a glance	265
13.3	SCP instances as MOCO models	267
13.4	Orders clustering for mix-planning	274
13.5	Variables clustering for the general SCP paradigm	280
13.6	Conclusions	284
	References	285

Index 287

Contributors

Prachi Agrawal
Department of Mathematics and Scientific Computing, National Institute of Technology Hamirpur, Himachal Pradesh, India

Rangel Arthur
Faculty of Technology (FT)-State University of Campinas (UNICAMP), Limeira, São Paulo, Brazil

E. Aslanzadeh
Department of Basic Sciences, Garmsar Branch, Islamic Azad University, Garmsar, Iran

A. Baghernejad
Department of Mechanical and Aerospace Engineering, College of Engineering, Garmsar Branch, Islamic Azad University, Garmsar, Iran

Emilio Bertolotti
Fast.square, Milan, Italy

Alberto Borghetti
Department of Electrical, Electronic, and Information Engineering, University of Bologna, Bologna, Italy

João P.S. Catalão
Institute for Systems and Computer Engineering, Technology and Science (INESC TEC), Porto, Portugal; Faculty of Engineering of the University of Porto, (FEUP), Porto, Portugal

Nisha Dayana
Dhanraj Baid Jain College, Chennai, Tamil Nadu, India.

A. Lenin Fred
Mar Ephraem College of Engineering and Technology, Elavuvilai, Tamil Nadu, India

Reinaldo Padilha França
School of Electrical Engineering and Computing (FEEC)-State University of Campinas (UNICAMP), Campinas, São, Paulo, Brazil

Amir H. Gandomi
Faculty of Engineering and Information Technology, University of Technology Sydney, Australia

Talari Ganesh
Department of Mathematics and Scientific Computing, National Institute of Technology Hamirpur, Himachal Pradesh, India

Shafieh Ghafori
Department of Computer Engineering, Urmia Branch, Islamic Azad University, Urmia, Iran

Farhad Soleimanian Gharehchopogh
Department of Computer Engineering, Urmia Branch, Islamic Azad University, Urmia, Iran

Amineh Ghazi
Department of Mathematics, Central Tehran Branch, Islamic Azad University, Tehran, Iran

Balázs Gulyás
Lee Kong Chian School of Medicine, Nanyang Technological University, Singapore

Ajay Kumar H
Mar Ephraem College of Engineering and Technology, Elavuvilai, Tamil Nadu, India

Ali Habibollahzade
School of Mechanical Engineering, College of Engineering, University of Tehran, Tehran, Iran

Said Ali Hassan
Department of Operations Research and Decision Support, Faculty of Computers and Artificial Intelligence, Cairo University, Giza, Egypt

Alireza Heidari
School of Electrical Engineering and Telecommunications (EE&T), The University of New South Wales (UNSW), Sydney NSW, Australia

Yuzo Iano
School of Electrical Engineering and Computing (FEEC)-State University of Campinas (UNICAMP), Campinas, São, Paulo, Brazil

Mohammad Sadegh Javadi
Institute for Systems and Computer Engineering, Technology and Science (INESC TEC), Porto, Portugal

Esmaeil Keshavarz
Department of Mathematics, Sirjan Branch, Islamic Azad University, Sirjan

S.N. Kumar
Amal Jyothi College of Engineering, Kerala, India

Farhad Hosseinzadeh Lotfi
Department of Mathematics, Science and Research Branch, Islamic Azad University, Tehran, Iran

Emmanuel Kwasi Mensah
Department of Economics, University of Insubria, Italy

L.R. Jonisha Miriam
Mar Ephraem College of Engineering and Technology, Elavuvilai, Tamil Nadu, India

Ali Wagdy Mohamed
Operations Research Department, Faculty of Graduate Studies for Statistical Research, Cairo University, Giza, Egypt; Wireless Intelligent Networks Center (WINC), School of Engineering and Applied Sciences, Nile University, Giza, Egypt

Ana Carolina Borges Monteiro
School of Electrical Engineering and Computing (FEEC)-State University of Campinas (UNICAMP), Campinas, São, Paulo, Brazil

A.F. Nematollahi
Electrical Engineering Department, Amirkabir University of Technology, Tehran, Iran

Ali Esmaeel Nezhad
Department of Electrical Engineering, School of Energy Systems, LUT University, Lappeenranta, Finland

Parasuraman Padmanabhan
Lee Kong Chian School of Medicine, Nanyang Technological University, Singapore

Iman Rahimi
University of Technology Sydney, Sydney, Australia

A. Rahiminejad
Department of Electrical and Computer Science, Esfarayen University of Technology, Esfarayen, North Khorasan, Iran

Abolfazl Ranjbar
Department of Civil Engineering, University of Tabriz, Tabriz, Iran; Department of GIS Engineering, Faculty of Surveying Engineering, Tehran University, Tehran, Iran

Morteza Taherkhani
Department of Electrical Engineering, West Tehran Branch, Islamic Azad University, Tehran, Iran

Babak Talatahari
Department of Civil Engineering, University of Tabriz, Tabriz, Iran

Siamak Talatahari
Department of Civil Engineering, University of Tabriz, Tabriz, Iran; School of Civil and Environment Engineering, University of New South Wales, Sydney, Australia

Mehdi Toloo
Department of Business Transformation, Surrey Business School, University of Surrey, Guildford, United Kingdom; Department of Systems Engineering, Faculty of Economics, Technical University of Ostrava, Ostrava, Czech Republic; Department of Operations Management & Business Statistics, College of Economics, and Political Science, Sultan Qaboos University, Muscat, Oman

Mohammad Tolouei
Department of Civil Engineering, University of Tabriz, Tabriz, Iran

B. Vahidi
Electrical Engineering Department, Amirkabir University of Technology, Tehran, Iran

Editors Biography

Prof. Mehdi Toloo, Ph.D.

 *Department of Systems Engineering, VŠB-Technical University of Ostrava, Sokolská třída 33, 70 200, Ostrava, Czech Republic*

 Department of Operations Management & Business Statistics, College of Economics and Political Science, Sultan Qaboos University, Al Khoudh, Muscat 123, Muscat, Oman

Mehdi Toloo is a Full Professor of Systems Engineering and Informatics in the Department of Systems Engineering and Informatics, Technical University of Ostrava, Czech Republic, Full Professor in the Department of Operations Management & Business Statistics, College of Economics and Political Science, Sultan Qaboos University, and Visiting Professor in the Department of Management and Economics, University of Turin, Italy. Areas of interest include operations research, decision analysis, performance evaluation, multiobjective programming, and mathematical modelling. He has contributed to numerous international conferences as a Chair, Keynote Speaker, Presenter, Track/Session Chair, Workshop Organizer, and member of the scientific committee. Mehdi has lots of experience in leading and collaboration in many successful research projects. He acts as an Editor for *Computers & Industrial Engineering* (ELSEVIER), *Decision Analytics* (ELSEVIER), *Healthcare Analytics* (ELSEVIER), *RAIRO-Operations Research*, *Mathematics*, and *Central European Journal of Operations Research* (SPRINGER). He has written 14 books and his research has been published in top-tier journals including the *European Journal of Operational Research,* and *OMEGA*, among others.

For more details visit http://homel.vsb.cz/~tol0013/

Siamak Talatahari after received the Ph.D. degree in Structural Engineering, Dr. Siamak Talatahari joined the University of Tabriz (one of the top 10 universities in Iran) where he is presently the Associate Professor. His main research interests include data science (DS), machine learning (ML), and artificial intelligence (AI) and their application on engineering. Thanks to introduce, improve, hybridize, and apply DS/AI/ML methods for engineering problems, he published over 120 refereed international journal papers, three edited books in Elsevier, eight chapters in international books. The citation of his works is over 8500. He has studied both how to present novel efficient DS/AI/ML methods and how to adaptive/improve the present methods to solve difficult problems. As instance, he presented Charged System Search (2010), Firefly Algorithm with Chaos (2013), Developed Swarm Optimizer (2016), Chaos Game Optimization (2020), Stochastic Paint Optimizer (2020), Crystal Structure Algorithm (2021), Material Generation Algorithm (2021), Social Network Search (2021) as some new metaheuristic algorithms. Multi Expression Programming (2014), Gene Expression Data Classification (2015), Linear Biogeography-based Programming (2018), and ANFIS-CSS (2021) are some other DS-based methods that are developed and applied by Dr. Talatahari for solving engineering problems. He is honored by many academic awards; he is recognized as the Top One Percent Scientist of the World in the field of Engineering and Computer Sciences for several years and as One of the 70 Most Influential Professors in The History of the Tabriz University. He was also recognized as Distinguished Scientist of Iranian Forefront of Sciences, Most Prominent Young Engineering Scientist, Distinguished Researcher,

Top Young Researcher, Most Acclaimed Professor, and Top Researcher and Teacher. In addition, he has been selected to receive the TWAS Young Affiliate-ship from the Central & South Asia Region and Elite Awards from the Iranian Elites Organization. Dr Talatahari worked as the Lead or Guest Editor of some special issues in different journals.

Iman Rahimi, Ph.D., received the BSc degree (Applied Mathematics) in 2009, MSc degree (Applied Mathematics—Operations Research) in 2011, and the Ph.D. degree from the Department of Mechanical and Manufacturing Engineering, Faculty of Engineering, Universiti Putra Malaysia, Malaysia, in 2017. After finishing his Postdoctoral program, he joined the Faculty of Engineering and Information Technology, University of Technology Sydney, Australia, as a Research Scholar. His research interests include machine learning and multiobjective optimization. He also has edited two books entitled Evolutionary Computation in Scheduling and Big Data Analytics in Supply Chain Management: Theory and Applications with Wiley and CRC Press (Taylor & Francis Group), respectively. He worked as an Editor for the following journals: *Computational Research Progress in Applied Science & Engineering* (CRPASE), *International Journal of Renewable Energy Technology* (IJRET), and *International Journal of Advanced Heuristic and Meta-Heuristic Algorithms*. Also, he has acted as a reviewer for the *International Journal of Production Research*. He also received several awards including an International Research Scholarship from the University of Technology Sydney and a Postdoctoral research grant from Iran National Science Foundation.

Preface

Combinatorial optimization problems appear in a wide range of applications in operations research (OR), engineering, biological sciences, computer science, as well as many others. Many combinatorial optimization approaches have been developed that link the discrete universe to the continuous one through geometric, analytic, and algebraic techniques. Optimization problems with multiobjective arise in a natural fashion in most disciplines and their solutions have been a challenge to researchers for a long time. Despite the considerable variety of techniques developed in OR and other disciplines to tackle these problems, the complexities of their solutions call for alternative approaches. Computational optimization has become increasingly well-studied in recent years and demand for optimal design and its applications in engineering and industry have become ever more important due to the need for more stringent designs in modern engineering practices. In addition, the interest and importance of optimization design problems have been much harder to be solved nowadays due to consider the difficulty of these problems containing realistic conditions, considering large-scale, and nonlinear problems while available resources, costs, and time are limited.

In this book, we will discuss the results of a recent multiobjective combinatorial optimization achievement considering metaheuristic, mathematical programming, and heuristic approaches. In other words, this book intends to show a diversity of various multiobjective combinatorial optimization issues that may benefit from different methods in theory and practice.

This edited book provides an indication of several of the state-of-the-art progresses in developing the field of multiobjective combinatorial optimization and solution finder techniques. This edited book will emphasize the audiences of engineers in industries, researchers, students, faculty members, and conducting research in OR and industrial engineering from academia, who work mainly on combinatorial optimization problems.

To facilitate this goal, Chapter 1 presents a brief analysis of most concepts and studies of solution approaches applied to multiobjective combinatorial optimization problems. A detailed scientometric analysis presents an influential tool for bibliometric analyses, which were performed on multiobjective combinatorial optimization problems and the solution approaches data from the Scopus databases.

The main contribution of Chapter 2 is to provide an updated review and overview of Heuristics and Metaheuristic methods, addressing its evolution and fundamental concepts, showing its relationship as well as approaching its success, with a concise bibliographic background, categorizing, and synthesizing the potential of the techniques. The purpose of Chapter 3 is on links between data envelopment analysis (DEA) and multiobjective decision making, and the robust presented papers in these fields are reviewed. Chapter 4 proposes a hybrid crow optimization algorithm for solving engineering problems. The hybrid crow optimization in this chapter is an improved version formulated by the incorporation of the arithmetic crossover concept of the genetic algorithm. Chapter 5 introduces the multiobjective version of a very recently proposed meta-heuristic optimization algorithm known as the Golden Ratio Optimization Method. The proposed method, named Multi-Objective Golden Ratio Optimization Method, uses an external repository matrix for storing the best obtained Pareto Front that is achieved by nondominated sorting of the whole population.

Chapter 6 focuses on determining the location of bank branches under competitive conditions considering different levels of customer attraction. The chapter applies Multi-Objective Charged System

Search for finding the optimum location of bank branches with conflicting purposes. Chapter 7 reviews the application of Multi-objective Grey Wolf Optimization in gasification-based problems. It provides a conventional waste-to-energy plant as well as simulating biomass gasification in a gasifier. Chapter 8 develops a dynamic biobjective model for the generation expansion planning together with the transmission system expansion planning. A Virtual Database Supported Non-dominated Sorting Genetic Algorithm-II (VDS-NSGA-II) is designed to tackle the above-mentioned problem. Chapter 9 proposes a new model based on the Multi-Objective Cuckoo Search Algorithm for community detection on social media. Chapter 10 presents a two-phase algorithm under variable return-to-scale and nonconvexity assumptions in DEA with the aim of solving the multicriteria assignment problem. Chapter 11 provides a comprehensive introduction to thermal system design and optimization from a contemporary perspective. Chapter 12 proposes a novel binary version of the recently developed gaining-sharing knowledge-based optimization algorithm to solve binary optimization problems. Finally, Chapter 13 outlines a new planning approach that reduces the search space for large supply chain planning instances using the Variables Clustering Method.

This book is proper to research students at all levels, and we hope it will be used as a supplementary textbook for several types of courses, including OR, computer science, statistics, and many fields of science and engineering related to multiobjective combinatorial optimization problems.

Mehdi Toloo
Technical University of Ostrava, Ostrava, Czech Republic &
Sultan Qaboos University, Muscat, Oman

Siamak Talatahari
University of Tabriz, Tabriz, Iran

Iman Rahimi
Faculty of Engineering and IT, University of Technology Sydney, Australia

Acknowledgments

The editors have to start by thanking each and every author who contributed to the edited book. The editors would also like to extend appreciation to all reviewers for their time, critical review of the chapters, and their insightful comments and constructive suggestions provided in several rounds. The editors of the book also would like to thank the project editors at Elsevier for their support and their cooperation.

Mehdi Toloo, PhD
Siamak Talatahari, PhD
Iman Rahimi, PhD

CHAPTER 1

Multiobjective combinatorial optimization problems: social, keywords, and journal maps

Mehdi Tolo[a,b,c], Siamak Talatahari[d,e], Amir H. Gandomi[f] and Iman Rahimi[g]

[a]*Department of Business Transformation, Surrey Business School, University of Surrey, Guildford, United Kingdom,* [b]*Department of Systems Engineering, Faculty of Economics, Technical University of Ostrava, Ostrava, Czech Republic,* [c]*Department of Operations Management & Business Statistics, College of Economics, and Political Science, Sultan Qaboos University, Muscat, Oman,* [d]*Department of Civil Engineering, University of Tabriz, Tabriz, Iran,* [e]*School of Civil and Environment Engineering, University of New South Wales, Sydney, Australia,* [f]*Faculty of Engineering and Information Technology, University of Technology Sydney, Australia,* [g]*University of Technology Sydney, Sydney, Australia*

1.1 Introduction

Combinatorial optimization problems possess a wide range of applications containing operations research/management, engineering, biological sciences, and computer science. Many combinatorial optimization approaches have been developed to link the discrete universe to the continuous one through geometric, analytic, and algebraic techniques [3,10].

Multiobjective optimization problems arise in a natural fashion in most disciplines and finding appropriate solutions has been known as a challenging issue for researchers. Although there are a variety of techniques developed in Operations Research and other disciplines to address this matter, it is urgently needed for alternative approaches, [1,2,7]. Computational optimization has become increasingly popular in recent years and design optimization and its applications in engineering and industry have become ever more important due to need more stringent designs in modern engineering practices. In addition, the interest and importance of optimum designing problems have been much harder to be solved nowadays due to the difficulty of these problem-containing realistic conditions, considering large-scale and nonlinear problems while, available resources, costs and time are limited.

This chapter presents some fundamental issues related to multiobjective combinatorial optimization problems and their solution methods. We consider 1321 published articles from the Scopus database for analyzing using scientometric analysis methods. We recognize the keywords, social and journal maps of publications dealing with multiobjective combinatorial optimization problems and solution methods.

1.2 Methodology

The research procedure in this chapter has been divided into four stages. Stage 1 gathers documents from the Scoups databases after restricting search to three keywords: "multi-objective," "optimization,"

Table 1.1 Document type.

Document type	Number of documents
Conference paper	1102
Published articles	1321
Editorial materials	2
Reviews	28
Book/ Book chapter	45

Table 1.2 Top 10 institutions ranked by number of documents published.

	Institution	Number of documents
1	Beihang University	25
2	Centre National de la Recherche Scientifique	23
3	Université de Lille	22
4	Centre de Recherche en Informatique, Signal et Automatique de Lille	22
5	Shanghai Jiao Tong University	21
6	Ministry of Education China	20
7	Universidade de Lisboa	17
8	Université libre de Bruxelles	16
9	Northwestern Polytechnical University	16
10	Instituto Superior Técnico	16

and "combinatorial." Furthermore, it is worthy to note that we consider almost all research articles and exclude books, book chapters, reviews, conference papers, and short communications/letters to provide high-quality research articles, which results in 1321 published research articles indexed in Scopus. Stage 2 employs Mendeley, as a powerful reference manager, that identifies and removes possible duplicate articles from the library. Stage 3 uses the well-known VOSviewer as social network analysis to provide a scientometric analysis for the selected documents [5,8,11,12]. This stage includes some steps namely: Co-occurrence Analysis, Co-authorship Analysis, Citation Analysis, Journal Map and Citation Network. Stage 4 provides conclusions and further discussions with the aim of coping with the above-mentioned research questions.

1.3 Data and basic statistics

As mentioned before, the Scopus as the largest abstract and citation database of peer-reviewed literature provided by Elsevier is used in this paper. The type of all used documents is clarified in Table 1.1.

Fig. 1.1 depicts the yearly distribution of publications (articles), showing that the majority were published between 1978 and December 2020. The two first articles were published in 1978 and 1989 [4,9], respectively.

Tables 1.2 presents the top 10 institutions ranked by a number of documents where the affiliation of the first author of documents is considered. The first top-ranked institutions are Beihang University

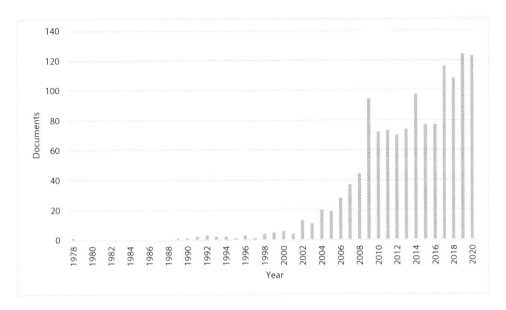

FIGURE 1.1

Distribution of publications per year.

(China), Centre National de la Recherche Scientifique (France), and Université de Lille (France), respectively.

1.4 Results and discussion

The following subsections present mapping the cognitive space including keywords analysis and journal maps and mapping the social space including coauthorship networks, respectively.

1.4.1 Mapping the cognitive space

The scientometric analysis scientifically measures and analyzes the literature [6]. Keywords present the basic parts of a certain field of research and could offer a sight of the knowledge organization. Fig. 1.2 depicts network visualization co-occurrence analysis by a map based on Scopus. Each node in the network displays a keyword and the link between the nodes illustrates the co-occurrence of the keywords. Referencing to Fig. 1.2 points out that combinatorial optimization, multiobjective combinatorial, genetic algorithm, structural optimization, process planning and scheduling, assembly, costs, multidisciplinary design optimization are among the top useful keywords. Table 1.3 presents the top 10 keywords for 1-, 2- and 3-word length. For 1-word length, optimization (461) followed by algorithms (208), scheduling (126), and integer programming(119); for 2-word length, combinatorial optimization owns the most frequency (833) followed by multiobjective optimization (529), genetic algorithms (291), and evolutionary algorithms (203); and for 3-word keyword length, combinatorial optimization problems possess the most frequency (169) followed by multiobjective optimization problem (79), particle swarm optimization (77), multiagent systems (60), and ant colony optimization (56). Moreover, Table 1.4 reports top 20 indexed keywords in the articles, which shows combinatorial

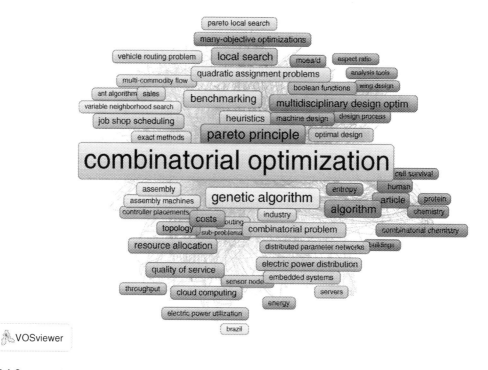

FIGURE 1.2

Network visualization.

optimization (793), multiobjective optimization (526), optimization (443), genetic algorithms (284), algorithms (208), evolutionary algorithms (195), and combinatorial optimization problem (168) are top indexed keywords with the most frequency.

Figs. 1.3 provides visualizations of the most active journals. Fig. 1.3 presents journals aggregated by density. The color of each item (journal) in the map is related to the density of the items at the point. The yellow area indicates the local high density of journals, and the blue area indicates low density. The densest area is on the right side of Fig. 1.3. It refers to the area occupied by *European Journal of Operational Research* and *Applied Soft Computing Journal*. The most cited journals are *European Journal of Operational Research* (3047), *Computers and Operations Research* (1128), and *Computers and Industrial Engineering* (1371). Fig. 1.3 also helps us to identify the least dense area of research, which is colored blue and small letters such as *WSEAS Transactions on Circuits and Systems*, *Transactions on Emerging Telecommunications Technologies*, and *Transactions of the Japanese Society for Artificial Intelligence*.

1.4.2 Mapping the social space

Analysis of the coauthorship networks shows the strong (number of co-authored works) and successful (highly cited works) connections among other collaborating researchers. The links across the networks in Figs. 1.4 and 1.5 are channels of knowledge, and the networks highlight the scientific communities engaged in research on multiobjective combinatorial optimization problems.

1.4 Results and discussion

Table 1.3 Top 10 keywords for 1-, 2- and 3-word length.

1-word		2-word		3-word	
Keyword	Frequency	Keyword	Frequency	Keyword	Frequency
Optimization	461	Combinatorial optimization	833	Combinatorial optimization problems	169
Algorithms	208	Multiobjective optimization	529	Multiobjective optimization problem	79
Scheduling	126	Genetic algorithms	291	Particle swarm optimization	77
Integer programming	119	Evolutionary algorithms	203	Pareto-optimal solutions	41
Design	68	Multiobjective optimization	208	Ant colony optimization	56
Algorithm	54	Multi-objective	154	Multi-agent systems	60
Benchmarking	50	Problem solving	162	Multidisciplinary design optimization	48
Non-dominated sorting genetic algorithm (NSGA)-II	42	Pareto principle	119	Local search optimization	48
Costs	44	Heuristic methods	102	Traveling salesman problem	41
Metaheuristics	40	Decision making	103	Multi-objective genetic algorithm	39

Table 1.4 The top 20 indexed keywords in the articles.

Indexed keywords	Occurrences
Combinatorial optimization	793
Multiobjective optimization	526
Optimization	443
Genetic algorithms	284
Algorithms	208
Evolutionary algorithms	198
Combinatorial optimization problems	168
Multi objective	154
Problem-solving	162
Pareto principle	119
Scheduling	116
Integer programming	115
Decision making	103
Heuristic methods	102
Multiobjective combinatorial optimization	95
Heuristic algorithms	83
Multiobjective optimization problem	79
Particle swarm optimization	77
Combinatorial mathematics	80
Simulated annealing	72

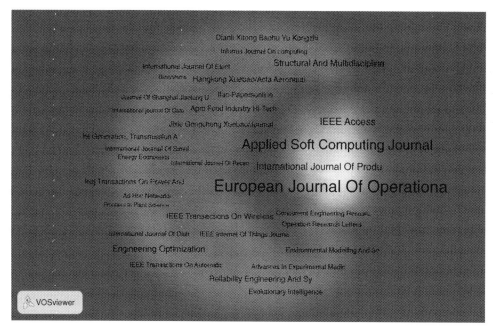

FIGURE 1.3

Journal map (density).

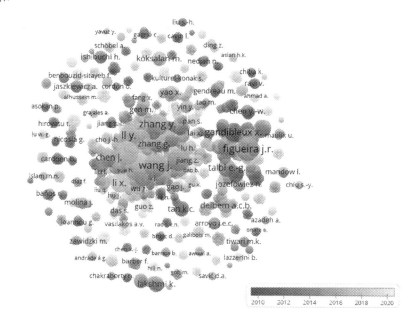

FIGURE 1.4

Scientific community (co-authorship) working on multiobjective combinatorial optimization problems.

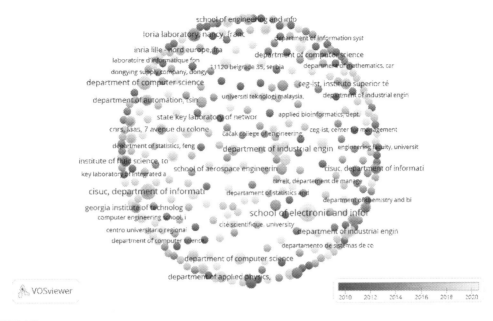

FIGURE 1.5

Scientific community (co-authorship) working from different organizations.

Fig. 1.4 shows the scientific community (coauthor) working on multiobjective combinatorial optimization problems over documents under investigation in the databases.

Fig. 1.4 illustrates overlay visualization and it can be observed that most collaborations such as Wang J., Zhang G., Chen J., Zhang Y., and Chen J., occurred before 2016 (dark blue) while the yellow color presents more recent collaborations including Cancela H., Hill N., Rentizelas A.

Fig. 1.5 refers to the scientific community (coauthorship) working from different organizations. Analogous to Fig. 1.4, the yellow color presents recent collaborations such as Center for Artificial Intelligent, Mount Royal University Bissett; and Department of Computer Science while the dark blue color shows collaboration before 2016 such as Department of Industrial Engineering and Department of System Engineering.

1.5 Conclusions and direction for future research

This chapter offered five informed perspectives on the emergence of multiobjective combinatorial optimization problems studies across social and cognitive spaces: (1) the main scientific disciplines involved in the emergence of multiobjective combinatorial optimization problems studies; (2) topics of interest in the scientific communities; (3) interdisciplinary collaboration among researchers and organization; and (4) the distribution of publications across journals; (5) in terms of keywords combinatorial optimization, multiobjective optimization, optimization, genetic algorithms, algorithms, and evolutionary algorithms are the most interesting keywords for scholars. The scientific communities presented that many researchers and institutions work on this matter and this shows the importance of this study.

As the main aim of this chapter was the keywords and scientometric analysis that considered the whole database, from which we chose some examples of published articles for review, a comprehensive review in the field can be considered as a future research direction. Moreover, comparison studies in the field using different evolutionary algorithms could be interested as future works.

References

[1] R. Behmanesh, I. Rahimi, A.H. Gandomi, Evolutionary many-objective algorithms for combinatorial optimization problems: a comparative study, Arch. Comput. Meth. Eng. (2020) 1–16.
[2] K. Deb, Multi-objective optimization, in: E.K. Burke, G. Kendall (Eds.), Search Methodologies, Springer,, Cham, 2014, pp. 403–449.
[3] M. Ehrgott, X. Gandibleux, A survey and annotated bibliography of multiobjective combinatorial optimization, OR-Spektrum 22 (4) (2000) 425–460.
[4] T.D. Fry, R.D. Armstrong, H. Lewis, A framework for single machine multiple objective sequencing research, Omega 17 (6) (1989) 595–607.
[5] A.H. Gandomi, A. Emrouznejad, I. Rahimi, Evolutionary computation in scheduling: a scientometric analysis, Evolutionary Computation in Scheduling (2020) 1–10.
[6] L. Leydesdorff, S. Milojević, Scientometrics, 2012 ArXiv Preprint ArXiv:1208.4566.
[7] R.T. Marler, J.S. Arora, Survey of multi-objective optimization methods for engineering, Struct. Multidiscip. Optim. 26 (6) (2004) 369–395.
[8] I. Rahimi, A. Ahmadi, A.F. Zobaa, A. Emrouznejad, S.H.E. Abdel Aleem, Big Data Optimization in Electric Power Systems: A Review, CRC Press,, FL, 2017.
[9] K. Scholz, On scheduling multi-processor systems with algebraic objectives, Computing 20 (3) (1978) 189–205.
[10] E.L. Ulungu, J. Teghem, Multi-objective combinatorial optimization problems: a survey, J. Multi-Criteria Decis. Anal. 3 (2) (1994) 83–104.
[11] N.J. Van Eck, L. Waltman, Software survey: VOSviewer, a computer program for bibliometric mapping, Scientometrics 84 (2) (2010) 523–538.
[12] N.J. Van Eck, L. Waltman, VOSviewer manual, Univeristeit Leiden, *Leiden*, 2013, pp. 1–53.

CHAPTER 2

The fundamentals and potential of heuristics and metaheuristics for multiobjective combinatorial optimization problems and solution methods

Ana Carolina Borges Monteiro[a], Reinaldo Padilha França[a], Rangel Arthur[b] and Yuzo Iano[a]

[a]*School of Electrical Engineering and Computing (FEEC)-State University of Campinas (UNICAMP), Campinas, São, Paulo, Brazil,* [b]*Faculty of Technology (FT)-State University of Campinas (UNICAMP), Limeira, São Paulo, Brazil*

2.1 Introduction

Combinatorial Optimization is closely linked to combined optimization problems, among which there is a finite (large) set of solutions, reduced to better. These problems are modeled as problems of maximizing (or minimizing) a function whose variables must obey certain restrictions, using the objective function or function used. These problems can be divided into three categories: those that assume real (or continuous) values, those that assume discrete (or integer) values, and those in which there are integer and continuous variables, classified, respectively, as continuous optimization problems, Combinatorial or Discrete Optimization, and Mixed Optimization [1,2].

Related computational challenges are faced in solving combined optimization problems belonging to the NP-hard or NP-complete class with little computational effort (execution time). Or even finding optimal, or even approximate, solutions to these types of problems is not always an easy challenge to be overcome. And in many practical problems, there is no need to find a "theoretically" optimal solution [1,2].

Heuristics have long been considered cognitive models par excellence, they are constituted as rules based on experience and planning, replacing the previous ones based on the algorithmic search that reaches the correct solutions after combining the problem with all possible solutions. Heuristic methods seek as great a degree of action as possible in a situation. Thus, it includes process, strategies, methods, procedures, of approaching trial/error, searching for the better mode to reach a certain end. Heuristic processes, in general, need less time than algorithmic methods, are closer to the mode human beings reason and obtain at problem resolutions, guaranteeing effective solutions [3].

In general, these techniques are strategies commonly used to solve problems that offer better solutions and generally with a shorter processing time than other types of techniques. Still considering that they employ conjunction of historical knowledge and random choices (from former results obtained by the process) guiding and conducting their searches within the research space through the research space in neighborhoods, preventing anticipated stops in great places [3,4].

Heuristics is a method or process created to find solutions to a problem. Heuristic research is research carried out through the quantification of proximity to a certain objective. It is said that it is possible to have a good (or high) heuristic if the object of evaluation is very close to the objective; it is said of bad (or low) heuristic if the object evaluated is too far from the objective [3,7].

An approximate algorithm (or approximation algorithm) is heuristic, that is, it uses information and intuition about the problem instance and its structure to solve it quickly. It is a simplifying procedure (although not simplistic) that, in the face of difficult questions, involves replacing them with others that are easier to solve to find viable, even if imperfect, answers. Such a procedure can be either a deliberate problem-solving technique, or an automatic, intuitive, and unconscious behavior operation [5].

However, not every heuristic algorithm is approximate, that is, not every heuristic has a mathematically proven quality ratio or formal proof of convergence. Heuristic algorithms are a good alternative when it is possible to have a very large instance of the problem or when there is little time available for resolution [6].

Heuristics and heuristic method are names for the algorithm that provides solutions not containing a quality limit, typically assessed empirically in terms of quality and complexity (average) of the solutions. These are those algorithms that have no guarantee of determining the optimal solution for the studied problem. Heuristics is a set of rules and methods that lead to discovery, invention, and problem solving. It is also an auxiliary science of history that studies the research of sources [7].

It should be noted that metaheuristics have also been largely employed for problems of continuous optimization, where the variables are real. For example, the use of the ant-colony optimization algorithm, with some modifications, to find minimum functions, or even considering a suitable metaheuristic for continuous optimization problems is derived by the behavior of a flock of birds during flight, called particle swarm optimization (PSO), among others [8].

In this sense, the objective of this study is to offer an introductory approach for people who want to start in the area of Combinatory Optimization, heuristics, and metaheuristics, highlighting and exemplifying some basic concepts, using problems, and illustrating the use of the technique for their solution.

Therefore, this chapter has the mission and objective of providing an updated review and overview of heuristics and metaheuristics, addressing its evolution and fundamental concepts, showing its relationship as well as approaching its success, with a concise bibliographic background, categorizing and synthesizing the potential of the technique.

2.2 Multiobjective combinatorial optimization

Optimization is related to any engineering process or managerial or technological decisions that need to be taken at different stages, consequently obtaining the best arrangement of these variables that minimizes or maximizes this merit function consists of an optimization process, reflecting on the cost or benefit of these decisions can be expressed through a function of certain decision variables [9,10].

Optimization means using computational methods to find the "best" way (represented by the best combination of values for the problem variables, i.e., optimal solution) to design or even operate a given system, considering the objectives and constraints of design or operation. Combinatorial problems are characterized by the possibility of combining several variables, as for optimization problems it is derived from Decision Space composing discrete variables, characterizing a multiobjective Combinatorial Optimization problem. In general, Objective Space makes up a finite set or it is possible to limit the search within a finite set [10,11].

With regard to multiobjective Combinatorial Optimization, it matches problems for which no exact, fast algorithm is known, that is, related to problems of the NP-hard class. In this context, the mathematical problem of multiobjective optimization consists of obtaining a set of variables (solution) satisfying the restrictions and optimizing a function consisting of several terms or functions involving two or more objective functions. In other words, they are a class of problems that can be computationally intractable due to their combinatorial nature and have an exponential dimension. Still considering that in general, they consist of problems that maximize or minimize an objective function involving several decision variables subject to equality/inequality restrictions and even completeness restrictions with respect to some or all variables [11,12].

Still reflecting on conflicting objectives, in general, there is no single solution that is optimal and optimizes all objective functions at the same time. In multiobjective problems, there is no optimum in the mono-objective sense, but a set of efficient or nondominated solutions. In that sense, no objective in a dominated solution can be improved without worsening at least one of the other objective functions. Therefore, the challenge of this type of problem is characterized by finding a solution or a set of feasible solutions, close to the Pareto-optimal frontier, without the need to explore the entire Decision Space [12,13].

The set of nondominated solutions is known as the Pareto-optimal set, and the image of a particular Pareto-optimal set, in the space of the objective values, is known as the Pareto boundary, that is, it is the set of feasible points that are not dominated by no other point [12,13].

Multiobjective optimization introduces certain flexibility to modeling; however, there is not a single defined solution, but a set of compromise solutions (Pareto solutions). Thus, solutions of this type consist of determining, in the feasible objective space, the efficient set, a subset of the efficient set or even, sets of solutions close to the Pareto-optimal frontier. Assessing the size and complexity of this method, most of the practical problems require the intervention of a decision maker [14,15].

In general, the result of multiobjective optimization problems leads to a decision problem involving one or more decision makers choosing from the set of alternatives the most interesting one [15,16].

Multiobjective decision methods can be classified (depending on the decision maker and the problem) as *a priori*, related to the tradeoff between the determined (modeled) objective functions before the execution of the optimization method, that is, transforming this type of problem into a mono-objective problem and a single solution is generated, which requires classic strategies of direct and accurate optimization. By changing the tradeoff between the objectives, a new solution can be generated, until the decision maker is satisfied. Or Progressive relating progressive optimization methods, consulting the decision-maker during the optimization, guiding the search toward the tradeoff solution that satisfies this. Or even *a posteriori* relating those methods designed to search for a set of representative solutions in the objective space. The result is a set of alternatives that are offered to the decision maker after optimization, without requiring preference modeling, but this process is more complex and computationally costly, as several solutions must be generated. In general, *a posteriori* methods are adopted in conjunction with multiobjective metaheuristics [16,17].

Another option with respect to the decision maker is to classify the objectives in order of priority, searching for the optimal solution in stages, starting the optimization of the first objective, without considering the others, and proceeding to the optimization of the following objectives, considering the previous optimal value, until reaching the last objective; however, this strategy does not guarantee the achievement of an efficient solution [17,18].

Another class of solution methods involves mapping feasible solutions (problems of an investigative nature), directing to efficient solutions, providing the decision maker with a set of approximate or Pareto-optimal solutions. In general, these methods require a great deal of computational effort, but it is appropriate where the decision maker does not have knowledge about the feasible objective space [17–19].

The search for efficient solutions in multiobjective combinatorial optimization problems is not trivial, in particular, for problems involving nonlinear functions and, even more so, discontinuous functions. In this context, a heuristic seeks to achieve a satisfactory solution, without having to scan the entire spectrum of solutions of exponential proportions (combinatorial nature) [17–19].

In this context, more flexible and easy to implement methods applicable to the search for solutions are also metaheuristic, they can solve multiobjective problems of combinatorial optimization and nonlinear optimization problems with flexibility. Describing a heuristic that is superimposed on another heuristic, constituting a more generic structure based on principles or concepts, superimposed on a specific heuristic of the problem under study [17–19].

2.3 Heuristics concepts

Heuristic methods are a procedure that, in the face of difficult questions, involves replacing answers for a project with others that are easier to solve to find viable, even if imperfect, answers. Often, this simple step can be sufficient to solve the problem. From the historical and etymological point of view, the word heuristic is derived from the Greek word "Heuriskein," with the meaning of discovering, just as curiously it also gave rise to the term Eureca [20].

A "heuristic" is a rule of thumb derived from experience, such a procedure being both a deliberate problem-solving technique and an automatic, intuitive, and unconscious behavior operation. There is no conclusive proof of its validity, and the heuristic technique is expected to work many times but not always. A heuristic optimization method can be deterministic or stochastic, depending on whether or not it will use randomly drawn numbers to execute its algorithm, that is, a heuristic helps to find good solutions, but not necessarily optimal ones [21].

A practical analogy, conceptualizing heuristics is in the aspect of a common everyday activity concerning the journey of a person leaving home and driving to work by car. Considering that she needs to choose the path to take, she evaluates the best path; however, this "definition of best" involves several factors such as travel time, pavement quality, safety issues, probability of bottling, slow traffic, among many others [21,22].

In this scenario, to optimize this problem related to the journey before leaving home, that person would need to connect to the Internet and search for various information online regarding the traffic conditions of their city at that time, checking the existence of works on their routes, news reporting accidents, among others. And based on this information, that individual evaluates pros and cons of each route, weighs the best options for each of the possible routes, financial costs, weighs bus or subway options, and then after analyzing the route with the best evaluation that day he would be chosen for the next day [23].

2.3 Heuristics concepts

This is obvious that people do not have this habit and do not practice this type of evaluation. What exists is based on individual experience, it is known that at some times some highways and routes must be avoided, and others favored. Which, as time goes by, people simply get in the car and go to work, using the route that, due to their accumulated experience, seems to be the best. In other words, most of what is done daily by people, whether professionally or personally, involves heuristic problem solving [23].

In this sense, the personal heuristics of each individual, in general, will probably work most of the time, but not always. Concerning this simple practical analogy of problems in the chosen route, sometimes this individual will result in being stuck in a traffic jam [24].

Heuristics have long been considered cognitive models par excellence, constituting rules based on experience and planning, replacing previous ones based on the algorithmic search that reaches the right solutions after combining the problem with all possible solutions. That is, heuristics is a method or process created with the aim of finding solutions to a problem, seeking as great a degree of action as possible in a situation. So, it involves tactics, means, procedures of trial/error approach, searching best mode to get a certain end. A heuristic is a simplifying procedure, although not simplistic in the face of questions [25].

Difficulties involve replacing them with others of easier resolution to find viable, though imperfect, answers. In this sense, heuristic processes require less time than algorithmic procedures, which are closer to the scheme human beings reason and get at problem resolutions, ensuring efficaciouspractical problems, there is no need to find a "theor solutions [26].

Heuristic searches are related to a search carried out by quantifying proximity to a given objective. What in this context, is evaluated as a good or high heuristic, if the object of evaluation is very close to the objective; on the other hand, it is rated as a bad or low heuristic if the object being evaluated is too far from the target [27].

Heuristic algorithms are a good alternative when it is possible to have a very large instance of the problem or when there is little time available for resolution. However, not all heuristic algorithms are approximate, that is, not all heuristics have a mathematically proven quality ratio or formal proof of convergence. Thus, a heuristic (it is a set of rules and methods that lead to discovery, invention, and problem solving) is related to the algorithm that provides solutions not needing a regular quality limit, typically analyzed empirically in practical terms of quality of solutions and complexity average [26,27].

In this respect, it is worth noting that an approximate algorithm is distinguished from heuristic. Regarding that "approximate" is respective to the algorithm that provides solutions within an absolute or asymptotic quality limit, as well as a polynomial asymptotic limit of complexity, that is, worst case, mathematically proven. An approximate (or approximation) algorithm can be considered heuristic by using information and intuition about the instance of the problem and its structure to solve it quickly [27].

Reflecting on this aspect, heuristic algorithms are those that have no guarantee of determining the optimal solution for the problem studied; however, the approximate methods may fall into this category, adding that, for these cases, properties are known to guarantee the worst case. And so, it is valid the statement that not every heuristic algorithm is approximate, that is, not every heuristic has a mathematically proven quality ratio or formal proof of convergence [28].

Heuristic methods are exploratory algorithms that seek to solve problems; many times these methods are classified as "blind search," as long as they do not involve the computational implementation of specialized knowledge, that is, a specific heuristic. In other words, to solve a second-degree equation, it would not necessarily use the basic formula, but would seek, by other methods, a solution that would meet the second-degree equation. So are the heuristic methods, they can be seen as a continuous and empirical search, with several great locations, whose result is the best that can be found under certain conditions [29].

An optimal solution to a problem is not always the target of heuristic methods, which in general find out the better possible solutions to problems, and not accurate, perfect, and definitive solutions; as, having as a starting point a feasible solution, they are inspired on consecutive approximations orientated to an optimum point. This "imprecision" of heuristic methods, or subjectivity, is not related to a deficiency, but a peculiarity analogous to human intelligence, as in general, people in everyday life solve several problems without knowing them precisely [30].

An illustration of the concept of heuristics comes from the mathematician George Pólya, concerning set up a schematic, if it is not possible to understand a problem; make a reverse mechanism to try to reach the solution (reverse engineering); if it is not possible to find out the solution; propose the same problem in a specific example; if the problem is abstract, to approach a more general problem first, concerning the analogy of the inventor's paradox concerning the fact that the most ambitious purpose is the most likely to succeed [30,31].

2.4 Metaheuristics concepts

Heuristics is a search by which various solutions of a given problem are analyzed to choose the one that appears to be the best solution according to a given evaluation function. Concerning metaheuristics, they consist of techniques referring to a wide class of algorithms for optimization and solving search problems and genetically solving problems, using combinatorial analysis, proving to be a powerful tool for resolving approximately combinatorial optimization problems in practice [32].

In this sense, *Constructive Methods* are those heuristics that gradually add individual components to the solution until a feasible solution is obtained. *Decomposition Methods* are those heuristics that fractionate the problem into small subproblems (divide and conquer), the output of a problem being the entry of the next problem, so that when solving, everyone gets a solution to the global problem. *Reduction Methods* are those heuristics that identify some characteristics that presumably should appear in the optimal solution, which represents that in this way it is possible to simplify the problem [32–34].

Still considering the *Model Manipulation Methods*, which are those heuristics that modify the structure of a given model to make it easier to solve, deducing, from its solution, the solution of the original problem. It consists of or reducing the solution space through the linearization of nonlinear functions, grouping variables to reduce their number, imposing new restrictions based on the expected behavior of the optimal solution, or even increasing the solution space, by eliminating restrictions [32,35].

Still reflecting on the *Neighborhood Search Methods*, which are those metaheuristics that start from an initial viable solution obtained by some other heuristic, choosing a set of solutions, the one that results in the better objective function value. And with changes in that solution, it is successively changed to a new solution, iteratively, until some stopping criterion is suited [36].

Metaheuristics can be considered as procedures that employ strategies to escape from local minimums in spaces of a search for complex solutions, they can be specified by controlling, at a higher level, this subspace, aiming to produce a satisfactory result for a given problem, but not guaranteeing the optimality. These techniques, in general, are applied to find answers to problems about which the brute force strategy is disregarded because the solution space is very large, still evaluating that there is little information about this problem [37].

Metaheuristics aim to improve the performance of a linear programming solver as a heuristic, considering the advantages related to *Ease of Implementation*, comparing that building a good traditional heuristic for a certain problem requires a minimal study of the characteristics of that problem, and in general, it is a difficult task; *Flexibility and Ease of Maintenance*, as when applying optimization to real complex problems, the exact definition of the problem is often changed several times, new constraints that were not foreseen initially appear, throughout a project. In this context, adapting a traditional heuristic incorporating a new constraint tends to be a difficult task. However, on the other hand, small changes in an entire programming formulation are usually easy, in the case of algorithms and data structures created to take advantage of each particular problem [38].

In this context, metaheuristics can be divided into different types of *Surrounding Search* (Simulated Annealing, Taboo Search), which traverse the search space taking into consideration, fundamentally, the "neighborhood" of the solution. As long as the set of solutions is defined, the application of an operator toward the current solution can be obtained from there. *Relaxation metaheuristics* are related to the simplification of the problem and use the solution found as a guide to the original problem. Concerning *Constructive types* (Greedy randomized adaptive search procedure, GRASP), which meticulously define the value of each component of the solution. And still reflecting on the *Evolutionary types* (Genetic Algorithms, GAs) that works with a great number of solutions, and through the interaction between its elements, it evolves. And even the *types of Decomposition*, which indicate how to decompose the problem into instances, using the solutions resulting in the construction of the solution to the original problem [33,39].

Thus, metaheuristics propose algorithms that exploit the search space effectively, aiming to find good quality solutions, which may even correspond to a certain optimal solution, comprising search methodologies in candidate solution spaces capable of managing computational search operators in local and global search. Solving combinatorial optimization problems that, for the most part, are specific and dedicated to a given problem, promoting robustness and efficiency in the search for solutions. Considering strategies to obtain the desired solution for the sake of feasibility, that is, there are no guarantees, after all, a "good" solution, not necessarily the optimal one, but it is better than no solution or a solution taken randomly from the search space [40].

In summary, the idea of implementing metaheuristics is based on heuristics that command other heuristics, which have a wide repertoire of possible applications, especially highlighting the scenarios in which the classic optimization methods are infactible or even inefficient. Among the metaheuristic techniques stand out: simulated annealing, GAs, taboo search, scatter search, grasp, ant colony, variable neighborhood search, and other techniques. It is also worth pointing out and emphasizing that metaheuristics, in general, do not have the properties related to the Guarantee of convergence, Guarantee of obtaining the optimal solution, or even Maximum cost guarantee to arrive at a solution [41].

Metaheuristics are derived from heuristics, which are heuristic methods that can deal with any optimization problem, that is, that are not linked to a specific problem. It is a set of concepts that is possible to be employed defining heuristic procedures applicable to a vast set of problems [42].

2.5 Heuristics and metaheuristics examples
2.5.1 Tabu search

Neighborhood search procedures are iterative process in which an $N(i)$ neighborhood is determined for each viable solution i, and the following solution j is searched for within the solutions in $N(i)$. The descent method clearly stops at a great location but not necessarily a great overall f [36,42].

Tabu Search is a metaheuristic, a very good approximation technique, competing with nearly all other known strategies and which, due to its flexibility, can overcome many classic procedures. The technique can be considered a more elaborate neighborhood search method than the descent method. This method can be evolved into a wide variety of levels of sophistication and application, considering the ease with which it is possible to adapt a rudimentary implementation to absorb several additional characteristics. This technique has important uses in areas such as evolutionary and genetic methods, improving neural network processes. Taboo search is an iterative method that consists of the general step of an iterative method of building, starting from a current solution i, the following solution j, and verifying whether to stop the process, or to execute a new step [43,44].

To improve the exploration process, it is an auxiliary adaptive procedure guiding a local search algorithm within a search space in a continuous exploration. The method avoids returning to a previously visited great place to overcome the optimality and achieve an optimal result or close to the global optimum [43,44].

The Tabu Search related to informed search strategies and local search is inspired by the use of adaptive memory used to drive the search process to leave great places and obtain solutions close to the global optimum, using a collection of intelligent solution principles problems, using flexible memory structures, an associated control mechanism to be used with the memory structures to define the Tabu restrictions, records of historical information that allow the use of strategies to intensify and diversify the search process [[43,44].

2.6 Evolutionary algorithms (EA)

Evolutionary algorithms (EA) are search strategies based on natural evolution, directed mainly by mutation and selection. It is a random search that employs variation and selection in a biological perspective these algorithms model the asexual reproduction of a population. In general, the algorithm is a metaheuristic, which adapts the number of variation by modifying the variance of the normal distribution, seeking to model this sexual reproduction (GAs) characterized by combining two parent strings in a descending string (called a crossover) [45,46].

They are based on natural evolution processes, consistent with a family of optimization algorithms inspired by biological evolution, such as natural selection, mutation, and recombination. Therefore, it is a process that follows a trajectory over time through a complex evolutionary search space (search space or state space) from an initial condition. They correspond to a class of algorithms of the type generate-and-test with the following characteristics: Population based; variational operators related to recombination and mutation; selection processes for reproduction and survival; and stochastics [45,46].

In the simplest case, genetic representation can be interpreted as a solution to the problem or individual, belonging to the phenotype space. With coding of the solution or chromosome, it belongs to the genotype space, considering only a string of size n, the chromosome. The positions of the string are known as the chromosome locus. The variable in a locus is known as a gene, and its value is an allele. The set of chromosomes is known as a genotype, with the following valid relationships: one for several: Phenotype = Genotype; one to one: Genotype = Phenotype. Thus, defining a phenotype (individual) with certain adaptability. Each subset of information encoded on a chromosome is called a gene. The mutation genetic operator modifies the value of each position of the string with a probability m, whereas the crossover operator works by combining two strings [47].

EAs play basic roles related to computational modeling to simulate (reproduce) natural evolutionary processes; in the same sense as considered as an optimization tool for problem solving. This type of algorithm is aimed at computable problems for which it is possible to provide an algorithm that leads to its solution. Considering the complexity of these computable problems, it is associated with the amount of memory and the processing time needed to reach the solution by exploring the search space. As regards, if one or both of these quantities are not computable, then the problem is considered to be infectible [48].

2.7 Genetic algorithms (GA)

GA are a special class of EA. A GA is an intelligent probabilistic search algorithm, it is an iterative computer simulation, simulating the process of evolution over a population of solutions by investing genetic operators in each reproduction. This technique is analogous to an evolutionary process of several generations of a population where each individual is an abstract representation of a solution to the problem, modeling the process of natural evolution. However, even though it is much faster than exhaustive search methods, they are still very slow compared to gradient methods, as they do not use any information related to the derivative of the objective function [49].

GA are a metaheuristic, which does not use any derivative information and, therefore, presents good chances of not being trapped in great places. That is, they require the definition of several parameters that affect the performance of the algorithm in various ways, and the meanings of the mutation, crossover, and other specific problem operators must be clearly defined, evaluating that a potential solution to the problem has representation from a set of parameters called genes, which concerns that these parameters are aggregated to form a string of values, that is, a chromosome [49,50].

The GAs, related to Population-Based Methods, have very peculiar characteristics concerning other search methods, relative to being based on a set of possible solutions; they do not involve modeling the problem, being restricted to solutions; resulting in a population of solutions (qualitatively classified by natural selection) and not just one; Parameters such as the probability of crossing, mutation rate, and mechanisms of mutation and crossing affect the performance of the algorithm less significantly; it is a probabilistic and nondeterministic method. That is, the same population will hardly present the same results for the same problem, considering that the size of the population, that is, the number of individuals, must be large enough to guarantee sufficient diversity and thus cover the space of solutions well [49,51].

2.8 Simulated annealing

The Simulated Annealing method is a type of local search method with extremely simple implementation, related to informed search and local search strategies, based on the process of annealing metals, bringing an analogy between the behavior of the combinatorial optimization problems and the large ones physical systems studied in statistical mechanics [52].

Local search only necessitates a method of assessing the cost and definition of a neighborhood scheme of a particular solution. Making the algorithm in the general end in an optimal solution, which depending on the initial solution can be very poor. What is respective that in classic local search methods, a new trial solution is created from the current solution, and it replaces the previous one if the value of the associated objective (energy) function is lower. Evaluating that as the method works with only one solution, it is suggested to run several copies of the algorithm in parallel, starting from different initial conditions [50,52].

The technique is a metaheuristic, which establishes a connection between the thermodynamic behavior and the search for the global maximum/minimum of a discrete optimization problem, wherein the technique transitions to higher energy solutions are allowed according to a certain probability, thus avoiding that the method gets stuck in the basin of attraction of a local minimum. At each iteration, the objective function generates values for two solutions that is, the current and the chosen one, which is compared, and then the solutions better than the current one are always accepted, while a fraction of the solutions worse than the current one is accepted in the hope of escaping from a local minimum/maximum [53,54].

2.9 Particle swarm optimization (PSO)

They are optimization algorithms, that is, swarm-based metaheuristics behave in such a way as to imitate models of nature to arrive at solutions close to an optimal model for a given solution. It is a population stochastic optimization strategy based on the collective behavior of animals, exploring the analogy with the social behavior of animals, such as bee swarms, flocks of birds, or even schools of fish. The PSO is based on social systems, based on the collective behavior of individuals interacting with each other and with the environment [55,56].

In this type of technique, related to Population-Based Methods, each individual of the group/gang that makes their own decisions, evaluated as a point in the n-dimensional space and the speed of this individual as the search direction to be employed in this solution candidate, but somehow based on the experience of the group leader. It is a technique for the optimization of nonlinear functions related to the particle swarm methodology. Mathematically, the search direction in each iteration is specified by weighing the experience of that solution, and the better solution ever found by the group (from the leading solution) [55,56].

Unlike other population-based methods, the PSO technique does not create new individuals, or even discard old ones, during iterations. On the other hand, a single static population is maintained, whose members have modified properties (such as position and speed) in response to new discoveries about space [55,56].

2.10 Scatter search (SS)

It is an evolutionary technique that assesses linear combinations of solution vectors, operating on a set of solutions known as a reference set, to produce new solution vectors through successive generations, combining them and generating new solutions to improve the solution originally. It is, therefore, an evolutionary method, considering the size of the reference set is relatively small, different from the population in GAs [57].

This metaheuristic consists of an initial phase (constituting a set of solutions, of which the better are chosen to be part of a reference set) and an evolutionary one (new solutions are created employing combinations of reference subsets that are strategically selected) differs in use deterministic rather than probabilistic strategies to achieve diversification and intensification, and population size. And from that, a set of the best solutions created is added in the reference set, evaluating that the procedures of the evolutionary phase are repeated until the process satisfies any stop criterion [57].

2.11 Greedy randomized adaptive search procedures (GRASP)

It is a metaheuristic related to an iterative process wherein each iteration resides of a construction phase and the second is a local search phase, corresponding to a Constructive Strategy. At the end of all iterations, the resulting solution is the better generated. The solution that proves to be the best, in general, is maintained as a result; however, it is not guaranteed that the solutions generated will not fall into optimal locations. Thus, it is significant to apply the local search phase in an attempt to improve such solutions [58,59].

With the use of customized data structures and careful implementation, as it is an adaptive heuristic, the benefits concomitant with the elements are updated with each construction iteration. So, that the last choice made, an efficient construction phase can produce good initial solutions, enabling an efficient local search [58,59].

At each construction iteration, a feasible solution is produced (the elements of the solution are chosen one by one), built using a random greedy function and the posterior solutions are obtained by applying, on the previous solution, a search algorithm site that provides a new best solution than the previous one. The choice of the next element to be included in the solution is settled by the ordering of all candidate elements, based on a function [58,59].

The local search in most cases (the maximum local is found in the vicinity of the solution obtained in the previous phase) as the algorithm does not guarantee that the solution of the first phase is locally optimal. Thus, the neighboring solutions are tested successively, until the best one is found. The technique still has a probabilistic component characterized by the random choice of one of the better candidates in the options of the construction phase, allowing distinct solutions to be achieved at each iteration of the method, but does not implicate its adaptive potential [58,59].

2.12 Ant-colony optimization

It is a population-based metaheuristic and derived by the behavior of an ant colony, of the finite size of ants that collectively look for better solutions in the search for food, corresponding to a Constructive

Strategy. Considering that each ant builds a solution from an initial state that is selected according to particular reestablished criteria for each problem. And so, an ant disperses pheromone in the soil, forming a trail at random, that when finding a trail of pheromone, the more ants follow this trail, the greater weight of attraction the trail will have due to the amount of pheromone [60–62].

The technique has as main objective the optimization and the search for solutions. As "artificial ants" and "artificial pheromone," as well as other variants, are probabilistic heuristics that construct information in two ways, corresponding to the pheromone trail and heuristic information. For the algorithm not to decorate or to tend to repeat and addict at optimal points, there is a constant evaporation factor. Assessing that once the ant has done its job, it dies, but the trail it left continues, in a way that contributes to the algorithm leaving its memory. Still considering that the parameters are defined at each iteration to minimize the possibility of converging to local solutions very quickly, ensuring greater exploration of the search space [60–62].

Stagnation is a situation, caused by excessive pheromone growth at the edges of a suboptimal journey, in which all "ants" always follow the same path. Still evaluating that despite the stochastic nature of the algorithm, the strong concentration of pheromone in the edges forces the "ant" to always follow the same route [60–62].

2.13 Clustering search

It is a hybrid metaheuristic that consists of a process of groupings of solutions to detect supposedly promising regions in the search space, and should be explored as soon as they are detected, using specific local search heuristics. It has the advantage of applying local search only in supposedly promising regions, detected through a process of grouping solutions. In order to improve the convergence process concomitant with a reduction in effort, considering that a region can be seen as a search subspace defined by a computational neighborhood relationship due to the more rational use of local search methods [63,64].

The technique is an iterative clustering process performed simultaneously with a metaheuristic identifying of groups of solutions that earn special interest. Create clusters randomly, considering the Maximum Diversity Problem, as given a set N with n elements, select a subset $M \subset N$ in such a way that the elements of M have the greatest possible diversity between them, that is, generating a large number of initial solutions at random. Considering its process of simple assimilation, by Path and by Recombination. And so, an improvement in the convergence process affiliated with a low in computational effort due to the more rational use of local search methods [63,64].

2.14 Hybrid metaheuristics

The development of a hybrid metaheuristic comes from the combination of different metaheuristics: metaheuristics with specific algorithms for the problem at hand; and metaheuristics with other operational research and/or artificial intelligence techniques [41,64].

Still considering that hybrid metaheuristics can be classified according to their respective characteristics at the level of hybridization, that is, respective high level to maintain the individual identity of the original algorithms that cooperate through a well-defined interface; and low-level, relating to algorithms that depend heavily on each other, the individual elements of the algorithms are exchanged [65].

Still evaluating its order of execution, whether Sequential, related to one algorithm is executed after the other, and information is only passed in one direction; Interleaved, making each iteration of an algorithm, the other is executed; Parallel concerning the algorithms are executed in parallel and the information can be passed in any direction. Or even the Control Strategy, be Integrated, where an algorithm is considered a subordinate, that is, embedded component of another algorithm; or Collaborated, related to algorithms exchange information, but are not part of each other [50,65].

2.15 Differential evolution (DE)

DE is a metaheuristic that uses NP D-dimensional parameter vectors, as a population in each G generation, generating new parameter vectors by adding the weighted difference between two parameter vectors to a third individual, that is, a mutation. Still considering that the initial set of vectors is generated randomly and should cover the entire search space, considering that in the absence of any knowledge about the search space, that is, with respect to promising regions or even partial solutions, a uniform distribution is used for the initial population [66,67].

So, these mutated parameter vectors are combined with other predetermined vectors (target vectors) to generate trial vectors, as this combination of parameters is known as DE crossover, emphasizing that each vector present in the current population must be used once as a trial vector. The operation corresponding to the selection is associated if this trial vector provides a higher fitness value (maximization) than that associated with the respective target vector, this higher fitness value will give way to the first in the next generation [67,68].

2.16 Teaching learning–based optimization (TLBO)

TLBO is an optimization metaheuristic inspired by nature based on the teacher-student relationship, considering the influence of a teacher (considered the aptest individual) on his students in the classroom. This approach is divided into two phases, the teacher phase, in which each iteration is selected a more apt individual (teacher). And the learner phase, in which the rest of the individuals learn and move toward the teacher, obeying a learning factor (teaching factor), still relating that each individual is compared to another randomly selected individual and a new individual is generated to move toward the best individual [69,70].

This approach is a very simple algorithm, in which the only parameters to be adjusted are the number of iterations and the size of the population of individuals. It has as a characteristic the good performance in the search for solutions, serving as a good option in the training of artificial neural networks [70,71].

2.17 Discussion

Two areas rely heavily on heuristic optimization techniques, Global Optimization, and Combinatorial Optimization.

Concerning Global Optimization in general, look for a local minimizer of the real function $F(x)$ where x is a vector of real variables. In other words, search for a vector x^* such that $F(x^*) \leq F(x)$ for every x next to x^*. The Global Optimization problem consists of finding an x^* that minimizes $F(x)$ to all possible values of x; however, there is no criterion to identify a global minimum. This is a much more difficult problem, and for most local applications it is already good or sufficient, considering that the user can use his experience to define a research region and/or provide a good starting point for a determining algorithm.

However, it is different from whether the algorithm achieves a better solution (local maximum) or not yet, if this algorithm continues to do only local searches for better solutions, but it will not be able to escape these "trap" for global optimization methods. For difficult problems such as Global Optimization, heuristic search methods are widely studied and used, but they cannot prove that these methods are an optimal solution or even a good one. However, in general, it is possible to find the best-known solution, which may be more than enough for the application in question.

Regarding Combinatorial Optimization, it is an area of application of heuristic methods, generally meaning that the state space is discrete, can be finite or just enumerable, that is, a space of types of molecules, or characteristics of a product. In this sense, any discrete problem can be seen as combinatorial. Perhaps the best known of the combinatorial problems is the Traveling Salesman Problem (determining the optimal sequence of cities to travel, so that the distance traveled is minimal), where a salesman is imagined who needs to visit a stated number of cities and then return to house. The cashier problem is a classic example of a Combinatorial Optimization problem.

However, reflecting on the aspects of Combinatorial Optimization, the great difficulty of combinatorial problems is that the number of possible solutions grows with a factorial speed, and quickly the computer becomes unable to enumerate all possible solutions to the problem. Since in order to solve problems like the clerk in polynomial time involving polynomial effort, despite numerous efforts, a method has not yet been found, which still represents one of the great open problems of Mathematics, related that this type of problem, is possible be reduced, in polynomial time. It is customary to summarize these properties of the cashier problem by saying that they belong to the category of complete polynomial nondeterministic problems, or NP-complete problems. Therefore, a polynomial algorithm is not known to find the optimal solution for this type of problem, but it is possible to build a polynomial algorithm to test the hypothesis of a set of solutions proposed as attempts to solve the problem.

In this sense, concerning Combinatorial Optimization problems, there are no algorithms that lead to the optimal solution in a viable time. And in this context, heuristic, deterministic, or random methods are used, that is, those used to obtain approximate solutions have the purpose that the solutions are found within a reasonable time for the practical purposes for which they are intended, and that their quality is the best possible given the time constraints. Used to achieve solutions that can be put into practice, even if they are not optimal (or there is no proof yet that they are).

Concerning Optimization, it consists of the process of choosing the best components in a set of available alternatives, related to an objective function, applied to the elements of the set, respective to types of optimization related to Linear Programming, Whole Programming, Combinatorial Optimization, Optimization Stochastic, and in this respect, in particular, metaheuristics. However, metaheuristics are related to process that coordinate local search methods with higher-level strategies, intending to generate a procedure capable of escaping local minimums and carrying out a robust search in the space of solutions, which are applied in solving problems about which there is little information, but which once offered a solution candidate can be tested, even if they do not present guarantees of optimality.

Thus, in this sense, the use of Exact Methods to find the solution to complex problems is often as efficient as using Iterative Methods. As the use of Gradient type method, which is a traditional algorithm of approximation of functions that are based on the slope of the function at a point, being a first-order optimization algorithm, where the choice of the direction of the minimum of the function 'f' is the direction opposite to the gradient, to define in which direction the function will seek a better solution, used when a matrix is symmetric and defined positive. Evaluating the great convergence time, since the gradient signal oscillates when it approaches a maximum/minimum, still considering that it can get stuck in local maximum/minimum or inflection points, in which the gradient is null. However, if a given problem can be solved by this type of method, this is probably the best alternative. However, not all problems can be treated by a gradient method

Faced with these cases, it is necessary to employ alternative strategies to solve these problems, which will often be heuristic optimization methods, such as those previously mentioned. Still evaluating that in general, heuristic methods are often called "methods of last resort," relative to the case in which the other methods have difficulty, that is, something to be used when everything else does not work. As these methods are usually slow, even though they are often the only ones available for a certain type of problem, and consequently they are the best. In the same sense, they are slower to solve the same problem than a method that uses derivatives.

At the same time as assessing the advancement of programming languages and the capabilities of today's computers, the computational cost associated with heuristic methods has become less costly, assessing that in general, any modern computer today has the capabilities to solve a reasonable problem. Still reflecting that although with different philosophies, metaheuristics (stochastic and inspired by natural phenomena) have in common characteristics that distinguish them from conventional heuristics, such as the inclusion of tools for ease of working in parallel environments and trying to escape the pitfalls of great places.

2.18 Conclusions

Reflecting on combinatorial problems, it is necessary to have an elementary procedure to find a desired optimal solution, which performs an exhaustive explosion of a given set of solutions, generally known as complete enumeration. In other words, all feasible solutions are generated and selected are those in which the presentation of the best value of an objective function is seen.

However, when the existence of many variables is very wide, the calculation time required to find a desired optimal solution increases exponentially with the number of variables in the problem. In this sense, given this practical difficulty of solving exactly a whole series of combinatorial problems, a solution is needed that provides feasible solutions, that is, the performance of heuristic and/or metaheuristic algorithms.

In general, situations that require heuristic algorithms are those in which there is no exact resolution method or even when the optimal solution is not required, or even the existence of precise resolution methods that require computational time or memory in addition to resources of the machine, or evaluating the existence of exact resolution methods that require computational time beyond the needs of the application.

The advantages that justify the use of this type of technique are derived from the greater flexibility in handling the characteristics of the problem, and the use of a fraction of the computational effort of an exact method. Still reflecting that this type of method offers more than one solution, which allows expanding the decision possibilities, especially when there are nonquantifiable factors that cannot be incorporated into the model, but that need to be considered.

In this context, heuristic techniques can be considered as simple procedures, they are useful tools for real problems, based on common sense, which are supposed to offer a good solution, although not necessarily optimal, related to difficult problems, in an easy and fast way. There is no guarantee that these solutions will optimize the objective function, but it is expected to produce suboptimal solutions of high quality in reasonable calculation time.

Concerning metaheuristics, these techniques can have the most unexpected and unusual origins, whether for applications in different areas, not just mathematical scaling, even considering that there is no supreme metaheuristic. Metaheuristics can also be classified as a single solution, concerning the types of simulated algorithms

Annealing, Tabu Search, GRASP, among others; or even a solution derived from Population, concerning the types of algorithms EAs, SS, Ant Colony, among others.

In this sense, it is worth noting that there is no optimization strategy to date which is globally better than all others, that is, considered as the best metaheuristic. In reality, what has currently considered a great effective option is the choice of an appropriate hybrid approach, which is decisive to achieve the best performance in solving the most difficult problems.

2.19 Future trends

Hyperheuristics has objective to resolve the conflict between ease of implementation and quality. The concept of hyperheuristics refers to heuristic algorithms that operate at a high level of generality, considering the domain referring to the composition by sets of heuristics (and not by the set of instances of an optimization problem). These techniques are implementations of metaheuristics and can be extremely powerful algorithms, finding solutions indirectly, using intelligently the heuristics and metaheuristics that have been provided, which in general, require substantial changes for each new problem to be addressed. In this sense, the idea is not necessarily to surpass the quality of other techniques [72,73].

Hyperheuristics have their potential advantages concerning robustness in the aspect of the algorithm is well defined, which presents efficiency and effectiveness in the treatment of several optimization problems. As your "intelligence" is associated with implementation at a level independent of the problem domain. Still evaluating that hyperheuristics are insensitive to these failures concerning the decision errors practiced by each of the heuristics, they tend to be poorly elaborated, that is, they are suppressed in a process of interleaving. And emphasizing the simplicity of adapting a hyperheuristic concerning a new problem, it is enough to change the set of simple low-level heuristics [73,74].

References

[1] K. Thulasiraman, S. Arumugam, T. Nishizeki, A. Brandstädt, Handbook of Graph Theory, Combinatorial Optimization, and Algorithms, Taylor & Francis, UK, 2016.
[2] F. Green, Review of Handbook of Graph Theory, Combinatorial Optimization, and Algorithms, ACM SIGACT News 50 (3) (2019) 6–11.

References

[3] K. Ali, M.I. Ghazaan. Meta-heuristic algorithms for optimal design of real-size structures. Springer, 2018.

[4] T.T. Mac, C. Copot, D.T. Tran, R De Keyser, Heuristic approaches in robot path planning: a survey, Rob. Autom. Syst. 86 (2016) 13–28.

[5] T. Stanley, D.B. Kirschbaum, A heuristic approach to global landslide susceptibility mapping, Nat. Hazards 87 (1) (2017) 145–164.

[6] M.S. Puga, J.S. Tancrez, A heuristic algorithm for solving large location–inventory problems with demand uncertainty, Eur. J. Oper. Res. 259 (2) (2017) 413–423.

[7] T. Rose, Heuristic research, What is Psychotherapeutic Research?, Routledge, UK, 2018, pp. 133–143.

[8] K.L. Du, M.N.S Swamy, Search and optimization by metaheuristics. Techniques and Algorithms Inspired by Nature, Birkhauser, Basel, Switzerland, 2016.

[9] E. Khalil, H. Dai, Y. Zhang, B. Dilkina, L. Song, Learning combinatorial optimization algorithms over graphs, in: M.I. Jordan, Y. LeCun, S.A. Solla (Eds.), Advances in Neural Information Processing Systems, The MIT Press, MA, 2017, pp. 6348–6358.

[10] L. Hulianytskyi, I. Riasna, Formalization and classification of combinatorial optimization problems, in: S. Butenko, P.M. Pardalos, V. Shylo (Eds.), Optimization Methods and Applications, Springer, Cham, 2017, pp. 239–250.

[11] M. Sánchez, J.M. Cruz-Duarte, J. carlos Ortíz-Bayliss, H. Ceballos, H. Terashima-Marin, I. Amaya, A systematic review of hyper-heuristics on combinatorial optimization problems, IEEE Access 8 (2020) 128068–128095.

[12] A.S. Hameed, B.M. Aboobaider, N.H. Choon, M.L. Mutar, W.H. Bilal, Review on the methods to solve combinatorial optimization problems particularly: quadratic assignment model, Int. J. Eng. Technol. 7 (3.20) (2018) 15–20.

[13] T. Huang, Y. Ma, Y. Zhou, H. Huang, D. Chen, Z. Gong, Y. Liu, A Review of combinatorial optimization with graph neural networks, in: Proc. 5th International Conference on Big Data and Information Analytics (BigDIA), IEEE, 2019, pp. 72–77.

[14] N. Benabbou, T. Lust, A general interactive approach for solving multi-objective combinatorial optimization problems with imprecise preferences, in: Proc. the International Symposium on Combinatorial Search (SoCS'19), 2019.

[15] N. Benabbou, C. Leroy, T. Lust, P. Perny, Combining local search and elicitation for multi-objective combinatorial optimization, in: Proc. International Conference on Algorithmic DecisionTheory, Springer, Cham, 2019, pp. 1–16.

[16] A. Raith, M. Schmidt, A. Schöbel, L. Thom, Multi-objective minmax robust combinatorial optimization with cardinality-constrained uncertainty, Eur. J. Oper. Res. 267 (2) (2018) 628–642.

[17] S. Martínez, C. González, A. Hospitaler, V. Albero, Sustainability assessment of constructive solutions for urban Spain: a multi-objective combinatorial optimization problem, Sustainability 11 (3) (2019) 839.

[18] J. Xu, C.C. Wu, Y. Yin, W.C. Lin, An iterated local search for the multi-objective permutation flowshop scheduling problem with sequence-dependent setup times, Appl. Soft Comput. 52 (2017) 39–47.

[19] Q. Liu, X. Li, H. Liu, Z. Guo, Multi-objective metaheuristics for discrete optimization problems: a review of the state-of-the-art, Appl. Soft Comput. 93 (2020) 106382.

[20] M. Fischer-Kowalski, K.H. Erb, Core concepts and heuristics, in: H. Haberl, M. Fischer-Kowalski, F. Krausmann, V. Winiwarter (Eds.), Social Ecology, Springer, Cham, 2016, pp. 29–61.

[21] Backhaus, J., de Carteret, R., Damerius, L., Huang, Y. Y., Pfisterer, A., Pöll, C., & Wallimann-Helmer, I. (2019). Introduction to critical systems heuristics In: Paschke, M. and Dahinden, M. eds. Applying Collective Inquiry. Engaging in the science-policy dialogue (Workbook 8). Zurich-Basel Plant Science Center, Zurich (pp. 42–45).

[22] E. Esposito, M. Santoro, D. Stark, Introduction to the symposium, heuristics of discovery., Sociologica 12 (1) (2018) 1.

[23] B.A. Kumar, M.S. Goundar, Usability heuristics for mobile learning applications, Educ. Inf. Technol. 24 (2) (2019) 1819–1833.
[24] Kilrea, K. A., & McCaslin, M. L. (2020). Deep heuristics: An emergent methodology for transformative, actualizing, & potentiating relationships (transformative inquiry in action).
[25] W. van den Bos, B. Eppinger, Developing developmental cognitive neuroscience: from agenda setting to hypothesis testing, Dev. Cogn. Neurosci. 17 (2016) 138.
[26] S. Aine, S. Swaminathan, V. Narayanan, V. Hwang, M. Likhachev, Multi-heuristic A, Int. J. Robot. Res. 35 (1–3) (2016) 224–243.
[27] V. Raman, N.S. Gill, Review of different heuristic algorithms for solving travelling salesman problem, Int. J. Adv. Res. Comput. Sci. 8 (5) (2017) 1–4.
[28] E. Khamehchi, M.R. Mahdiani, Optimization algorithms, in: E. Khamehchi, M.R. Mahdiani (Eds.), Gas Allocation Optimization Methods in Artificial Gas Lift, Springer, Cham, 2017, pp. 35–46.
[29] Barrett, T. D., Clements, W. R., Foerster, J. N., & Lvovsky, A. I. (2019). Exploratory combinatorial optimization with reinforcement learning. arXiv preprint arXiv:1909.04063.
[30] B.A. Conway, Evolutionary and heuristic methods applied to problems in optimal control, in: A. Frediani, B. Mohammadi, O. Pironneau, V. Cipolla (Eds.), Variational Analysis and Aerospace Engineering, Springer, Cham, 2016, pp. 117–143.
[31] G. Gigerenzer, The Heuristics Revolution: Rethinking the Role of Uncertainty in Finance. On The Behavioural Finance Revolution, Edward Elgar Publishing, 2018.
[32] P. Stender, S. Stuhlmann, (2018). Fostering Heuristic Strategies in Mathematics Teacher Education, INDRUM, 2018 (2018).
[33] A. Duarte, M. Laguna, R. Martí, Metaheuristics for Business Analytics, Springer, Cham, 2018, pp. 29–55.
[34] U. Schroeders, O. Wilhelm, G. Olaru, Meta-heuristics in short scale construction: ant colony optimization and genetic algorithm, PLoS One 11 (11) (2016) e0167110.
[35] G.R. Raidl, J. Puchinger, C. Blum, Metaheuristic hybrids, in: A. Duarte, M. Laguna, R. Marti (Eds.), Handbook of Metaheuristics, Springer, Cham, 2019, pp. 385–417.
[36] K. Hussain, M.N.M. Salleh, S. Cheng, Y Shi, Metaheuristic research: a comprehensive survey, Artif. Intell. Rev. 52 (4) (2019) 2191–2233.
[37] D. Pisinger, S. Ropke, Large neighborhood search, in: A. Duarte, M. Laguna, R. Marti (Eds.), Handbook of Metaheuristics, Springer, Cham, 2019, pp. 99–127.
[38] Sorensen, K., Sevaux, M., & Glover, F. (2017). A history of metaheuristics. arXiv preprint arXiv:1704.00853.
[39] N. Dey, (Ed.). Advancements in Applied Metaheuristic Computing. IGI Global, PA, 2017.
[40] J. Silberholz, B. Golden, S. Gupta, X. Wang, Computational comparison of metaheuristics, in: M. Gendreau, J-Y Potvin (Eds.), Handbook of Metaheuristics, Springer, Cham, 2019, pp. 581–604.
[41] E. Cuevas, J. Gálvez, O. Avalos, Recent Metaheuristics Algorithms for Parameter Identification, Springer, Cham, 2020.
[42] T. Ganesan, P. Vasant, I. Elamvazuthi, Advances in Metaheuristics: Applications in Engineering Systems, CRC Press, Boca Raton, FL, 2016.
[43] L. Deroussi, An Introduction to Metaheuristics, Metaheuristics for Logistics, 4, Wiley, NJ, 2016, pp. 37–55.
[44] A. Amuthan, K.D. Thilak, Survey on Tabu search meta-heuristic optimization, in: Proc. 2016 International Conference on Signal Processing, Communication, Power and Embedded System (SCOPES), IEEE, 2016, pp. 1539–1543.
[45] K.Z. Zamli, B.Y. Alkazemi, G. Kendall, A Tabu search hyper-heuristic strategy for T-way test suite generation, Appl. Soft Comput. 44 (2016) 57–74.
[46] O. Bozorg-Haddad, M. Solgi, H.A. Loáiciga, Meta-heuristic and Evolutionary Algorithms for Engineering Optimization, John Wiley & Sons, NJ, 2017.
[47] T.S. Bindiya, E. Elias, Meta-heuristic evolutionary algorithms for the design of optimal multiplier-less recombination filter banks, Inform. Sci. 339 (2016) 31–52.

[48] A.C.B. Monteiro, Proposta de uma metodologia de segmentação de imagens para detecção e contagem de hemácias e leucócitos através do algoritmo WT-MO, 2019,.
[49] K. De Jong, Evolutionary computation: a unified approach, in: Proc. 2016 on Genetic and Evolutionary Computation Conference Companion, 2016, pp. 185–199.
[50] O. Kramer, Genetic algorithms. In: Genetic Algorithm Essentials, Springer, Cham, 2017, pp. 11–19.
[51] P. Siarry (Ed.), Metaheuristics, Springer International Publishing, Cham, 2016.
[52] G. Wu, R. Mallipeddi, P.N. Suganthan, Ensemble strategies for population-based optimization algorithms–a survey, Swarm Evol. Comput. 44 (2019) 695–711.
[53] F.Y. Vincent, A.P. Redi, Y.A. Hidayat, O.J. Wibowo, A simulated annealing heuristic for the hybrid vehicle routing problem, Appl. Soft Comput. 53 (2017) 119–132.
[54] K.L. Du, M.N.S Swamy, Simulated annealing, in: K-L Du, M.N.S. Swamy (Eds.), Search and Optimization by Metaheuristics, Springer, Cham, 2016, pp. 29–36.
[55] D. Delahaye, S. Chaimatanan, M. Mongeau, Simulated annealing: from basics to applications, in: M. Gendreau, J-Y Potvin (Eds.), Handbook of Metaheuristics, Springer, Cham, 2019, pp. 1–35.
[56] D. Wang, D. Tan, L. Liu, Particle swarm optimization algorithm: an overview, Soft Comput. 22 (2) (2018) 387–408.
[57] M.R. Bonyadi, & Z. Michalewicz, Particle swarm optimization for single-objective continuous space problems: a review. Evol. Comput., 25 (1), (2017). 1–54
[58] M. Laguna, Scatter search, Search Methodologies, Springer, Boston, MA, 2014, pp. 119–141.
[59] M.G. Resende, C.C. Ribeiro, GRASP: Greedy randomized adaptive search procedures, Search Methodologies, Springer, Boston, MA, 2014, pp. 287–312.
[60] M.G. Resende, C.C. Ribeiro, Greedy randomized adaptive search procedures: Advances and applications, in: M. Gendreau, J-Y Potvin (Eds.), Handbook of Metaheuristics, 146, Springer, Cham, 2010, p. 281-317),.
[61] M. Dorigo, T. Stützle, Ant colony optimization: overview and recent advances, in: M. Gendreau, J-Y Potvin (Eds.), Handbook of Metaheuristics, Springer, Cham, 2019, pp. 311–351.
[62] S. Mirjalili, J.S. Dong, A. Lewis, Ant Colony optimizer: theory, literature review, and application in AUV path planning, in: S. Mirjalili, J.S. Dong, A. Lewis (Eds.), Nature-Inspired Optimizers, Springer, Cham, 2020, pp. 7–21.
[63] A. Akhtar, "Evolution of ant colony optimization algorithm – a brief literature review." arXiv: 1908.08007. 2019.
[64] W.A. Altoé, D.D.C. Bissoli, G.R. Mauri, A.R Amaral, A clustering search metaheuristic for the bi-objective flexible job shop scheduling problem, in: Proc. 2018 XLIV Latin American computer conference (CLEI), IEEE, 2018, pp. 158–166.
[65] N. Tremblay, A. Loukas, Approximating spectral clustering via sampling: a review, in: F. Ros, S. Guillaume (Eds.), Sampling Techniques for Supervised or Unsupervised Tasks, Springer, Cham, 2020, pp. 129–183.
[66] K.R. Opara, J. Arabas, Differential evolution: a survey of theoretical analyses, Swarm Evol. Comput. 44 (2019) 546–558.
[67] T. Eltaeib, A. Mahmood, Differential evolution: a survey and analysis, Appl. Sci. 8 (10) (2018) 1945.
[68] N. Javaid, Differential evolution: An updated survey, in: L. Barolli, K. Yim, T. Enokido (Eds.), Conference on Complex, Intelligent, and Software Intensive Systems, Springer, Cham, 2018, pp. 681–691.
[69] F. Zou, D. Chen, Q. Xu, A survey of teaching–learning-based optimization, Neurocomputing 335 (2019) 366–383.
[70] R. Xue, Z. Wu, A survey of application and classification on teaching-learning-based optimization algorithm, IEEE Access 8 (2019) 1062–1079.
[71] P. Sarzaeim, O. Bozorg-Haddad, X. Chu, Teaching-learning-based optimization (TLBO) algorithm, in: O. BozorgHaddad (Ed.), Advanced Optimization by Nature-Inspired Algorithms, Springer, Singapore, 2018, pp. 51–58.

[72] F.D. Mohammed, M.F. Ramli, M.Z. Zakaria, A review on hybrid metaheuristics in solving assembly line balancing problem, Proc. AIP Conference, 2138, AIP Publishing LLC, 2019.

[73] J.H. Drake, A. Kheiri, E. Özcan, E.K. Burke, Recent advances in selection hyper-heuristics, Eur. J. Oper. Res 285 (3) (2019) 405–428.

[74] T. Macias-Escobar, B. Dorronsoro, L. Cruz-Reyes, N. Rangel-Valdez, C. Gómez-Santillán, A survey of hyper-heuristics for dynamic optimization problems, in: O. Castillo, P. Melin, J. Kacprzyk (Eds.), Intuitionistic and Type-2 Fuzzy Logic Enhancements in Neural and Optimization Algorithms: Theory and Applications, Springer, Cham, 2020, pp. 463–477.

CHAPTER 3

A survey on links between multiple objective decision making and data envelopment analysis

Amineh Ghazi[a] and Farhad Hosseinzadeh Lotfi[b]

[a]*Department of Mathematics, Central Tehran Branch, Islamic Azad University, Tehran, Iran,*
[b]*Department of Mathematics, Science and Research Branch, Islamic Azad University, Tehran, Iran*

3.1 Introduction

Multiple criteria decision making (MCDM) is an essential methodology in operations research. The purpose of this methodology is the selection of best solution in the feasible region of the decision space (continuous or discrete), according to the multiple objective functions or multiple criteria. Generally, the MCDM methodology is divided into two subsets, multiple objective decision making (MODM) and multiple attribute decision making (MADM) [83]. In MODM, the objective is the designation of the best alternative in the feasible region of the decision space so that all objectives are satisfied. In MADM, there are multiple alternatives and according to considered criteria the best alternative is selected from available alternatives. Note that in MADM, there are some predetermined alternatives; yet, in MODM, there are not predetermined alternatives. Multiple objective linear programming (MOLP) problem plays a key role in MODM [154]. It is of interest to note that, in MOLP problems, there does not exist a solution in the feasible region of decision space which optimizes all objective functions at the same time; therefore, instead of the optimal solution, the notion efficient, Pareto Koopmans, or nondominated solution is utilized. In fact, the important task of an MOLP problem is to obtain nondominated solutions to select the most preferred solution (MPS) using a large number of approaches [154].

Data envelopment analysis (DEA) is a robust mathematical methodology to determine the relative efficiency of a set of peer entities, decision-making units (DMU) with common inputs and outputs, such as bank branches, schools, firms, hospitals, etc. [27]. Initially, DEA was introduced by Charnes et al. [27] based on the work of Farrell [54] in obtaining nonparametric productivity function. Essentially, DEA is a nonparametric estimation of frontier methodology based on linear programming to measure the relative efficiency of DMUs. In evaluation of DMUs, this methodology focuses on each individual DMU to select the best weights for inputs and outputs to locate the envelopment-efficient frontier (surfaces) of the production possibility set (PPS). If a DMU lies on the efficient frontier, it is referred to as an efficient DMU, otherwise an inefficient one [113]. DEA has two different kind models, radial and nonradial. In the literature of DEA, some traditional (standard) models are CCR [27], BCC [17], Additive [26], SBM [180], and ERM [136]. The radial DEA models have an input or output orientation; however, nonradial DEA models fail to have any orientation. Moreover, the traditional DEA models have two forms, envelopment and multiplier, obtained from viewpoints of constructing PPS and defining relative efficiency, respectively. In decision making, DEA has had dramatic increase in the past decade and become a robust method to analyze the efficiency of various organizational DMUs. The popularity of DEA is reflected by a wide range of successful applications of this methodology [13].

MCDM and DEA are two powerful and significant methodologies in operations research and management science [156]. A large number of researchers in these two fields have in-depth attention in MCDM and DEA methodologies. MCDM and DEA are structurally very similar to each other, and there exist some similarities between them. In a large amount of research, the similarity of these methodologies has been discussed from several point of views. The objective of both methodologies is to identify efficient points, units, or alternatives and obtain the projection of inefficient points [102,107,108,156]. Researchers have argued that MCDM is a tool to improve the theory of DEA problems; in other words, they have believed that MCDM provides essential theories for DEA problems. In addition, they have argued that DEA is an applied area of MCDM [44,102]. Consequently, DEA is considered as an application of MCDM. In DEA, if inputs and outputs are considered as criteria and DMUs are considered as alternatives, the formulation of DEA is converted to one similar to that of MCDM, although the objective of these two methodologies might be different. Hence, DEA and MCDM are complements, and should never be viewed as substitutes [44,107,108]. In recent years, a large number of researchers have made excellent efforts to integrate of these two methodologies, including the capability of MCDM in DEA and the capability of DEA in MCDM [44,102,156].

Both the MCDM and DEA methodologies were taken from the work of Koopmans [110] in the activity analysis. Charnes and Cooper worked on the MCDM methodology before introducing the DEA methodology. They played a key role in connecting these methodologies. In the first presented DEA paper by Charnes et al. [27], they indirectly pointed out the similarity between these two methodologies, but did not make an attempt to directly tie these two fields together. Then, this task was carried out by other researchers who worked on both methodologies.

In general, classifying the relationships between MCDM and DEA can be divided into two principal parts. The first category is the application of the MCDM concepts in the DEA methodology. It was offered by Charnes et al. [27] using the exploratory comparison between the concept of Pareto Koopmans in MOLP problems and conditions of the efficient DMU in DEA. The second class is the application of the DEA concepts in the MCDM methodology. It was suggested by Doyle and Green [44] using the exploratory application of the cross-efficiency DEA approach [149] in ranking predetermined alternatives. On the other hand, MCDM is divided into two subsets, and thus, relationships between DEA and MCDM are precisely divided into four classes. However, seeing that the aim of this research is classifying the links between MODM and DEA, only two classes are investigated. The first class is the usage of the MODM concepts in the DEA methodology. In this part, MODM techniques are applied to solve the DEA problems, including the formulation of classical (traditional) DEA models, target setting, value efficiency, common set of weights (CSW), secondary goal, discriminant analysis (DA), and specification of the efficient frontier for a PPS. The traditional nonradial DEA models are constructed by multiple objective programming (MOP) problems [25,26,136,180]. On the other hand, some traditional radial models can be reformulated as MOP problems [38,203]. To overcome the problem of inappropriate projection for inefficient DMUs onto the efficient frontier, researchers have proposed many MOP problems in the DEA methodology. To set targets, Golany [68] primarily implemented an interactive technique to solve the MOP problem in DEA. The value efficiency is an efficiency that incorporates decision maker's preference information in DEA models; it was preliminary presented by Halme et al. [72]. To solve the problem of alternative optimal solutions in DEA models, Doyle and Green [45] offered the exploratory incorporation of the secondary goal formulation. There are several methods for facing optimistic DEA models; CSW methods is one of them, introduced by Charnes et al. [30] and Kornbluth [111] via cone restrictions in the MOP problem. It seems that Retzlaff-Roberts [144] first investigated the relationships between two unrelated linear programming techniques, DEA and DA. It is of interest to note that the concept of the goal programming method is common for both of them. In recent years, some DEA researchers modified the proposed algorithms for determining efficient faces of MOLP problems into the DEA methodology for specifying efficient hyperplanes of a PPS, for example, Hosseinzadeh Lotfi et al. [82]. Second part is the usage of DEA concepts for MODM problems. In this part, the DEA techniques are applied to obtain efficient points in MODM problems. This ability is often utilized for multiobjective integer linear programming problems. Liu et al. [58] and Yougbare and Teghem [206] developed exploratory research methods to show the usage of DEA techniques for solving 0-1 MOLP problems.

According to the above discussions, there exist tightly coupled relationships between the DEA and MCDM methodologies. There are a large number of studies where the MCDM concepts were applied in the DEA methodology and vice versa. Yet, none of them classified the issues related to the connections between them. The purpose of current study is to explore the relationship between DEA and MODM methodologies and review on some available papers in these fields. Therefore, this study attempts to collect almost all of the presented DEA-MODM papers, extracted from online databases such as Science Direct, Springer, and Jstore. Over the last years, there exist a large volume of published papers in this field. Therefore, the authors prefer to eliminate many papers for the reasons such as their low scientific level, publication in low level journals, having application rather than theoretical point of view, and having similar topics.

This paper is organized as follows: In Section 3.2, the basic implications of DEA and MODM are discussed. In Section 3.3, the application of MODM concepts in the DEA methodology is reviewed, and the application of DEA concepts in the MODM methodology is reviewed in Section 3.4. At last, a summary and conclusion of the classification in the relationships between DEA and MODM methodologies is presented.

3.2 Preliminary discussion

3.2.1 Multiple objective decision making

The single-objective decision-making methods reflect an earlier revolution. However, in the real word, the problems have become more complex and involve more than one objective. Therefore, multiple objective problems are required. MOLP problems are one kind of the most popular models utilized in MODM. An MOLP problem is constructed to maximize or minimize several linear objective functions subject to a set of linear constraints as follows [83,154]:

$$
\begin{aligned}
&\text{Max} \quad \{c^1 X = z_1\} \\
&\text{Max} \quad \{c^2 X = z_2\} \\
&\qquad \vdots \\
&\text{Max} \quad \{c^s X = z_s\} \\
&\text{s.t.} \quad X \in S,
\end{aligned}
\tag{3.1}
$$

TABLE 3.1 Classification of solution methods for MOLP problems.

Methods for no preference information given:	Global criterion methods,...
Methods for no preference information given:	Global criterion methods,...
Methods for a priori preference information given:	Goal programming method, Lexicographic method,...
Methods for a progressive preference information given:	Steuer method, Zionts–Wallenius method,...
Methods for a posteriori preference information given:	Adaptive search method, Parametric methods,...

where $S = \{X \in \Re^n \mid AX = b, X \geq 0, b \in \Re^m\}$ and $C = (c^1, c^2, \ldots, c^s)^T \in \Re^{s \times n} (c^r \in \Re^n, r = 1, \ldots, s)$. The purpose of solving MOLP (3.1) is to find a point or points in the feasible region of decision space (S) that maximize the criteria vector ($z_1, \ldots, z_s$). Yet, there exists a problem that in each MOLP problem there is not only one point in S that maximizes all objectives, simultaneously. Therefore, instead of the optimal solution, the concept of efficient solution is adopted [154].

Definition 1. $x^* \in S$ is called an efficient (strong efficient) solution if and only if there is no $x \in S$ such that $C^T x \geq C^T x^*$ and $C^T x \neq C^T x^*$.

To identify the efficient solutions of MOLP problems, some preference information from the decision maker may be required. The type of these information and corresponding methods are classified in four states as shown in Table 3.1.

A 0-1 MOLP problem is considered as follows:

$$\begin{aligned}
\text{Max} \quad & \{c^1 W, \ldots, c^s W\} \\
\text{s.t.} \quad & A_i W \leq b_i, \quad i = 1, \ldots, m, \\
& w_j \in \{0, 1\}, \quad j = 1, \ldots, n,
\end{aligned} \quad (3.2)$$

where $c^r = (c^{1r}, c^{2r}, \ldots, c^{nr})(r = 1, \ldots, s)$, $A_i = (a_{i1}, a_{i2}, \ldots, a_{in})(i = 1, \ldots, m)$ and $W = (w_1, w_2, \ldots, w_n)^T$. Corresponding to each vector W in the feasible region of decision space there exists a vector $Y = (y_1, \ldots, y_s) = (c^1 W_1, \ldots, c^s W_s)$ in the feasible region of criterion space.

3.2.2 Data envelopment analysis

Consider a set of n homogeneous DMUs, each consuming m inputs to produce s outputs. Let x_{ij} and y_{rj} denote the ith and rth input and output for DMU_j, respectively. The input vectors $x_j = (x_{1j}, \ldots, x_{mj})$ and the output vectors $y_j = (y_{1j}, \ldots, y_{sj})$ are semipositive ($x_j \geq 0, y_j \geq 0, j = 1, \ldots, n$). The general PPS in the DEA methodology is defined as follows [17,27]:

$$T = \left\{ (x, y) \mid \sum_{j=1}^n \lambda_j x_{ij} \leq x_i, \ i = 1, \ldots, m, \ \sum_{j=1}^n \lambda_j y_{rj} \geq y_r, \ r = 1, \ldots, s, \ \lambda \in \Lambda \right\} \quad (3.3)$$

where Λ is one of the following:

$$\Lambda_C = \{\lambda_j \mid \lambda_j \geq 0, \ j = 1, \ldots, n\}, \quad (3.4)$$

$$\Lambda_V = \left\{ \lambda_j \mid \sum_{j=1}^n \lambda_j = 1, \ \lambda_j \geq 0, \ j = 1, \ldots, n \right\}. \quad (3.5)$$

T_C and T_V are PPSs with constant returns to scale (CRS) and variable returns to scale (VRS) technologies, respectively. The input-oriented DEA model to measure the relative efficiency of DMU_o (the DMU under evaluation) is constructed as follows:

$$
\begin{aligned}
\text{Min} \quad & \theta - \epsilon \left(\sum_{i=1}^{m} s_i^- + \sum_{r=1}^{s} s_r^+ \right) \\
\text{s.t.} \quad & \sum_{j=1}^{n} \lambda_j x_{ij} + s_i^- = \theta x_{io}, & i = 1, \ldots, m, \\
& \sum_{j=1}^{n} \lambda_j y_{rj} - s_r^+ = y_{ro}, & r = 1, \ldots, s, \\
& s_i^- \geq 0, \; s_r^+ \geq 0, & i = 1, \ldots, m, \; r = 1, \ldots, s, \\
& \lambda_j \in \Lambda_C, & j = 1, \ldots, n.
\end{aligned}
\tag{3.6}
$$

Model (3.6) is the well-known CCR model in the envelopment form.

Definition 2. DMU_o is called CCR efficient if and only if an optimal solution $(\theta^*, \lambda^*, s^{-*}, s^{+*})$ of Model (3.6) satisfies $\theta^* = 1$ and zero slacks ($s^{-*} = 0, s^{+*} = 0$).

Otherwise DMU_o is called inefficient. The dual of Model (3.6) is

$$
\begin{aligned}
\text{Max} \quad & \sum_{r=1}^{s} u_r y_{ro} \\
\text{s.t.} \quad & \sum_{i=1}^{m} v_i x_{io} = 1, \\
& \sum_{r=1}^{s} u_r y_{rj} - \sum_{i=1}^{m} v_i x_{ij} \leq 0, & j = 1, \ldots, n, \\
& u_r \geq \epsilon, \; v_i \geq \epsilon, & r = 1, \ldots, s, \; i = 1, \ldots, m.
\end{aligned}
\tag{3.7}
$$

Model (3.7) is called the CCR model in the multiplier form.

Definition 3. DMU_o is called CCR efficient if and only if in Model (3.7), $\sum_{r=1}^{s} u_r^* y_{ro} = 1$ and there exists at least one optimal solution (u^*, v^*) with $u^* > 0$ and $v^* > 0$.

Similar to Models (3.6) and (3.7), the output-oriented DEA models to measure the relative efficiency DMU_o can be constructed.

Chapter 3 A survey on links between multiple objective decision making

Now, both orientations are combined in a single model, called Additive DEA model. There are several types of Additive models, from which we select following [26]:

$$\text{Max} \quad \varrho = \sum_{i=1}^{m} s_i^- + \sum_{r=1}^{s} s_r^+$$

$$\text{s.t.} \quad \sum_{j=1}^{n} \lambda_j x_{ij} + s_i^- = x_{io}, \quad i = 1, \ldots, m,$$

$$\sum_{j=1}^{n} \lambda_j y_{rj} - s_r^+ = y_{ro}, \quad r = 1, \ldots, s,$$

$$s_i^- \geq 0, \ s_r^+ \geq 0, \quad i = 1, \ldots, m, \ r = 1, \ldots, s,$$

$$\lambda_j \in \Lambda_C, \quad j = 1, \ldots, n.$$

(3.8)

Definition 4. DMU_o is called ADD efficient if and only if in Model (3.8), $\sum_{i=1}^{m} s_i^{-*} + \sum_{r=1}^{s} s_r^{+*} = 0$.

A measure for Additive models in the form of a single scalar called Slack-Based Measure (SBM) [180]. To estimate the efficiency of DMU_o, it is formulated as following fractional program:

$$\text{Min} \quad \rho = \frac{1 - \frac{1}{m}\sum_{i=1}^{m}\frac{s_i^-}{x_{io}}}{1 + \frac{1}{s}\sum_{r=1}^{s}\frac{s_r^+}{y_{ro}}}$$

$$\text{s.t.} \quad \sum_{j=1}^{n} \lambda_j x_{ij} + s_i^- = x_{io}, \quad i = 1, \ldots, m,$$

$$\sum_{j=1}^{n} \lambda_j y_{rj} - s_r^+ = y_{ro}, \quad r = 1, \ldots, s,$$

$$s_i^- \geq 0, \ s_r^+ \geq 0, \quad i = 1, \ldots, m, \ r = 1, \ldots, s,$$

$$\lambda_j \in \Lambda_C, \quad j = 1, \ldots, n.$$

(3.9)

Definition 5. DMU_o is called SBM efficient if and only if in Model (3.9), $\rho^* = 1$.

The general combined-oriented model is constructed as follows [24,72]:

$$\text{Max} \quad \theta$$

$$\text{s.t.} \quad \sum_{j=1}^{n} \lambda_j x_{ij} \leq x_{io} - \theta d_{ix}, \quad i = 1, \ldots, m,$$

$$\sum_{j=1}^{n} \lambda_j y_{rj} \geq y_{ro} + \theta d_{ry}, \quad r = 1, \ldots, s,$$

$$\lambda_j \in \Lambda_C, \quad j = 1, \ldots, n,$$

(3.10)

where the direction vector $d = (d_x, d_y) = (d_{1x}, \ldots, d_{mx}, d_{1y}, \ldots, d_{sy})$ shows that the direction which DMU_o can move to lie on the efficient frontier.

Definition 6. In Model (3.10), DMU_o is efficient if and only if $\theta^* = 0$ and all optimal slacks of the constraints are zero.

The BCC models based on the VRS technology were introduced by Banker et al. [17]. Note that the BCC models are the extension of the CCR model where the VRS technology is adopted instead of the CRS technology.

3.3 Application of MODM concepts in the DEA methodology

In this section, the MODM techniques are applied as tools to solve the DEA problems. The application of MODM in DEA is classified into seven groups: the classical models, target setting, value efficiency, CSW, secondary goal, DA, and constructing efficient frontier of PPS. In the classical DEA models, some techniques of MODM such as lexicographic, goal programming, minimax, maximin, and weighted-sums are utilized. In the DEA problems, the preference information using the interactive techniques of MODM is used to provide a more meaningful efficiency score referred to value efficiency and to provide more effective targets called target setting. To remove the problem of alternative solution in DEA models, the lexicographic method is used to create secondary goal DEA models. Usually, in constructing the CSW DEA models, MOP problems are used, and then, the goal programming method is implemented to solve these problems. Also, the goal programming method can be applied in DA; accordingly, DEA and MODM are integrated to formulate the nonparametric DEA-DA models. The multiple simplex method (MSM) [208] and existing algorithms to obtain the efficient faces of MOLP problems can be utilized to construct the efficient hyperplanes of PPS.

3.3.1 Classical DEA models

The initial presented CCR model is radial, hence, it is more restrictive. Then, to overcome this problem the MODM techniques are applied to propose the nonradial DEA models. On the other hand, the proposed radial DEA models can be expressed as MOLP problems [203]. As the MCDM methodology has powerful theory, the classical DEA models as MOLP problems are more preferred. The two-phase process approach is the initial application of the lexicographic and weighted-sums methods in constructing the classical DEA models.

The objective of Model (3.6) attempts to find an optimal solution that first minimizes θ then maximizes the sum of input excess and output shortfalls while fixing $\theta = \theta^*$. This approach is called the two-phase process. Totally, this model has the two-phase process. Also, the two-phase process method was developed into the three-phase process [38]. Banker and Morey [18] extended the two-phase process in the presence of discretionary and nondiscretionary inputs.

In most of the research in DEA, the input-oriented CCR model is constructed using two famous viewpoints in Models (3.6) and (3.7). Still, another method to present the CCR model is the implementation of the MOP problem as follows [38]:

$$\text{Min} \quad \left\{ \frac{\sum_{j=1}^{n} \lambda_j x_{ij}}{x_{1o}}, \ldots, \frac{\sum_{j=1}^{n} \lambda_j x_{mj}}{x_{mo}} \right\} \quad (3.11)$$

$$\text{s.t.} \quad \sum_{j=1}^{n} \lambda_j y_{rj} \geq y_{ro}, \quad r = 1, \ldots, s$$

$$\lambda_j \geq 0, \quad j = 1, \ldots, n.$$

To utilize Model (3.11), input vectors must be positive, and the minimax method is applied to solving. Then, the objective function of MOLP (3.11) is converted to Min θ, and $\theta \geq \frac{\sum_{j=1}^{n} \lambda_j x_{ij}}{x_{io}}$ ($i = 1, \ldots, m$) is added to the constraints of the model. MOLP (3.11) is input-oriented; similarly, the output-oriented CCR model is constructed using an MOLP problem with s objective functions, and for solving the maximin method is applied.

With regard to the Pareto Koopmans efficiency test for $(x_o, y_o) \in \Re^{m+s}$, Cherens et al. [26] proposed the Additive DEA model (3.8). Essentially, this model is the usage of the weighted-sums method in the DEA methodology. It is worth mentioning that the Additive DEA model is similar to the dual of the multiplicative efficiency DEA model in a linear form, presented by Charnes et al. [28], and similar to the model proposed by Charnes et al. [29] for estimating the piecewise Cobb–Douglas production function. However, Additive DEA model was not offered in the mentioned times.

To obtain the Russell measure, several researchers presented some methods to solve the following MOLP problem:

$$\begin{aligned}
\text{Min} \quad & \{\theta_1, \ldots, \theta_m, \phi_1, \ldots, \phi_s\} \\
\text{s.t.} \quad & \sum_{j=1}^{n} \lambda_j x_{ij} \leq \theta_i x_{io}, & i = 1, \ldots, m, \\
& \sum_{j=1}^{n} \lambda_j y_{rj} \geq \phi_r y_{ro}, & r = 1, \ldots, s, \\
& \theta_i \leq 1, & i = 1, \ldots, m, \\
& \phi_r \geq 1, & r = 1, \ldots, s, \\
& \lambda_j \geq 0, & j = 1, \ldots, n.
\end{aligned} \qquad (3.12)$$

Pastor et al. [136] is one of the researchers who determined the nonradial enhanced Russell model (ERM) as the ratio between the average of decreasing input coefficients and the average of increasing output coefficients. As a result, the objective function of MOLP (3.12) in ERM is converted to Min $\frac{1/m \sum_{i=1}^{m} \theta_i}{1/s \sum_{r=1}^{s} \phi_r}$ and the constraints remain unchanged. Regarding the objective function in ERM and substitutes

$$\theta_i = \frac{x_{io} - s_i^-}{x_{io}} = 1 - \frac{s_i^-}{x_{io}}, \quad i = 1, \ldots, m,$$

$$\phi_r = \frac{y_{ro} - s_r^+}{y_{ro}} = 1 + \frac{s_r^+}{y_{ro}}, \quad r = 1, \ldots, s, \qquad (3.13)$$

we obtain

$$\frac{\sum_{i=1}^{m} \theta_i}{\frac{s}{m} \sum_{r=1}^{s} \phi_r} = \frac{\sum_{i=1}^{m} \left(1 - \frac{s_i^-}{x_{io}}\right)}{\frac{s}{m} \sum_{r=1}^{s} \left(1 + \frac{s_r^+}{y_{ro}}\right)} = \frac{1 - \frac{1}{m} \sum_{i=1}^{m} \frac{s_i^-}{x_{io}}}{1 + \frac{1}{s} \sum_{r=1}^{s} \frac{s_r^+}{y_{ro}}}, \qquad (3.14)$$

which is the same as the objective function for the SBM model (3.9). We may therefore apply ERM to solve SBM or vice versa.

The generalized DEA model using an MOLP problem was proposed as follows [189,190,207]:

$$\text{Max} \quad \{y_1, y_2, \ldots, y_s, -x_1, -x_2, \ldots, -x_m\}$$

$$\text{s.t.} \quad \sum_{j=1}^{n} \lambda_j \bar{x}_{ij} - x_i \leq 0, \qquad i = 1, \ldots, m,$$

$$\sum_{j=1}^{n} \lambda_j \bar{y}_{rj} - y_r \geq 0, \qquad r = 1, \ldots, s, \qquad (3.15)$$

$$\delta_1 \left(\sum_{j=1}^{n} \lambda_j + \delta_2 (-1)^{\delta_3} \lambda_{n+1} \right) = \delta_1,$$

$$\lambda_j \geq 0, \qquad j = 1, \ldots, n,$$
$$x_i, y_r \geq 0, \qquad i = 1, \ldots, m, \ r = 1, \ldots, s,$$

where $(\bar{x}_{1j}, \ldots, \bar{x}_{mj})$ and $(\bar{y}_{1j}, \ldots, \bar{y}_{sj})$ are vectors of input and output, respectively, for DMU_j ($j = 1, \ldots, n$). Also, δ_1, δ_2, and δ_3 are $\{0, 1\}$ binary parameters ($*$ indicates either 0 or 1). Different values of parameters and variable changes lead to basic DEA models as follows:

1. **Input-oriented BCC model**
   When $(\delta_1, \delta_2, \delta_3) = (1, 0, *)$ and $(x_i, y_r) = (\theta \bar{x}_{io}, \bar{y}_{ro})$ for all i, r, then the values $\bar{x}_{io}, \bar{y}_{ro}$ are constant, and also the maximization of $-\theta \bar{x}_{io}$ and $\bar{y}_{ro}$ is equivalent to minimization of θ.
2. **Output-oriented BCC model**
   When $(\delta_1, \delta_2, \delta_3) = (1, 0, *)$ and $(x_i, y_r) = (\bar{x}_{io}, \varphi \bar{y}_{ro})$ for all i, r, then the values $\bar{x}_{io}, \bar{y}_{ro}$ are constant, and also the maximization of $\bar{x}_{io}$ and $\varphi \bar{y}_{ro}$ is equivalent to maximization of φ.
3. **Additive BCC DEA model**
   When $(\delta_1, \delta_2, \delta_3) = (1, 0, *)$ and $(x_i, y_r) = (\bar{x}_{io} - s_i^-, \bar{y}_{ro} + s_r^+)$ for all i, r, then the values $\bar{x}_{io}, \bar{y}_{ro}$ are constant, and also $-s_i^-$ and s_r^+ are maximized.
4. **Russell DEA model**
   When $(\delta_1, \delta_2, \delta_3) = (1, 0, *)$ and $(x_i, y_r) = (\theta_i \bar{x}_{io}, \phi_r \bar{y}_{ro})$ for all i, r, then the values $\bar{x}_{io}, \bar{y}_{ro}$ are constant, and also $-\theta_i \bar{x}_{io}$ and $\phi_r \bar{y}_{ro}$ are maximized.

3.3.2 Target setting

The problem of target setting is one of the classical applications in management science. The objective of target setting methods is not to measure efficiency, but they make an effort to determine an appropriate projection point on the efficient frontier. On the grounds that the closer projection point to the MPS is more maximized than the decision maker's implicitly utility function, the MPS is a more effective and important point. Setting the future targets of an inefficient DMU drawing on radial projection being too restrictive, the traditional DEA models inappropriately project the inefficient DMUs onto the efficient frontier, which is a disadvantage in DEA. In recent years, to overcome this weakness a great number of researchers have offered MOLP problems instead of the traditional DEA models for target setting. Also, in most cases, to solve the proposed MOLP problems, the interactive techniques are implemented to reflect the decision maker's preference. The target obtained through these techniques is closer to the MPS compared to the targets obtained through traditional DEA models. Furthermore, the use of interactive methods for target setting is easier and more rational. Golany [68] proposed the preliminarily usage of an interactive technique in the DEA context.

Joro et al. [102] argued that the output-oriented two-phase process of the CCR model in the envelopment form tends to be structurally similar to the MOLP problem drawing on the reference point model proposed by Wierzbicki [192]. By changing variables, the reference point model is reformulated as follows:

$$
\begin{aligned}
\text{Max} \quad & \sigma + \delta 1^T s^+ \\
\text{s.t.} \quad & Y\lambda - \sigma W - s^+ = g, \\
& X\lambda + s^- = b, \\
& \lambda \geq 0, s^- \geq 0, s^+ \geq 0, \\
& \delta > 0.
\end{aligned}
\qquad (3.16)
$$

By comparing the reference point model (3.16) and the output-oriented CCR model, some differences between them are obtained. For example, in Model (3.16), W, g, and b are replaced by y_o, 0, and x_o in the output-oriented two-phase process of the CCR model, respectively. Also, in the objective function of Model (3.16), there exist slacks only in output shortfalls, but in the objective function of the CCR model there exist slacks in the input excesses and output shortfalls. Therefore, if DMU_o is weakly efficient, Model (3.16) may evaluate it as efficient. Korhonen et al. [113] proposed a nonradial method in which DMUs are free to choose their own target on the efficient frontier. Their method is based on an MOLP problem, and in solving the Pareto Race Software [112,114] is applied to find the MPS. Based on REM, Thanassoulis and Dyson [167] proposed a nonradial DEA model to reach the target value. In addition, Zhu [213] presented another nonradial DEA model to determine targets. Unlike Thanassoulis and Dyson's model [167], the observed DMU in Zhu's model is not necessarily dominated by its own target. In both the mentioned models, the weights of increasing and decreasing factors are chosen arbitrarily before testing the model, which is a weakness in the DEA methodology. Then, Estellita Lins et al. [53] eliminated this weakness by employing a multiobjective model. They used *a posteriori* preference incorporation method to determine nonradial projections of inefficient DMUs onto the efficient frontier. Also, Estellita Lins et al. [53] proposed another multiple objective target optimization model where the interactive technique is used to directly optimize the target values. Moreover, Quariguasi Frota Neto and Angulo-Meza [138] compared the proposed models by Thanassoulis and Dyson [167], Zhu [213], alternative multiobjective formulation proposed by Angullo Meza [10], and the traditional radial DEA models. As a result, the traditional radial projection models provide useless targets, and they preferred multiobjective formulation despite its high complexity. Lozano and Villa [127] proposed two target-setting DEA methods. The first one is based on the interactive multiobjective method, and the second one is based on the lexicographic method. In these methods, the analytical hierarchy process (AHP) is used to compute the weights of input and output factors. Also, Azadi et al. [14] suggested two target-setting procedures to identify targets for the two-stage network problems. The proposed models extended the work of Stewart [157] and Du et al. [46].

Yang et al. [203] investigated an equivalence between the output-oriented CCR model in the envelopment form and an MOLP problem drawing on the minimax formulation. This MOLP problem is

$$
\begin{aligned}
\text{Max} \quad & \left\{ \sum_{j=1}^{n} \lambda_j y_{1j}, \ldots, \sum_{j=1}^{n} \lambda_j y_{sj} \right\} \\
\text{s.t.} \quad & \sum_{j=1}^{n} \lambda_j x_{ij} \leq x_{io}, && i = 1, \ldots, m \\
& \lambda_j \geq 0, && j = 1, \ldots, n.
\end{aligned}
\qquad (3.17)
$$

The purpose of their study is to implement MOLP (3.17) instead of the output-oriented CCR model and apply an interactive technique for target setting. They used the interactive tradeoff analysis procedure, the gradient projection, and local region search method [200,201,202]. In later years, some researchers applied Yang et al.'s idea [203] for establishing an equivalence between the classical DEA models and MOLP problems, to set target values for the observed DMU using interactive techniques. By drawing on this idea, Malekmohammadi et al. [128] improved the work of Yang et al. [203] using the decrease in the total input consumption and the increase in the total output production. Similar to their work [203], this MOLP problem was solved using the gradient projection method. More decrease in the consumption of total input than obtained results by Yang et al.'s method [203] is advantage of Malekmohammadi et al.'s method [128]. Wong et al. [193] solved MOLP (3.17) using five interactive methods to compare and analyze. These techniques are G-D-F [60], Wierzbicki [192], STEM [22], Tchebychev [153,155], and STOM [130]. As a result, the Tchebychev method is preferred because it is a user friendly and easy-to-understand solution process. Note that the number of the iterations of this approach is more much than others. Still, they [193] argued that the accuracy of the results is more important than the length of the solution process. Moreover, Yang et al. [204] have done a detailed investigation on the features of integrated efficiency and tradeoff analysis in [203]. This research led to the definition of the new efficiency measures which helps to achieve a more realistic target setting. Hosseinzadeh Lotfi et al. [80] established the equivalence between the output-oriented CCR model in the envelopment form and a multiple linear programming drawing on the min ordering formulation. They used Zionts–Wallenius' method [215] to reflect the decision maker's preference to search the MPS in the provided MOLP problem. In addition, Hosseinzadeh Lotfi et al. [77] established an equivalence between the general combined-oriented CCR model and an MOLP problem based on the min ordering formulation. Similar to their own previous work in [80], they [77] applied Zionts–Wallenius' method [215] to design an interactive procedure to locate the MPS on the efficient frontier. Ebrahimnejad and Hosseinzadeh Lotfi [48] provided equivalence conditions between the general combined-oriented the CCR model and a multiobjective linear programming problem using the minimax formulation. They [48] used Zionts–Wallenius' method [215] to solve the MOLP problem and target setting. Note that in this method [48] and Hosseinzadeh Lotfi et al.'s method [77], there exists a tradeoff analysis on both inputs and outputs, but Yang et al.'s method [203] and Hosseinzadeh Lotfi et al.'s method [80] implemented tradeoff only in outputs. Similar to the previous studies on target setting, Ebrahimnejad and Tavana [49] established an equivalence between the DEA model in the presence of undesirable outputs proposed by Seiford and Zhu [148] and an MOLP problem which increases the total desirable outputs and decreases the total undesirable outputs with respect to the minimax formulation. They [49] used the interactive satisfying tradeoff method [129] to search the MPS and target setting.

An et al. [12] introduced a nonoriented network DEA model with considering fairness between two stages for setting target intermediate products and determining frontier projections for a two-stage system. In this regard, an MOLP problem is proposed to analyze the properties of target efficiencies. Then, a Nash bargaining game model was formed for setting the fair intermediate products of the system. Dehnokhalaji et al. [40] presented two multiobjective programming problems to allocate some inputs among DMUs so that their cost efficiencies improve or stay unchanged after resource allocation. Sharahi and Khalili-Damghani [150] developed a decision support system to allocate resources and setting targets across DMUs with network structures in an equitable manner in the presence of uncertainty. Chen and Wang [32] expressed the limitations of traditional DEA in target setting, and developed some new alternative target-setting approaches within the cross-efficiency framework to provide different improvement directions of cross-efficiency. Lozano et al. [126] suggested a target-setting approach that uses the compromise programming method of multiobjective optimization. This method computes the closest target (using l_p metric) to the ideal point.

There is a great deal of research in the DEA methodology about the problems that attempt to determine the best possible output for a given input and the efficiency score of DMUs remains unchanged at the same time. Researchers called this problem as the inverse DEA. In recent years, a large number of DEA models based on an MOLP structure have been presented to solve such problems. The basic idea of the inverse DEA is extending the concepts of the inverse optimization problem [211] to the DEA context. Preliminary work on the inverse DEA is carried out by Golany [68].

Golany [68] suggested an algorithm based on an MOLP problem to help decision makers to set up their own desired outputs. He solved the proposed MOLP problem using the interactive technique. It is worth mentioning that this is the initial usage of the interactive method in the DEA methodology. The algorithm has five steps. Steps 1 and 2 are similar to the first step of the STEM algorithm [22] and using the results of MOLP (3.18)

$$\begin{aligned} \text{Max} \quad & y_{ro}, \quad r = 1, \ldots, s \\ \text{s.t.} \quad & Y\lambda - y_o = 0, \\ & X\lambda \leqslant x_o, \\ & e^T \lambda = 1, \\ & \lambda \geqslant 0, \end{aligned} \qquad (3.18)$$

the payoff matrix is constructed. Step 3 finds an average point by Model (3.19),

$$\begin{aligned} \text{Max} \quad & \sum_{r=1}^{s} (y_{ro}/\bar{Y}_r) \\ \text{s.t.} \quad & Y\lambda - y_o = 0 \\ & X\lambda \leqslant x_o \\ & e^T \lambda = 1 \\ & \lambda \geqslant 0, \end{aligned} \qquad (3.19)$$

where $\bar{Y}_r = (1/n) \sum_{j=1}^{n} y_{rj}$. Step 4 solves a sequence of programs, one for each y_{ro}, to generate a set of efficient points for the given resource vector x_o. Step 5 aids the decision maker to focus on the efficient frontier to make their final selection. The new information provided by the decision maker in Step 5 helps to correct the objective function.

Wei et al. [191] offered an inverse DEA model as an MOLP problem with purpose of determining how much more outputs/inputs level need to be produced when some inputs/outputs level of a DMU are increased and its efficiency level will not change. This research is the first serious discussion and analysis of the inverse DEA problem. Their idea was based on the work of Zhang and Cui [210]. Then, Yan et al. [197] proposed an inverse DEA model by introducing additional preference cones to Wei et al.'s model [191]. Also, they showed that under special conditions, their own model is converted to the original inverse DEA model presented in [191]. Moreover, Jahanshahloo et al. [94] developed the presented method by Yan et al. [197] to identify how much outputs level are changed when some or all inputs level of a DMU increase and its efficiency level is improved. Jahanshahloo et al. [95] estimated the best level of inputs when a DMU changes some or all of its outputs level, while preserving or increasing its efficiency level. Furthermore, they introduced another method to identify extra inputs when outputs are estimated drawing on the models proposed by Yan et al. [197] and Jahanshahloo et al. [94]. Jahanshahloo et al. [87] suggested an inverse DEA model in the presence of undesirable factors to control the changes of inputs/outputs level of a DMU when some of outputs/inputs level are changed and the efficiency level is preserved. Based on Yan et al.'s model [197], Jahanshahloo et al. [96] investigated the sensitivity of efficiency classification of efficient and inefficient DMUs when all or some inputs/outputs level are changed. Also, by nondominated solutions of the presented models, the lower bounds and upper bounds of inputs and outputs variations range are obtained. Hadi-vincheh and Foroughi [70] suggested a generalized model to control the changes of inputs/outputs level of a DMU, and at the same time its efficiency level is unchanged. Their proposed MOLP is able to estimate inputs/outputs level, regardless of the efficiency or inefficiency of DMUs. Therefore, they [70] claimed that their method has some advantage in comparison with the other previous methods. Hadi-vincheh et al. [71] using a contravention example showed that if the weak efficient solution of the proposed MOLP problem by Wei et al. [191] is not strongly efficient, then this method is not correct. Also, they proposed sufficient conditions for inputs estimation when outputs level increase. These proposed conditions are based on the strongly efficient solution and certain weakly efficient solution of the MOLP problems. Lertworasirikul et al. [118] suggested an inverse BCC model. This model is different from the previous ones because it takes into account the relative efficiency of all DMUs. The proposed inverse BCC model is used to determine the best possible inputs level of a DMU so that, by changing outputs level, the relative efficiency level of all DMUs remains unchanged. However, Ghiyasi [65] pointed out the proof of the proposed theorem in [118] has some mistakes. He corrected the problem by a simpler proof. Ghobadi and Jahangiri [67] reviewed the theoretical results and applications of the inverse DEA models, and extended these models in the presence of fuzzy data. Jahanshahloo et al. [100] addressed the inverse version of the dynamic DEA model proposed by Emrouznejad and Thanassoulis [51] and Jahanshahloo et al. [99], and provided necessary and sufficient conditions for input-estimation and output-estimation under intertemporally dependence assumption. Also, they introduced a new optimality notion (periodic weak Pareto optimality) for MOLP problems. Ghiyasi [66] proposed inverse DEA models when price information is available. In fact, these models guarantee not only the fixed technical efficiency but also the unchanged cost (revenue) efficiency of all DMUs. Amin et al. [5] proposed a method to deal with target setting by considering the inverse DEA model as a multiobjective programming problem, and then, utilized the goal programming approach for solving the problem when there exists a preference for saving specific resources. Wegener and Amin [188] developed an inverse DEA model to optimize greenhouse gases emissions. This model minimizes the overall emissions generated by DMUs to produce a certain level of outputs, given that the firms' efficiency will not be deteriorated.

3.3.3 Value efficiency

The classical DEA models are value free. In other words, they do not have any assumption that inputs or outputs are more important than one another. However, each efficient DMU is more important than an inefficient one in evaluating by the traditional DEA, but in the real world this case is not established because all factors may not have the same value. Hence, researchers make an attempt to incorporate preference information in DEA. The value efficiency analysis is one of the available approaches based on the MODM methodology to incorporate preference information of the decision maker in the DEA models that were initially presented by Halme et al. [72]. Seeing that the value function fails to have an explicit form, it is approximated by an indifference curve at the MPS. Therefore, the MPS plays a key role in the value efficiency analysis. The value efficiency score is calculated for each DMU in comparison with DMUs that have the same value as the MPS. Thus, it determines how much the input and output values need to be improved to reach a point on the indifference contour of the value function that goes through the MPS. Consequently, researchers proposed different approaches to approximate the value efficiency score.

Halme et al. [72] proposed a method to calculate the value efficiency analysis of DMUs. They approximated the contour using the possible tangents of a value function at the MPS. This method is based on these assumptions: the value function must be pseudoconcave, strictly increasing in outputs, strictly decreasing in inputs, and reaching its maximum at the MPS on the efficient frontier. With the value function being pseudoconcave, the value of units in PPS is less or equally preferred than those in the MPS. In other words, the obtained value efficiency scores are optimistic estimates of the true ones. Halme et al. [72] provided the following valuable theorem to determine the value efficiency score.

Theorem 1. Let $u^* = (-x^*, y^*) \in PPS$ be the decision maker's MPS. Then $u = (-x, y)$, an arbitrary unit in PPS is value inefficient with respect to any strictly increasing pseudoconcave value function $v(u)$, $u = (-x, y)$ with a maximum at point u^*, if the optimum value Z^* of the following problem is strictly positive:

$$\begin{aligned}
\text{Max} \quad & Z = \sigma + \epsilon(1^T s^+ + 1^T s^-) \\
\text{s.t.} \quad & Y\lambda - \sigma w^y - s^+ = y, \\
& X\lambda + \sigma w^x + s^- = x, \\
& A\lambda + \delta = b, \\
& \epsilon > 0, \quad (\text{"Non-Archimedean"}) \\
& s^+ \geq 0, \, s^- \geq 0, \\
& \lambda_j \geq 0 \text{ if } \lambda_j^* = 0 \;\; j = 1, \ldots, n, \\
& \delta_j \geq 0 \text{ if } \delta_j^* = 0 \;\; j = 1, \ldots, k,
\end{aligned} \quad (3.20)$$

where $\lambda^* \in \Lambda = \{\lambda \mid \lambda \in \mathfrak{R}_+^n, A\lambda \leq b\}$, μ^* corresponds to the MPS: $y^* = Y\lambda^*$, $x^* = X\lambda^*$.

Proof: See [72].

Model (3.20), which shows the simple modification of the classical DEA model, makes it possible to take into account the value judgment according to the MPS.

Joro et al. [103] proposed a model to estimate lower bounds for the value efficiency score. Therefore, a set of bounds on the true value efficiency score is obtained. Furthermore, they presented an interactive technique to find the true value efficiency score. Halme and Korhonen [73] proposed two methods to incorporate preference information into the value efficiency. First, they augmented additional preference information using weight restrictions in the dual of Model (3.20) to improve the estimation of the value efficiency score. Note that the restricted value efficiency model may be infeasible because the weight constraints are in conflict with the MPS or the basic value efficiency model. Second, they used the set of optimal weights for the MPS. For this reason, they introduced the optimality conditions for the MPS. Both monotonicity and pseudoconcavity assumptions fail to produce enough information to evaluate the change in input and output values. Therefore, Korhonen and Syrjanen [115] suggested two sets of additional assumptions, homogeneity and linearity assumptions, to produce a function form of the value function. Zohrebandian [216] developed Halme et al.'s approach [72] to measure the value efficiency score more precisely. He introduced an MOLP problem and solved it drawing on Zionts–Wallenius' method [215] to identify the MPS. Also, without solving any linear programming problem, the value efficiency score of each *DMU* is obtained. Jahanshahloo et al. [91] extended the work of Halme et al. [72] to the interval value efficiency analysis. Hence, they suggested two models to determine the upper and lower bounds of the value efficiency score to classify DMUs into three classes.

3.3.4 Secondary goal models

In the cross-efficiency approach proposed by Sexton et al. [149], there are some basic advantages such as it provides a unique ordering of DMUs and eliminates unrealistic weights without requiring any weight restrictions. However, it has a principal disadvantage when the CCR model in the multiplier form has alternative optimal solutions, which results in a different rank ordering. To overcome this problem, Doyle and Green [45] initially suggested the incorporation of the secondary objectives in the multiplier form of DEA models. Afterwards, several researchers proposed a large number of the secondary goal formulation with various purposes to determine unique optimal solutions, but none of them guarantee that the obtained optimal solution is unique. It is of interest to note that the most common usage of the lexicographic method in the DEA context is the secondary goal DEA models to avoid the alternative optimal solutions.

Doyle and Green [45] presented three alternative ways to formulate the secondary goal models by aggressively minimizing the objective of all other DMUs. The first one is

$$\text{Min} \quad A_o = \sum_{j=1, j \neq o}^{n} \frac{\sum_{r=1}^{s} u_r y_{rj}}{\sum_{i=1}^{m} v_i x_{ij}}$$

$$\text{s.t.} \quad \sum_{r=1}^{s} u_r y_{rj} - \sum_{i=1}^{m} v_i x_{ij} \leq 0, \quad j = 1, \ldots, n, \ j \neq o \quad (3.21)$$

$$\sum_{r=1}^{s} u_r y_{ro} - E_o^* \sum_{i=1}^{m} v_i x_{io} = 0$$

$$u_r \geq 0, \ v_i \geq 0, \quad r = 1, \ldots, s, \ i = 1, \ldots, m,$$

where E_o^* is the efficiency score DMU_o obtained by the CCR model. Model (3.21) leads to nonlinear fractional programming that is not converted to linear programming. In the second secondary goal model, instead of the fractional form, they adopted the indifference between the numerator and denominator of the objective function of Model (3.21) as follows:

$$\text{Min} \quad B_o = \sum_{r=1}^{s} \left(u_r \sum_{j=1, j \neq o}^{n} y_{rj} \right) - \sum_{i=1}^{m} \left(v_i \sum_{j=1, j \neq o}^{n} x_{ij} \right)$$

$$\text{s.t.} \quad \sum_{i=1}^{m} v_i x_{io} = 1,$$

$$\sum_{r=1}^{s} u_r y_{rj} - \sum_{i=1}^{m} v_i x_{ij} \leq 0, \quad j = 1, \ldots, n, \ j \neq o, \quad (3.22)$$

$$\sum_{r=1}^{s} u_r y_{ro} - E_o^* \sum_{i=1}^{m} v_i x_{io} = 0,$$

$$u_r \geq 0, \ v_i \geq 0, \quad r = 1, \ldots, s, \ i = 1, \ldots, m.$$

This model is a linear programming. Finally, they considered the fractional form of the secondary goal model as

$$\text{Min} \quad C_o = \frac{\sum_{r=1}^{s} \left(u_r \sum_{j=1, j \neq o}^{n} y_{rj} \right)}{\sum_{i=1}^{m} \left(v_i \sum_{j=1, j \neq o}^{n} x_{ij} \right)}$$

$$\text{s.t.} \quad \sum_{r=1}^{s} u_r y_{rj} - \sum_{i=1}^{m} v_i x_{ij} \leq 0, \quad j = 1, \ldots, j \neq o$$

$$\sum_{r=1}^{s} u_r y_{ro} - E_o^* \sum_{i=1}^{m} v_i x_{io} = 0$$

$$u_r \geq 0, \ v_i \geq 0, \qquad r = 1, \ldots, s, \ i = 1, \ldots, m.$$

(3.23)

Model (3.23) can be converted to linear programming using the Charnes and Cooper transformation [27]. The benevolent formulation for Models (3.21), (3.22), and (3.23) are determined by maximizing the secondary goals A_o, B_o, and C_o, respectively.

Liang et al. [120] extended Doyle and Green's secondary goal models [45] by introducing different objective functions including minimizing the total deviation from the ideal point, minimizing the maximum efficiency score, and minimizing the mean absolute deviation. Wang and Chin [183] proposed four secondary goal models that were minimizing or maximizing the total deviation from the ideal point, minimizing or maximizing the squared sum of deviations from the ideal point, minimizing the maximum deviation from the ideal point for the benevolent form or maximizing the minimum deviation from the ideal point for the aggressive form, and minimizing the mean absolute deviation from the ideal point. These models being based on the ideal point are more realistic than Liang et al.'s models [120]. Another problem of the aggressive and benevolent models is that these two models may lead to two different rank priority results; therefore, selecting one of them is a difficult task. To eliminate this problem, Wang and Chin [182] proposed a neutral secondary goal model for the cross-efficiency evaluation, in which each DMU determines the weights purely from its own point of view. Another approach to remove the above-mentioned problem was proposed by Yang et al. [199] which composed an interval cross-efficiency matrix using two new secondary goal models. To rank all DMUs by this matrix, they considered it as a stochastic MCDM problem and then solved it drawing on the stochastic multicriteria acceptability analysis method [116]. Oral et al. [132] presented an improvement to the secondary goal model using the maximum resonated appreciative model for the cross-efficiency. Also, they showed the proposed model is better than the benevolent formulation. Another one of the drawbacks in the previous aggressive and benevolent secondary goal models is that the achievement of a better ranking is less important than the maximization of the individual score. To tackle this problem, Wu et al. [195] proposed a mixed integer linear programming DEA model where there principle of rank priority has more preference than the efficiency score. Furthermore, Contreras [36] improved the model proposed by Wu et al. [195], in which the attention is not only focused on determining the best individual efficiency score but it also minimizes the efficiency score of the closest DMU to the DMU under evaluation to optimize its rank position. Whashio and Yamada [187] proposed a model called the rank-based measure to evaluate DMUs. They determined the individual maximum score after keeping the best ranking, using some mathematical programming. The proposed model can be considered as a secondary goal model in the cross-efficiency approach. Wang et al. [185] proposed some secondary goal models based on a neutral way to minimize the virtual disparity in the cross-efficiency evaluation. Also, they showed that the proposed DEA models reduce the number of zero weights. Alcaraz et al. [2] proposed two secondary goal models to obtain the best and the worst rank of each DMU. In this method, instead of determining a single rank for each DMU, a range is determined for ranking. The neutral DEA model proposed by Wang and Chin [182] maximizes the relative efficiency of each output; therefore, only the number of zero weights for outputs is reduced, and it cannot reduce the number of zero weights among inputs. For removing this drawback, Wang et al. [184] extended Wang and Chin's model [182] to reduce the number of zero weights for both inputs and outputs using a simultaneously input- and output-oriented weight determination secondary goal model. According to the work of Wang and Chin [182] and Orkcu and Bal [134] in which each DMU make an effort on its own to choose of weights without considering the effects on the other DMUs, Wu et al. [196] presented a weight-balanced secondary goal model. This model fails to guarantee to maximize the individual efficiency of the DMU under evaluation but also reduces the differences in the weighted inputs and weighted outputs and reduces the number of zero weights as well. To avoid unrealistic and unreasonable weight in the cross-efficiency evaluation, Ramon et al. [141] extended the multiplier bound approach in [140] into the secondary goal model. This model is similar to Wang and Chin's model [182] in choice of weights without considering the effects on the other DMUs. Indeed, the model makes an attempt to avoid the large differences in weights and avoid the zero weights. Lam [117] suggested a model for ranking DMUs based on DA, super efficiency, and a mixed integer linear programming to obtain an optimal weight set and implied it in cross-efficiency evaluation. In this model, the obtained weight set can reflect the capabilities of the efficient DMU under the consideration. Moreover, this method makes an attempt to preserve the original classificatory results of DEA and also it produces much fewer zero weights than the classical DEA models. Orkcu and Bal [133] modified the works of Bal and Orkcu [15] and Bal et al. [16] into the secondary goal models based on the goal programming method. The proposed two secondary goal models [133] are better than previous mentioned models because they are based on the multicriteria DEA model and have three different efficiency concepts [119]. Lim [121] extended the aggregative model to include the benevolent case in addition to it. Then, he developed his previous work [121] to present another type of the aggressive and benevolent secondary goal models based on minimax formulation. The proposed models determine the optimal weights that maximize the efficiency score of the individual DMU under the evaluation and subsequently minimize (or maximize) the cross-efficiency of the best (or worst) peer DMU. To select the symmetric weights [42] in a secondary goal model, Jahanshahloo et al. [88] added the symmetric constraints into output weights to reward symmetric output as long as linearity is preserved. Also, Guo and Wu [69] extended it to the secondary goal model in the presence of undesirable outputs by restrictions presented to realize a unique ranking of DMUs through the new Maximal Balance Index based on the optimal shadow prices. Wu et al. [194] addressed the deficiencies of the traditional benevolent and aggressive models and proposed appropriate secondary goal models. To do it, they gave a target identification model to get desirable and undesirable cross-efficiency targets for all DMUs. Then, several secondary goal models with different purposes were presented considering both desirable and undesirable identified targets for all DMUs. Liu et al. [125] introduced a technique for the cross-efficiency evaluation considering undesirable outputs. An aggressive model was given to guarantee the uniqueness of the optimal solution. This approach can not only perform the cross-efficiency evaluation for DMUs with undesirable outputs but also pays attention to each DMU's ranking preference. Zero weights make much information on inputs and outputs be ignored in the cross-efficiency evaluation. Therefore, Lin et al. [122] presented an interval method for determining weights in the cross-efficiency evaluation, which not only ensures a unique weight set for positive input and output data but also reduces the number of zero weights maximally without imposing any prior weight restriction. Carrillo and Joro [23] suggested a secondary goal from a neutral perspective. The proposed formulation considers two hypothetical units, ideal and anti-ideal, and, respectively, forces the efficiency score of DMUs to be as small and large as possible in a simultaneous way. Ang et al. [11] developed group efficiency and group cross-efficiency models to evaluate Taiwan hotel chains. To overcome the nonuniqueness of the cross-efficiency evaluation, aggressive secondary formulations are built on the basis that groups compete with each other. Orkcu et al. [135] presented a neutral cross-efficiency model to measure the efficiency of the two-stage network systems. This model is able to rank DMUs in substages and decompose the cross-efficiency measure of the system into the product of those of stages.

3.3.5 Common set of weights

The traditional DEA models are optimistic. To deal with this difficulty, the concepts of the weight restriction and value judgment were introduced. Also, another method to deal with this problem is the CSW methods, which was initially introduced by Charnes et al. [30] and Kornbluth [111] by including cone restrictions to MOLP problems. In common weight methods, only one hyperplane is generated as the frontier to evaluate all DMUs. Hence, these methods are utilized to rank all DMUs on a unified scale. The advantages are (1) common weights models reduce the computation in comparison with the standard DEA models, and (2) common weights models have higher discrimination power than the classical DEA models. Almost all common weights DEA models are based on MOP problems which have a strong theoretical background.

Charnes et al.'s model [30] was a developed form of the CCR model by introducing the additional weight restrictions in predetermined spaces. Afterwards, based on this model [30], Kornbluth [111] suggested an MOLP problem including cone restrictions to reduce weight flexibility in the MOLP problem. The restricted MOLP problem is

$$\text{Max} \quad \left\{ \sum_{r=1}^{s} u_r y_{r1}, \ldots, \sum_{r=1}^{s} u_r y_{rn}, -\sum_{i=1}^{m} v_i x_{i1}, \ldots, -\sum_{i=1}^{m} v_i x_{in} \right\}$$

$$\text{s.t.} \quad \sum_{r=1}^{s} u_r y_{rj} - \sum_{i=1}^{m} v_i x_{ij} \leqslant 0, \qquad j = 1, \ldots, n,$$

$$\sum_{i=1}^{m} v_i x_{ij} \geqslant q_{lim}, \qquad j = 1, \ldots, n, \qquad (3.24)$$

$$u \in \bar{U}, v \in \bar{V},$$

q_{lim} some arbitrary positive lower limit,

where $\bar{U}$ and $\bar{V}$ are cone restrictions. The cone restrictions are more general than the bound restrictions proposed by Dyson and Thanassoulis [47]. Note that MOLP (3.24) gives more information than the classical DEA models because these models are optimistic and their objectives attempt to find only the maximum efficiency score for the DMU under evaluation but do not guarantee to give the best efficiency for other DMUs.

Belton and Vickers [19] proposed a mathematical linear programming based on the 1-norm to test the Pareto Koopmans efficiency of the DMU under evaluation. The obtained model is the dual of the proposed model by Charens et al. [26]. Moreover, Stewart [156] changed the linear formulation to test the Pareto Koopmans efficiency in Belton and Vickers's paper [19] by applying the infinity norm instead of the 1-norm in calculating deviations; as a result, the dual of this linear programming is equal to the input-oriented BCC model in the envelopment form. This indicates that the original DEA ratio model and testing the linear formulation for the Pareto Koopmans efficiency based on the distance form are equal. Roll and Golany [145] suggested a CSW model based on an MOLP problem. For solving the problem, the sum weighted method is used. Their model provides a true common weights vector because there is no need to predict either the reference point or the preference weights. Doyle [43], using the Doyle and Green's method [45], presented a new CSW model. The model [43] involves the weakness of the predetermination of the reference point and the weights. Thus, this model is less convenient than Roll and Golany's model [145]. Sinuany-Stern et al. [152] proposed a two-stage ranking method based on linear DA. Furthermore, Sinuany-Stern and Friedman [151] developed a nonlinear DEA-DA model for ranking DMUs. This model is based on ratios and provides CSW. Hosseinzadeh Lotfi et al. [78] presented an MOP problem to find CSW in DEA. In this method, instead of solving n linear programming problem to obtain efficiency scores, only one nonlinear problem is solved. Similarly, Chiang and Tzeng [35] solved the mentioned MOP problem in [78] by the maximin formulation and then obtained a nonlinear programming problem. Also, for solving the MOP problem in [78] and [35], Jahanshaloo et al. [98] introduced another approach that used only one nonlinear programming problem instead of n linear DEA models to obtain the efficiencies all DMUs based on common weights. Then, they suggested a method to rank all efficient DMUs by solving only two problems. Furthermore, Davoodi and Zhiani Rezai [39] used the MOP problem addressed in [78,35], and [98] to provide common weights. Unlike models in [78,35], and [98], the MOP problem in their paper [39] is converted into the linear problem drawing on the goal programming method.

Kao and Hung [105] presented a true common weights model based on the compromise solution approach and fractional form of efficiency to rank all DMUs. The three important drawbacks of this method are, solving $n + 1$ optimization problems, determining the best norm, and obtaining nonlinear programming. Also, an advantage of this model is that no additional prior information is needed. The proposed MOLP problem in [105] was derived from the classic fractional DEA model; as a result, a nonlinear model is constructed. To deal with this problem, Chen et al. [33] suggested a CSW model through the difference between inputs and outputs instead of the ratio form, using different norms. From a computational point of view, this model has an important advantage over the Kao and Hung's model [105]. Also, Chen et al.'s model [33] does not need to predetermine the reference point and weights that are the advantages of this method. Zohrebandian et al. [217] extended Kao and Hun's model [105] to a linear model. Similarly, the compromise solution approach is used to solve the proposed MOLP problem. In the previous common weights DEA models, it was assumed that there is no uncertainty in the input and output data. To develop the previous models, Omrani [131] introduced a CSW DEA model in the presence of uncertain data. To find out common weights, they applied Kao and Hung's model [105] and the goal programming approach. Salehi et al. [146] developed Kao and Hung's model [105] to find CSW of DMUs under interval uncertainties on inputs and outputs data. To avoid nonlinearity, they used the 1-norm, for the distance between actual efficiency and ideal efficiency scores of each DMU.

The purpose of Kao's work [104] attempts to design a Malmquist productivity index based on Kao and Hung's model [105]. In this method, the productivity change of all DMUs over time is compared on a common basis. Using a numerical example, he [104] showed that the Malmquist productivity index using the CSW model produced more reliable results than the conventional Malmquist productivity index models. Then, Yang et al. [198] improved Kao's model [104].

Liu and Peng [123] introduced a CSW model to maximize the group's efficiency score. Yet, the proposed model has two difficulties. First, it may have alternative solutions. Second, more than one DMU may be efficient. In addition, Liu and Peng [124] proposed a systematic procedure based on Liu and Peng's method [123] to obtain a preferable and robust ranking method. Also, Payan [137] and Ramezani-Tarkhorani et al. [139] showed that the proposed model by Liu and Peng [123] is not generally correct; then, they [137,139] improved it. Moreover, Jahanshahloo et al. [89] used the opinion of Liu and Peng [123] to present two common weights ranking methods. In the first and second methods, an ideal line and a special line are, respectively, defined. All DMUs are compared with these lines to obtain their efficiency score and rank them with these results. Jahanshahloo et al. [101] introduced a ranking method drawing on a CSW model based on an MOLP problem that is solved by Zionts–Wallenius' method [215]. The interactive method incorporates preference structures of the decision maker into the common weights model. Therefore, it has an advantage over the previous approaches in the CSW methods. Chiang et al. [34] implemented the MOLP problem mentioned in Roll and Golany [145], Hosseinzadeh Lotfi et al. [78], Chiang and Tzeng [35], and Jahanshaloo et al. [98] to obtain CSW in a DEA model using a separation vector. Unlike the previously mentioned nonlinear common weights models, the separation vector model is linear. Furthermore, in comparison with Liu and Peng's model [123], this model utilized fewer auxiliary variables, and thus, it is applicable.

Hatefi and Torabi [75] proposed an MCDA-DEA model based on the CSW method, which has more discriminating power than the classical DEA models to construct composite indicators. Although this model is capable of ranking efficient DMUs, it does not require adjusting any parameter; that is, they removed the drawbacks of Zhou et al.'s model [212]. Cook and Zhu [37] proposed a CSW model based on the goal programming method. Ramon et al. [142] proposed another CSW model by minimizing deviations of the common weights from the weights of the efficient DMUs provided by the CCR model. They used several norms to measure these differences; therefore, DMUs get several ranks. Sun et al. [165] suggested two approaches to generate a set of common weights by introducing two virtual DMUs, the ideal DMU and the anti-ideal DMU. The first one determines the weights by minimizing the distance of all DMU from the ideal DMU, and the second one determines the weights by maximizing the distance of all DMUs from the anti-ideal DMU. Note that the two different sets of optimal weights may generate different efficiency for a DMU, and consequently, they would have two different rank; however, Sun et al. [165] found that two kinds of ranking orders are the same. Hosseinzadeh Lotfi et al. [76] proposed a mathematical programming based on a CSW method in resource allocation. The proposed model is solved using the goal programming approach. Also, Hatami-Marbini et al. [74] proposed a common weights model to optimize the resource contraction so that the efficiency of all DMUs is improved or equal to the efficiency prior to change. Similar to most of the CSW models, this proposed model utilized the goal programming approach to specify a common reduction of resources. Also, the proposed model has several advantages over Amirteimoori and Emrouznejad's model [8]. Ghazi and Hosseinzadeh Lotfi [62] suggested a CSW DEA model to a budget allocation problem, in which the available budget is reallocated among all DMUs and at the same time the central authority aims to maximize the total efficiency scores of DMUs. To avoid an inappropriate reallocation, the model is restricted by incorporating the opinion of the decision maker to model using budgetary constraints. Kiani Mavi et al. [109] introduced an approach to find CSW in a two-stage network model regarding goal programming, considering the undesirable inputs, intermediate products, and outputs in the context of big data. Razavi Hajiagha et al. [143] suggested a method to determine CSW in a multiperiod DEA. First, a CSW problem is formulated as a multiobjective fractional programming problem. Then, a multiperiod form of the problem is formulated and the mean efficiency of DMUs is maximized while their efficiency variances are minimized.

Jafarian-Moghaddam and Ghoseiri [85] introduced a fuzzy dynamic multiobjective DEA model in which data are changed in T periods. Therefore, the proposed MOLP problem will have $n \times T$ objectives. This MOLP problem is solved using Zimmerman's method [214]. As a result, the obtained model is nonlinear, and using optimal common weights, the efficiency score of all DMUs is calculated at the same time. Then, Jafarian-Moghaddam and Ghoseiri [84] extended the above-mentioned method to fuzzy data by implementing the original fuzzy dynamic multiobjective DEA model twice. Also, Wang et al. [181] extended Jafarian-Moghaddam and Ghoseiri's model [85] in the two-stage fuzzy dynamic multiobjective DEA model.

The classical DEA models classify all DMUs into two sets, efficient and inefficient. However, in these models, usually more than one DMU is recognized as efficient. Hence, the models fail to provide more information about the efficient DMUs. To resolve this problem, ranking [1,38] and weight restrictions [38,168,169] methods are introduced. It is of interest to note that CSW models decrease a number of efficient DMUs; therefore, they can be considered as ranking models. In some studies, the main concern is to find a single efficient DMU, called the most efficient DMU, based on CSW models, instead of ranking all efficient DMUs. The most efficient DMU is the best one that can be determined by ranking methods. In many applications, finding the most efficient DMU is required, particularly when the decision maker wants to select only one DMU among the available DMUs. Yet, similar to ranking models, in the proposed models for finding most efficient DMU, there is not a common criterion to specify the best one, which is the weakness of these models.

Li and Reeves [119] presented a multiple criteria DEA model without prior information about weights. The proposed MOLP problem utilized three objectives: minimization of deviations, minimax, and minsum formulations under the same constraints of the classical DEA models. Therefore, the discrimination power of the cited model is improved in comparison with classical DEA models. Furthermore, to improve the discrimination power of DEA models, Despotis [41] introduced the global efficiency approach that is the natural extension of the original DEA model in the view of multiobjective mathematical programming context. Also, he presented a global index to rank efficient DMUs and compared the global efficiency approach with the multicriteria DEA model in [119] and the cross-efficiency approach. Li and Reeves's model [119] fails to optimize all objectives, simultaneously. To tackle this problem, Bal et al. [16] proposed a model by integrating the goal programming method and the CCR model. However, their model has some drawbacks such as the issue of zero weights for variables in DMUs, not achieving the similar results with Li and Reeves's model [119], and investigating the case of the VRS technology. To deal with these problems and to improve its discrimination power, Ghasemi et al. [61] presented a biobjective weighted model. The proposed model is almost similar to Li and Reeves' model but only considered the second and third objectives in it.

Karask and Ahiska [106] suggested an algorithm based on a parametric CSW model to rank DMUs. They claimed that the presented algorithm is convergent to a single efficient DMU and saves a great amount of computation in comparison with the cross-efficiency approach. Then, Toloo [171] illustrated that, in some cases, Karsak and Ahishka's method [106] finds an inefficient unit as the most efficient DMU, incorrectly. To deal with this issue, he [171] proposed an integrated mixed integer linear model with more discriminating power without explicit inputs to obtain the most CCR-efficient DMU. Using a numerical example, Amin et al. [7] showed that Karask and Ahiska's ranking algorithm [106] is not always convergent. Then, they suggested an improved model to specify the best efficient DMUs without considering any parameter. The method presented in [7] is capable of ranking all DMUs by removing the most efficient DMU in each iteration. Also, Amin [3] showed that the proposed method in [7] may obtain more than one efficient DMU. Then, he modified it and proved a theorem to indicate his proposed model gives only one efficient DMU. Ertay et al. [52] in the last step of the methodology utilized a trial and error model to obtain the single most efficient DMU. Remember that this method needs to solve n linear programming; in other words, one linear programming for each DMU. To determine the most CCR-efficient DMU, Amin and Toloo [6] introduced a common weights DEA model that has less computation than the previously cited methods. Unlike Ertay et al.'s model [52], their model [6] has no parameters. However, the model is useful only for the CRS technology. Then, You et al. [205] showed that there are two mistakes in the work of Amin and Toloo [6]. First, the process of proving the theorem is incorrect, and second, the proposed model was incorrectly built. Also, Amin [4] exhibited that the proposed DEA model in Amin and Toloo [6] may obtain more than one efficient DMU, and he then presented an improved integrated mixed integer nonlinear programming DEA model to eliminate this problem. To overcome the problems in [6] and [4], Foroughi [56] proposed a mixed integer linear programming. This model is more discriminative than the mentioned models, and also, it is always feasible. The results of this model are similar to the super efficiency DEA model (AP) [9] in ranking method and has some advantages over the super efficiency model. To fill the gap of Amin and Toloo's model [6] which finds only the CCR-efficient DMU, Toloo and Nalchigar [178] proposed a mixed integer linear programming based on a common weights DEA model to find the most BCC-efficient DMU. On the grounds that this model is employed in the VRS technology, it has a wide range of applications. Also, Toloo [170] showed that Toloo and Nalchigar's model [178] has two drawbacks. First, it fails to guarantee the production of a single BCC-efficient DMU, and second, a binary variable cannot be considered as a deviation from the efficiency. Therefore, a new integrated mixed integer linear DEA model based on a CSW model was introduced to determine the most BCC-efficient DMU. At last, he [170] proved that the new model specifies only one single BCC-efficient DMU. In addition, Foroughi [57] proposed a generalized model to remove the infeasibility problem of Toloo and Nalchigar's model [178]. The generalized new model is always feasible with a finite optimal solution. Also, the proposed models in Amin and Toloo [6], Toloo and Nalchigar [178], and Foroughi [56] (when it is feasible) are specific cases of this generalized model. He [57] offered an algorithm considering the problem of alternative optimal solutions and determining all DMUs that can be considered as the most efficient DMU. Asosheh et al. [13] combined two famous managerial methodologies, balanced scorecard (BSC) and DEA, and introduced a method for IT project selection. Hence, they proposed an integrated DEA model based on a CSW model to determine the most efficient IT project in the presence of cardinal and ordinal data. This method is less computational in comparison to other ranking methods, which is an advantage of this method. Toloo and Nalchigar [179] improved the work of Farzipoor Saen [55] and proposed a new integrated DEA model capable of determine the most efficient supplier in the presence of both cardinal and ordinal data. As an advantage, their method specifies the best supplier by solving only one mixed integer linear programming problem. Moreover, they introduced another DEA model capable of recognizing the most BCC-efficient DMU in the presence of imprecise data. Furthermore, Toloo [174] illustrated the drawbacks of Toloo and Nalchigar's model [179] and suggested an improved mixed integer linear programming DEA model to identify the most efficient DMU in the presence of both cardinal and ordinal data. This model is more effective than the previously proposed models, computationally. Wang and Jiang [186] proposed three mixed integer linear programming problems to determine the most efficient DMU under different returns to scale technologies. These models have some advantages over the previous DEA models, including feasibility, fewer variables and constraints, more reliability, more practicality, not requiring the specification of any assurance region for input and output weights, and correct selection of the most efficient DMU. Toloo and Kresta [177] presented a mixed integer linear programming problem to find the most efficient DMU based on common weights without explicit outputs and only with pure input data set. Toloo and Ertay [176] suggested a new approach to find the most cost-efficient DMU and applied it on the automotive industry in Turkey. From the theoretical point of view, the non-Archimedean infinitesimal provides a lower bound to keep input and output weights as positive. Remember that if the non-Archimedean epsilon is ignored, the proposed models are not able to determine either the most efficient DMU or their rank. Based on this note, to determine the non-Archimedean epsilon for Amin's model [4] and Foroughi's model [56], Toloo ([175] presented new mixed integer linear models to assess the lower bound weights. Based on the previous work in [170], Toloo [173] proposed an integrated linear programming problem to specify candidate DMUs to be the most efficient DMU and an integrated mixed integer linear programming problem to determine the most efficient DMU. These models exclude the non-Archimedean epsilon, and the optimal solution is conveniently obtained. Toloo [172] introduced a minimax mixed integer linear programming model based on Wang and Jiang's model [186] to obtain the most efficient DMU. In the presented model, the input and output weights are strictly positive and no lower bound is necessary to be calculated; therefore, the problem of non-Archimedean epsilon is eliminated.

3.3.6 DEA-discriminant analysis

DA is originally a statistical tool that can predict group membership of a newly sampled observation. Several researchers proposed some nonparametric DA approaches that specify the situation of new observation. For the first time, Retzlaff-Roberts [144] investigated relationships between two unrelated linear programming techniques, DEA and DA. DEA is a nonparametric method to evaluate relative efficiencies of a group of DMUs and classify them into two sets, efficient and inefficient, and DA is a method to predict group membership of new observation. In DA membership, all observations $z_j = (x_j, y_j)$ $(j = 1, \ldots, n)$ are specified and classified into two groups (G_1 and G_2) which have n_1 and n_2 members ($n = n_1 + n_2$). A discriminant function estimates weights or parameters by minimizing their group misclassification. Then, by using them, new observations are classified. There are several techniques to classify observations in DA, one of them is linear programming, which has some advantages over other methods. The two mathematical techniques DA and DEA share some significant properties, although they are impertinent. With the concept of goal programming being common for both DEA and DA methodologies, most of the proposed nonparametric DEA-DA models are based on the goal programming method. In fact, DEA-DA models can be considered as CSW models.

Retzlaff-Roberts [144] discussed some similarities and differences between DA and DEA methodologies. He showed that when there is no prior knowledge of group membership, the basic DA model is converted to the CCR model in the multiplier form. Also, in this research, another nonparametric DA model was proposed based on the DEA methodology when there is a prior knowledge of group membership. The group membership of the new observation will be predicated by the obtained optimal weights. Note that these nonparametric models are more restrictive than basic DA models because they are based on a ratio form of DEA models. To overcome the problem of the parametric goal programming DA models, Sueyoshi and Kirihara [164] offered a new type of nonparametric goal programming DA model.

Because of the similarities between the additive model and the minimized sum of deviations model in [59], Sueyoshi [158] proposed a two-stage method based on Additive DEA model, referred to DEA-DA, as follows:

Stage 1: Classification and overlap identification
Proposed DEA-DA model in Stage 1 is

$$
\begin{aligned}
\text{Min} \quad & \sum_{j \in G_1} s_{1j}^+ + \sum_{j \in G_2} s_{2j}^- \\
\text{s.t.} \quad & \sum_{i=1}^{k} \alpha_i z_{ij} + s_{1j}^+ - s_{1j}^- = d, \quad j \in G_1, \\
& \sum_{i=1}^{k} \beta_i z_{ij} + s_{2j}^+ - s_{2j}^- = d - \eta, \quad j \in G_2, \\
& \sum_{i=1}^{k} \alpha_i = 1, \\
& \sum_{i=1}^{k} \beta_i = 1, \\
& s_{1j}^+, s_{1j}^-, s_{2j}^+, s_{2j}^- \geq 0, \quad j = 1, \ldots, n, \\
& \alpha_i, \beta_i \geq 0, \quad i = 1, \ldots, k, \\
& d \text{ free.}
\end{aligned}
\tag{3.25}
$$

Considering α^*, β^*, and d^* are the optimal solutions of Model (3.25). In this stage, the classification and overlap identification are specified as follows:

a) If $\sum_{i=1}^{k} \alpha_i^* z_{im} > d^* \geq \sum_{i=1}^{k} \beta_i^* z_{im}$ or $\sum_{i=1}^{k} \alpha_i^* z_{im} \leq d^* < \sum_{i=1}^{k} \beta_i^* z_{im}$ is identified for the newly sampled mth observation (DMU), then DEA-DA concludes that there is an overlap and the observation belongs to $G_1 \cap G_2$.

b) If $\sum_{i=1}^{k} \alpha_i^* z_{im} \geq d^*$ and $\sum_{i=1}^{k} \beta_i^* z_{im} \geq d^*$ including $\sum_{i=1}^{k} \alpha_i^* z_{im} = \sum_{i=1}^{k} \beta_i^* z_{im} = d^*$ is identified for the new observation, then DEA-DA concludes that there is no overlap and the observation belongs to G_1.

c) If $\sum_{i=1}^{k} \alpha_i^* z_{im} < d^*$ and $\sum_{i=1}^{k} \beta_i^* z_{im} < d^*$ is identified for the new observation, then DEA-DA concludes that there is no overlap and the observation belongs to G_2.

Stage 2: Handling overlap
If there exists an overlap, for handling, the following linear programming problem is employed:

$$\text{Min} \quad \sum_{j \in G_1} s_{1j}^+ + \sum_{j \in G_2} s_{2j}^-$$

$$\text{s.t.} \quad \sum_{i=1}^{k} \alpha_i z_{ij} + s_{1j}^+ - s_{1j}^- = d, \quad j \in G_1,$$

$$\sum_{i=1}^{k} \alpha_i z_{ij} + s_{2j}^+ - s_{2j}^- = d - \eta, \quad j \in G_2, \quad (3.26)$$

$$\sum_{i=1}^{k} \alpha_i = 1,$$

$$s_{1j}^+, s_{1j}^-, s_{2j}^+, s_{2j}^- \geq 0, \quad j = 1, \ldots, n,$$

$$\alpha_i, \geq 0, \quad i = 1, \ldots, k,$$

$$d \text{ free.}$$

Let α^* and d^* be the optimum solutions for Model (3.26). Consequently, the following cases are obtained:

a) If $\sum_{i=1}^{k} \alpha_i^* z_{ij} \geq d^* \quad j \in G_1 \cap G_2$, then $j \in G_1$,
or
b) If $\sum_{i=1}^{k} \alpha_i^* z_{ij} < d^* \quad j \in G_1 \cap G_2$, then $j \in G_2$.

Unfortunately, this DEA-DA method has three major drawbacks: cannot deal with a negative data, produces two separate hyperplanes for discrimination, and does not implement any large scale simulation study.

To remove some drawbacks in [158], Sueyoshi [159] proposed a new nonparametric method, called the extended DEA-DA in the two-stage classification process. The extended DEA-DA technique is introduced to produce a piece wise linear discriminant function for a single separated hyperplane at each stage and to deal with negative data. However, this method has two methodological shortcomings: it is formulated to minimize the total distance of misclassified observations instead of minimizing the number of the misclassified observations, and it does not record mathematical conditions to formulate the model. To overcome the mentioned drawbacks, Sueyoshi [161] reformulated the extended DEA-DA model by two mixed integer linear programming models. Note that each mixed integer programming problem needs more computational time than the original linear programming problems. Also, mixed integer programming problems fail to guarantee the global optimal solution for weight estimates of a discriminant function. Furthermore, different selections of the small number ϵ and the large number M produce different solutions. These are three problems of Sueyoshi's model [161]. Then, Sueyoshi [160] and Sueyoshi and Hwang [163] eliminated the first stage of the two-stage mixed integer programming approach to simplify the estimation process. This one-stage mixed integer programming approach is useful when an overlap between two groups is not a serious problem. A major shortcoming of the previous methods is their inability to handle the classification of more than two groups. For the extension of DEA-DA models to deal with more than two groups, Sueyoshi [162] considered the classification of three groups, using a mixed integer programming model, and compared the previously proposed DEA-DA models with DA models from different viewpoints. As mentioned, one of the drawbacks of Sueyoshi's model [158] is its inability to assess the overlap between two groups. To deal with this problem, Jahanshahloo et al. [90] revised the model in [158], and extended it for interval data using the two-step method, including classification and overlap identification and then use of the Monte Carlo method [97] for overlap identification.

Boudaghi and Farzipoor Saen [21] introduced a DEA-DA model based on BCC model for predicting group membership of suppliers in the sustainable supply chain context. This model can predict group membership of suppliers regarding the nature of factors including inputs, outputs, and efficiency scores. Tavassoli and Farzipoor Saen [166] developed a stochastic two-stage DEA-DA model that, unlike the conventional DEA-DA models, includes random data and accordingly predicts supplier's group membership with high precision.

3.3.7 Efficient units and efficient hyperplanes

Employing concepts of MOLP problems is one of the approaches to find efficient hyperplanes of a PPS in DEA. To do it, DEA researchers modify the presented algorithms for obtaining efficient faces in an MOLP problem to the DEA methodology for obtaining efficient hyperplanes of a PPS, for example, the work of Hosseinzadeh Lotfi et al. [79]. It is of interest to note that specifying efficient hyperplanes of the PPS, using the MODM methodology, was discussed only in T_V, yet.

To determine all extreme efficient DMUs of T_V, Hosseinzadeh Lotfi et al. [82] proposed a method based on the following MOLP problem:

$$\begin{aligned} \text{Max} \quad & \{-x_1, \ldots, -x_m, y_1, \ldots, y_s\} \\ \text{s.t.} \quad & \sum_{j=1}^n \lambda_j \bar{x}_{ij} \leqslant x_i, && i = 1, \ldots, m, \\ & \sum_{j=1}^n \lambda_j \bar{y}_{rj} \geqslant y_r, && r = 1, \ldots, s, \quad (3.27) \\ & \sum_{j=1}^n \lambda_j = 1, \\ & x_i \geqslant 0, \ y_r \geqslant 0, && i = 1, \ldots, m, \ r = 1, \ldots, s, \end{aligned}$$

where $(\bar{x}_j, \bar{y}_j)$ $(j = 1, \ldots, n)$ are input and output vectors. For solving MOLP (3.27), they used the MSM. To obtain efficient units, the following valuable theorem is proved.

Theorem 2. Each Pareto-optimal solution of MOLP (3.27) is corresponding to an efficient production possibility in T_V and vice versa.

Proof: See [82]

Note that the generation of all extreme efficient DMUs using this method has less computation than the classical DEA models. Moreover, they extended the above method to generate efficient hyperplanes of T_V. For this purpose, they defined the set of weights to solve MOLP (3.27) using the weighted-sums method. Assume

$$W = \{(w_1, \ldots, w_m, \acute{w}_1, \ldots, \acute{w}_s) \mid \sum_{i=1}^{m} w_i + \sum_{r=1}^{s} \acute{w}_r = 1, \ w_i > 0, \ \acute{w}_r > 0\}, \quad (3.28)$$

according to the weights W, MOLP (3.27) is converted to the following linear programming:

$$\begin{aligned}
\text{Min} \quad & \sum_{i=1}^{m} \omega_i x_i - \sum_{r=1}^{s} \acute{\omega}_r y_r \\
\text{s.t.} \quad & -\sum_{j=1}^{n} \lambda_j \bar{x}_{ij} + x_i \geq 0, \quad i = 1, \ldots, m, \\
& \sum_{j=1}^{n} \lambda_j \bar{y}_{rj} - y_r \geq 0, \quad r = 1, \ldots, s, \\
& \sum_{j=1}^{n} \lambda_j = 1, \\
& \lambda_j \geq 0, \quad j = 1, \ldots, n, \\
& x_i \geq 0, \ y_r \geq 0, \quad i = 1, \ldots, m, \ r = 1, \ldots, s, \\
& (\omega_i, \acute{\omega}_r) \in W, \quad i = 1, \ldots, m, \ r = 1, \ldots, s.
\end{aligned} \quad (3.29)$$

Gradients of efficient hyperplanes for T_V can be obtained from the optimal solution of the dual of Model (3.29). To show that this method characterizes all efficient hyperplanes of T_V, Theorem 3 is proved.

Theorem 3. Suppose all components of the input and output vectors are positive, then, the weights chosen from W are gradient vectors of the efficient hyperplane in T_V.

Proof: See [82]

Hosseinzadeh Lotfi et al. [81] suggested another method to find efficient DMUs of T_V using MOLP (3.27); for solving, they used the weighted-sums method, similarly. Using a finite number of weights that are extreme rays of the generated cone by the efficient solutions, all weak efficient DMUs in T_V are provided. It is worth mentioning that if the number of inputs and outputs are smaller than the number of DMUs, this method will be useful. Based on Sayin's approach [147], for finding efficient faces of an MOLP problem, Hosseinzadeh Lotfi et al. [79] developed an MOLP problem to find all efficient hyperplanes of T_V in the DEA methodology. An important task of this approach is that without finding the extreme efficient points of the proposed MOLP, the efficient faces of T_V are identified. Zamani and Hosseinzadeh Lotfi [209] obtained all efficient units of T_V by solving a parametric liner programming problem. It should be noted that solving a parametric linear programming problem requires high amounts of calculation. However, for large amounts of n (the number of DMUs), solving the parametric linear programming is better than solving n linear programming problems. Therefore, this parametric linear program is utilized to specify the efficient units in T_v with only one input and output, and the number of DMUs is arbitrary. Based on solving an MOLP problem in the DEA methodology, Ghazi et al. [63] obtained explicit form equations of strong and weak defining hyperplanes of the PPS with negative data. Regarding these equations, they determined efficiency scores and target units for inefficient DMUs. Moreover, Ghazi et al. [64] developed an algorithm, drawing on the theory of MODM, to specify strong efficient DMUs, the strong efficient frontier, and strong defining hyperplanes for T_V. Essentially, this algorithm was constructed based on solving the MOLP problem by the MSM.

Table 3.2 lists classifying the application of MODM techniques in the DEA methodology and the several researchers who worked in this field.

3.4 Classification of usage of DEA in MODM

In this section, the DEA technologies are applied as tools to solve MODM problems. The classification of usage of DEA in MODM is categorized into only one group, which is obtaining efficient points.

3.4.1 Efficient points

There are numerous algorithms for solving MOLP problems, but less attention has been given for solving integer MOLP problems. The seminal paper presented by Liu et al. [58] was developed to show the usage of DEA techniques for solving 0-1 MOLP problems. Then, other researchers extended this work to solve integer MOLP problems.

Liu et al. [58] proposed a method for solving multiobjective 0-1 linear programming problems using DEA techniques. In the 0-1 MOLP problem, if there are Q activities, then the total possible number of alternatives would be 2^Q. In this method, each DMU presents the following relationship between resources and objectives:

$$0 \leqslant x_{id} = A_i.W_d = \sum_{q=1}^{Q} a_{iq}w_{qd} \leqslant b_i,$$
$$0 \leqslant y_{rd} = C_r.W_d = \sum_{q=1}^{Q} c_{rq}w_{qd} \leqslant g_r. \tag{3.30}$$

In vector $W_d = (w_{1d}, \ldots, w_{Qd})$, $w_{qd} = 1$ if the qth activity is performed; otherwise, $w_{qd} = 0$. A PPS is built using the DMUs obtained from all alternatives (feasible solutions) of the 0-1 MOLP problem. Hence, according to the number of alternatives, the number of DMUs would be 2^Q. The following property, which forms the basis of their method, is proposed to solve the 0-1 MOLP problem.

Property 1. Let $D_d = (x_d, y_d)$ be the vector DMU_d resulting from alternative-d and be BCC efficient, then alternative-d with vector $W_d = [w_{1d}, w_{2d}, \ldots, w_{Qd}]^T$ is an efficient solution.

Proof: See [58].

The three steps algorithm to acquire efficient solutions of the 0-1 MOLP problem is suggested as follows:

Step 1: Initialization

First choose arbitrary feasible alternatives. Then, the BCC model in the multiplier form is used to evaluate the considered DMUs from the mentioned alternatives. If there does not exist any efficient DMU, choose other feasible alternatives; otherwise, construct the convex hull of efficient DMUs, and go to Step 2.

TABLE 3.2 Applying the MODM concepts for DEA problems.

Classification	Authors
Standard DEA models	Banker and Morey (1986); Charnes and Cooper (1985); Charnes, Cooper, and Rhodes (1978); Charnes, Cooper, Golany, Seiford, and Stutz (1985); Pastor, Ruiz, and Sirvent (1999)
Target setting	Angullo Meza (2002); Ebrahimnejad and Hosseinzadeh Lotfi (2012); Ebrahimnejad and Tavana (2014); Estellita Lins, Angulo-Meza, and Moreila da Silva (2004); Ghiyasi (2015); Golany (1988); Hadi-Vencheh and Foroughi (2006); Hadi-Vencheh, Foroughi, and Soleimani-Damaneh (2008); Jahanshahloo, Hosseinzadeh Lotfi, Shoja, Tohidi, and Razavyan (2004); Joro, Korhonen, and Wallenius (1998); Korhonen and Syrjan (2004); Korhonen, Stenfors, and Syrjanen (2003); Lertworasirikul, Charnsethikul, and Fang (2011); Lozano and Villa (2009); Thanassoulis and Dyson (1992); Wei, Zhang, and Zhang (2000); Yang, Wong, Xu, and Stewart (2009)
Value efficiency	Halme and Koehonen (2000); Halme, Joro, Korhonen, Salo, and Wallenius (1999); Joro, Korhonen, and Zionts (2003); Korhonen and Syrjan (2005); Zohrebandian (2011)
Secondary goal	Alcaraz, Ramon, Ruiz, and Sirvent (2013); Contreras (2012); Doyle and Green (1994); Guo and Wu (2013); Lam (2010); Liang, Wu, Cook, and Zhu (2008); Lim (2008); Oral, Amin, and Oukil (2015); Orkcu and Bal (2011); Orken and Bal (2012); Ramon, Ruiz, and Sirvent (2010); Wang and Chin (2010); Wang, Chin, and Jiang (2011); Washio and Yamada (2013); Wu, Liang, Zha, and Yang (2009); Yang, Ang, Xia, and Yang (2012)
Common set of weights	Amin (2008); Amin (2009); Amin and Toloo (2007); Bal, Orkcu, and Celebioglu (2010); Belton and Vicker (1993); Charens, Cooper, Wei, and Huang (1989); Chiang, Hwang, and Liu (2011); Doyle (1995); Foroughi (2011); Hatami-Marbini, Tavana, Agrell, Hosseinzadeh Lotfi, and Ghelej Beigi (2015); Hatefi and Torabi (2010); Hosseinzadeh Lotfi, Hatami-Marbini, Agrell, Aghayi, and Gholami (2013); Jahanshahloo, Hosseinzadeh Lotfi, Khanmohammadi, Kazemimanesh, and Rezaie (2010); Kao (2010); Liu and Peng (2008); Kao and Hung (2009); Karask and Ahiska (2005); Kornbluth (1991); Lin and Reeves (1999); Omrani (2013); Ramon, Ruiz, and Sirvent (2012); Roll and Golany (1993); Sinuany-Stern and Friedman (1998); Stewart (1996); Toloo (2012); Toloo and Ertay (2014); Toloo and Nalchigar (2011); Wang and Jiang (2012); Yang, Zhang, Zhang, Zhang, and Xu (2016); You, Jie, and Xin, (2013); Zohrehbandian, Makui, and Alinezhad (2010)
DEA-discriminant analysis	Jahanshahloo, Hosseinzadeh Lotfi, Rezai Balf, and Zhiani Rezai (2007); Retzlaff-Roberts (1996); Sueyoshi (1999);
Efficient hyperplanes	Hosseinzadeh Lotfi, Jahanshahloo, Mozzaffari, and Gerami (2011); Hosseinzadeh Lotfi, Noora, Jahanshahloo, Jablonsky, Mozaffari, and Gerami (2009); Ghazi, Hosseinzadeh Lotfi, and Sanei (2020)

Step 2: Generation
Solve Model (3.31) to find an initial alternative $-d$,

$$\text{Max} \quad \sum_{q=1}^{Q} w_{qd}$$

$$\text{s.t.} \quad \sum_{q=1}^{Q} \left[\sum_{r=1}^{s} u_{rj}^* c_{rq} - \sum_{i=1}^{m} v_{ij}^* a_{iq} \right] w_{qd} > u_{0j}^*, \quad \forall j \in \acute{G}_k,$$

$$0 \leqslant \sum_{q=1}^{Q} c_{rq} w_{qd} \leqslant g_r, \qquad r = 1, \ldots, s, \qquad (3.31)$$

$$0 \leqslant \sum_{q=1}^{Q} a_{iq} w_{qd} \leqslant b_i, \qquad i = 1, \ldots, m,$$

$$w_{qd} \in \{0, 1\},$$

where g_r and b_i are, respectively, bounded constraints of objectives and resources, and (u^*, v^*, u_0^*) is the optimal solution of the BCC DEA model. The convex hull of all the efficient DMUs in $\acute{G}_k$ (a subset of G_k, denotes the efficient DMUs) is constructed by the associated supporting hyperplanes $u_j y_j - v_j x_j \leqslant u_{0j}^*$ ($j \in \acute{G}_k$). For any DMU that is not contained in $H(\acute{G}_k)$ and results from an alternative, say, d, its x_d and y_d must satisfy the set of constraints $u_j y_j - v_j x_j > u_{0j}^*$ ($j \in \acute{G}_k$) which imply the first group of constraints in Model (3.31). If there exists an optimal solution then obtain a more distant DMU from the convex hull, and go to Step 3. Otherwise, stop.

Step 3: Evaluation
Evaluate the efficiency of the new obtained DMUs using the BCC model in the multiplier form and construct the new convex hull of the efficient DMUs, and go to Step 2.

According to Step 2, when Model (3.31) is infeasible then the algorithm stops because there is not another alternative to investigative whether it is efficient or not. The related alternatives of efficient DMUs in the last iteration are efficient solutions of the 0–1 MOLP.

Liu et al.'s method [58] has some drawbacks, such as existence of the convexity constraint in the BCC model, which may eliminate some efficient DMUs and also its inability to consider a feasible solution corresponding to the DMUs with negative inputs or outputs. For these reasons, Liu et al.'s method [58] lost some efficient solutions. To eliminate the second drawback, Jahanshahloo et al. [92] proposed a two-stage algorithm in which Additive DEA model without input is used. In comparison between this method and the proposed method in [58], this method is more computationally efficient. Yet, there exists the convexity constraint in Additive DEA model without input; therefore, this may eliminate some efficient solutions in the 0-1 MOLP problem. Also, the suggested method in Jahanshahloo et al. [93] obtained all efficient solutions of the 0-1 MOLP problem using a one-stage algorithm and without generating all feasible solutions. In this algorithm, with each iteration, at least one efficient solution is found. Still, this method has a significant drawback; in each iteration of the proposed algorithm, some constraints must be added to the constructed model in the previous iteration. Therefore, solving the mentioned model needs more computation. Yougbare and Teghem [206] argued that Property 1 in [58] is false, saying that they forgot to consider inputs in their proof and the criteria space of the 0-1 MOLP problem is constructed using only outputs. Then, they [206] established the correct relations

TABLE 3.3 Applying the DEA concepts for MODM problems.

Classification	Authors
Efficient points	Franklin Liu, Huang, and Yen (2000); Jahanshahloo, Hosseinzadeh Lotfi, Shoja, and Tohidi (2003);
	Jahanshahloo, Hosseinzadeh Lotfi, Shoja, and Tohidi (2005); Keshavarz and Toloo (2014); Youghbare and Teghem (2007)

between efficient points in the 0-1 MOLP problem and efficient DMUs in DEA. Hence, all the previous methods in this field must be revised. Jahanshahloo and Foroughi [86] implemented the concept of DEA in the MOLP problem to obtain efficient points. Also, using a simple example, they illustrated that their own proposed method can find extremely efficient solution, but Ecker-Kouada and Benson's methods [20,50] had no solution for this case. Keshavarz and Toloo [107] presented an approach based on the DEA methodology to test the efficiency status of feasible solutions in multiobjective integer linear programming problems. In the proposed test, Additive FDH DEA model is applied. Note that the test is similar to the problem of Ecker and Kouade [50] to obtain efficient solutions of an MOLP problem. Then, they proposed an algorithm based on Additive BCC model in the multiplier form to check a given efficient solution is either supported or nonsupported. Remember that these approaches cannot be used for negative data because FDH and BCC models are implemented. At last, by an example, the drawback of Chen and Lu's method [31] is illustrated. Moreover, Keshavarz and Toloo [108] designed another two-phase approach based on the DEA methodology to find and classify a set of efficient points of multiobjective integer linear programming problems. In Phase 1, a mixed integer linear programming problem based on Additive FDH model is formulated to obtain efficient points. They [108] extended it to an algorithm that determines a minimal complete set of efficient solutions related to nondominated points in the criteria space. In Phase 2, the BCC model is used to classify all efficient solutions obtained from Phase 1 as supported and nonsupported.

Table 3.3 lists classifying the application of DEA techniques in the MODM methodology and the several researchers who worked in this field.

3.5 Discussion and conclusion

Several researchers have made attempt to integrate the DEA and MODM methodologies. They applied the MODM techniques as tools to solve the DEA problems and inversely the DEA techniques as tools to solve MODM problems. In a large amount of research in operations research and management science, the links between DEA and MODM were discussed from several points of view, but none of them classified the relationships between these methodologies. In light of the mentioned discussion, the current study classified the relationships between the DEA and MODM methodologies and reviewed the main seminal papers in each area. The link started since 1978, when the first DEA paper was published. As a result of the previous subsections, the usage of the MODM techniques in DEA is more frequent than other situation because the preliminary DEA researchers worked on the MODM methodology before introducing DEA and played the key role in developing DEA. In Section 3.3, some MODM techniques were used to solve DEA problems. This part includes seven categories, formulating the traditional DEA models, target setting, value efficiency, CSW, secondary goal, DA, and specifying the efficient frontier of a PPS. The researchers who worked in these groups presented more robust DEA research in the theoretical point of view. Also, in Section 3.4, some DEA techniques were suggested to solve MOP problems. This part includes only one category, solving 0-1 MOLP problems. The researchers who worked in this group presented applied MODM studies. It is recommended that future research be undertaken in classifying the relationships between the DEA and MADM methodologies. In addition, examining fields of the application for each category is suggested for future studies.

References

[1] N. Adler, L. Friedman, Z.S. Stern, Review of ranking methods in data envelopment analysis context, Eur. J. Oper. Res. 140 (2002) 249–265.
[2] J. Alcaraz, N. Ramon, J.L. Ruiz, I. Sirvent, Ranking ranges in cross efficiency evaluations, Eur. J. Oper. Res. 226 (2013) 516–521.
[3] G.R. Amin, A note on "an improved MCDM DEA model for technology selection", Int. J. Prod. Res. 46 (2008) 7073–7075.
[4] G.R. Amin, Comments on finding the most efficient DMUs in DEA: an improved integrated model, Comput. Ind. Eng. 56 (2009) 1701–1702.
[5] G.R. Amin, S. Al-Muharrami, M. Toloo, A combined goal programming and inverse DEA method for target setting in mergers, Expert Syst. Appl. 115 (2019) 412–417.
[6] G.R. Amin, M. Toloo, Finding the most efficient DMUs in DEA: an improved integrated model, Comput. Ind. Eng. 52 (2007) 71–77.
[7] G.R. Amin, M. Toloo, B. Sohrabi, An improved MCDM DEA model for technology selection, Int. J. Prod. Res. 44 (2006) 2681–2686.
[8] A. Amirteimoori, A. Emrouznejad, Input/output deterioration in production processes, Expert Syst. Appl. 38 (2011) 5822–5825.
[9] A. Andersen, N.C. Petersen, A procedure for ranking efficient units in data envelopment analysis, Manage. Sci. 39 (1993) 1261–1264.
[10] L. Angulo Meza, Um enfoque multiobjectivo para determinagdo de alvos de andlise envoltdria de dados (DEA), Coordenadio dos Programas de Pos-Graduaaio em Engenharia. Universidade Federal do Rio de Janeiro (COPPE/UFRJ), 2002 Ph.D. thesis.
[11] S. Ang, M. Chen, F. Yang, Group cross-efficiency evaluation in data envelopment analysis: an application to Taiwan hotels, Comput. Ind. Eng. 125 (2018) 190–199.
[12] Q. An, H. Chen, B. Xiong, J. Wu, L. Liang, Target intermediate products setting in a two-stage system with fairness concern, Omega 73 (2017) 49–59.
[13] A. Asosheh, S. Nalchigar, M. Jamporazmey, Information technology project evaluation: an integrated data envelopment analysis and balanced scorecard approach, Expert Syst. Appl. 37 (2010) 5931–5938.
[14] M. Azadi, A. Shabani, M. Khodakarami, R. Farzipoor Saen, Reprint of "planning in feasible region by two-stage target setting DEA methods: an application in green supply chain management of public transportation service providers", Transport. Res. E Log. 74 (2015) 22–36.
[15] H. Bal, H.H. Orkcu, A goal programming approach to weight dispersion in data envelopment analysis, Gazi Univ. J. Sci. 20 (2007) 117–125.
[16] H. Bal, H.H. Orkcu, S. Celebioglu, Improving the discrimination power and weights dispersion in the data envelopment analysis, Comput. Oper. Res. 37 (2010) 99–107.
[17] R.D. Banker, A. Charnes, W.W. Cooper, Some models for estimating technical and scale inefficiency in data envelopment analysis, Manage. Sci. 30 (1984) 1078–1092.

[18] R.D. Banker, R.C. Morey, Efficiency analysis for exogenously fixed inputs and outputs, Oper. Res. 34 (1986) 513–521.
[19] V. Belton, Demystifyin DEA-a visual interactive approach based on multiple criteria analysis, J. Oper. Res. Soc. 44 (1993) 883–896.
[20] H.P. Benson, Finding an initial efficient extreme point for a linear multiple objective program, J. Oper. Res. Soc. 32 (1981) 495–498.
[21] E. Boudaghi, R. Farzipoor Saen, Developing a novel model of data envelopment analysis-discriminant analysis for predicting group membership of suppliers in sustainable supply chain, Comput. Oper. Res. 89 (2017) 348–359.
[22] R. Benayoun, J. Montgolfier, J. Tergny, O. Laritchev, Linear programming with multiple objective functions: step method (STEM), Math. Program. 1 (1971) 366–375.
[23] M. Carrillo, J.M. Jorge, An alternative neutral approach for cross-efficiency evaluation, Comput. Ind. Eng. 120 (2018) 137–145.
[24] R.G. Chambers, Y. Chung, R. Fare, Profit, directional distance functions, and nerlovian efficiency, J. Optimiz. Theory Appl. 98 (1998) 1078–1092.
[25] A. Charnes, W.W. Cooper, Preference to topics in data envelopment analysis, Ann. Oper. Res. 2 (1985) 59–94.
[26] A. Charnes, W.W. Cooper, B. Golany, L. Seiford, J. Stutz, Foundation of data envelopment analysis for Pareto-Koopmans efficient empirical production function, J. Econ. 30 (1985) 91–107.
[27] A. Charnes, W.W. Cooper, E. Rhodes, Measuring the efficiency of decision making unit, Eur. J. Oper. Res. 2 (1978) 429–444.
[28] A. Charnes, W.W. Cooper, L. Seiford, J. Stutz, Multiplicative model for efficiency analysis, Socio-Econ. Plan. Sci. 16 (1982) 223–224.
[29] A. Charnes, W.W. Cooper, L. Seiford, J. Stutz, Interval multiplicative efficiency and piecewise Cobb-Douglas envelopments, Oper. Res. Lett. 2 (1983) 101–103.
[30] A. Charens, W.W. Cooper, Q.L. Wei, Z.M. Huang, Cone ratio data envelopment analysis and multi objective programming, Int. J. Syst. Sci. 20 (1989) 1099–1118.
[31] L.H. Chen, H.W. Lu, An extended assignment problem considering multiple inputs and outputs, Appl. Math. Model. 31 (2007) 2239–2248.
[32] L. Chen, Y.M. Wang, DEA target setting approach within the cross efficiency framework, Omega 96 (2020) 1–9.
[33] Y.W. Chen, M. Larbani, Y.P. Chang, Multiobjective data envelopment analysis, J. Oper. Res. Soc. 60 (2009) 1556–1566.
[34] C.I. Chiang, M.J. Hwang, Y.H. Liu, Determining a common set of weights in a DEA problem using a separation vector, Math. Comput. Model. 54 (2011) 2464–2470.
[35] C.I. Chiang, G.H. Tzeng, A new efficiency measure for DEA: efficiency achievement measure established on fuzzy multiple objectives programming, J. Manage. 17 (2000) 369–388. (in Chinese)
[36] I. Contreras, Optimizing the rank position of the DMU as secondary goal in DEA cross-evaluation, Appl. Math. Model. 36 (2012) 2642–2648.
[37] W.D. Cook, Z. J., Within-group common weights in DEA: an analysis of power plant efficiency, Eur. J. Oper. Res. 178 (2007) 207–216.
[38] W.W. Cooper, L.M. Seiford, K. Tone, Introduction to Data Envelopment Analysis and Its Uses with DEA-Solver Software and References, Springer, Cham, 2007.
[39] A. Davoodi, H. Zhiani Rezai, Common set of weights in data envelopment analysis: a linear programming problem, Cent. Eur. J. Oper. Res. 20 (2012) 355–365.
[40] A. Dehnokhalaji, M. Ghiyasi, P. Korhonen, Resource allocation based on cost efficiency, J. Oper. Res. Soc. 68 (2017) 1279–1289.

[41] D.K. Despotis, Improving the discriminating power of DEA: focus on globally efficient units, J. Oper. Res. Soc. 53 (2002) 314–323.
[42] S. Dimitrov, W. Sutton, Promoting symmetric weight selection in data envelopment analysis: a penalty function approach, Eur. J. Oper. Res. 200 (2010) 281–288.
[43] J. Doyle, Multiattribute choice for the lazy decision maker: Let the alternatives decides, Organ. Behav. Hum. Decis. Process. 62 (1995) 87–100.
[44] J. Doyle, R. Green, Data envelopment analysis and multiple criteria decision making, Omega 21 (1993) 713–715.
[45] J. Doyle, R. Green, Efficiency and cross-efficiency in DEA: derivations, meanings and uses, J. Oper. Res. Soc. 45 (1994) 567–578.
[46] D. J., L. L., C. Yao, G.B. Bi, DEA based production planning, Omega 38 (2010) 105–112.
[47] R.G. Dyson, E. Thanassoulis, Reducing weight flexibility in data envelopment analysis, J. Oper. Res. Soc. 39 (1988) 563–576.
[48] A. Ebrahimnejad, F. Hosseinzadeh Lotfi, Equivalence relationship between the general combined-oriented CCR model and the weighted minimax MOLP formulation, J. King Saudi Univ. Sci. 24 (2012) 47–54.
[49] A. Ebrahimnejad, M. Tavana, An interactive MOLP method for identifying target unit in output-oriented DEA models: the NATO exlargement problem, Measurement 52 (2014) 124–134.
[50] J.G. Ecker, I.A. Kouada, Finding efficient points for linear multiple objective programs, Math. Program. 8 (1975) 375–377.
[51] A. Emrouznejad, E. Thanassoulis, A mathematical model for dynamic efficiency using data envelopment analysis, Appl. Math. Comput. 160 (2005) 363–378.
[52] T. Ertay, D. Ruan b, U.R. Tuzkaya, Integrating data envelopment analysis and analytic hierarchy for the facility layout design in manufacturing systems, Inf. Sci. 176 (2006) 237–262.
[53] M.P. Estellita Lins, L. Angulo-Meza, A.C. Moreila daSilva, A multi-objective approach to determine alternative targets in data envelopment analysis, J. Oper. Res. Soc. 55 (2004) 1090–1101.
[54] M.J. Farrell, The measurement of productive efficiency, J. R. Stat. Soc. 120 (1957) 253–281.
[55] R. Farzipoor Saen, Supplier's selection in the presence of both cardinal and ordinal data, Eur. J. Oper. Res. 183 (2007) 741–747.
[56] A.A. Foroughi, A new mixed integer linear model for selecting the best decision making units in data envelopment analysis, Comput. Ind. Eng. 60 (2011) 550–554.
[57] A.A. Foroughi, A revised and generalized model with improved discrimination for finding most efficient DMUs in DEA, Appl. Math. Model. 37 (2013) 4067–4074.
[58] F.H. Franklin Liu, C.C. Huang, Y.L. Yen, Using DEA to obtain efficient solutions for multi objective $0-1$ linear programs, Eur. J. Oper. Res. 126 (2000) 51–68.
[59] N. Freed, F. Glover, Evaluating alternative linear programming models to solve the two-group discriminant problem, Decis. Sci. 17 (1986) 151–162.
[60] A.M. Geoffrion, J.S. Dyer, A. Feinberg, An interactive approach for multi-criterion optimization with an application to the operation of an academic department, Manage. Sci. 19 (1972) 357–368.
[61] M.R. Ghasemi, J. Ignatius, A. Emrouznejad, A bi-objective weighted model for improving the discrimination power in MCDEA, Eur. J. Oper. Res. 233 (2014) 640–650.
[62] A. Ghazi, F. Hosseinzadeh Lotfi, Assessment and budget allocation of Iranian natural gas distribution company- a CSW DEA based model, Socio-Econ. Plann. Sci. 66 (2019) 112–118.
[63] A. Ghazi, F. Hosseinzadeh Lotfi, M. Sanei, Hybrid efficiency measurement and target setting based on identifying defining hyperplanes of the PPS with negative data, Oper. Res. 20 (2020) 1055–1092.
[64] A. Ghazi, F. Hosseinzadeh Lotfi, M. Sanei, Finding the strong efficient frontier and strong defining hyperplanes of production possibility set using multiple objective linear programming, Oper. Res. (2020) 1–34.

[65] M. Ghiyasi, On inverse DEA model: the case of variable returns to scale, Comput. Ind. Eng. 87 (2015) 407–409.
[66] M. Ghiyasi, Inverse DEA based on cost and revenue efficiency, Comput. Ind. Eng. 114 (2017) 258–263.
[67] S. Ghobadi, S. Jahangiri, Inverse DEA: review, extension and application, Int. J. Inf. Technol. Decis. Mak. 13 (2014).
[68] B. Golany, An interactive MOLP procedure for the extension of DEA to effectiveness analysis, J. Oper. Res. Soc. 39 (1988) 725–734.
[69] D. Guo, J. Wu, A complete ranking of DMUs with undesirable outputs using restrictions in DEA models, Math. Comput. Model. 58 (2013) 1102–1109.
[70] A. Hadi-Vencheh, A.A. Foroughi, A generalized DEA model for inputs/outputs estimation, Math. Comput. Model. 43 (2006) 447–457.
[71] A. Hadi-Vencheh, A.A. Foroughi, M. Soleimani-damaneh, A DEA model for resource allocation, Econ. Model. 25 (2008) 983–993.
[72] M. Halme, T. Joro, P. Korhonen, S. Salo, J. Wallenius, A value efficiency approach to incorporating performance information in data envelopment analysis, Manage. Sci. 45 (1999) 103–115.
[73] M. Halme, P. Korhonen, Restricting weights in value efficiency analysis, Eur. J. Oper. Res. 126 (2000) 175–188.
[74] A. Hatami-Marbini, M. Tavana, P.J. Agrell, F. HosseinzadehLotfi, Z. Ghelej Beigi, A common-weights DEA model for centralized resource reduction and target setting, Comput. Ind. Eng. 79 (2015) 195–203.
[75] S.M. Hatefi, S.A. Torabi, A common weight MCDA-DEA approach to construct composite indicators, Ecol. Econ. 70 (2010) 114–120.
[76] F. Hosseinzadeh Lotfi, A. Hatami-Marbini, P.J. Agrell, N. Aghayi, K. Gholami, Allocating fixed resources and setting targets using a common-weights DEA approach, Comput. Ind. Eng. 64 (2013) 631–640.
[77] F. Hosseinzadeh Lotfi, G.R. Jahanshahloo, A. Ebrahimnejad, M. Soltanifar, S.M. Mansourzadeh, Target setting in the general combined-oriented CCR model using an interactive MOLP method, J. Comput. Appl. Math. 234 (2010) 1–9.
[78] F. Hosseinzadeh Lotfi, G.R. Jahanshahloo, A. Memariani, A method for finding common set of weights by multiple objective programming in data envelopment analysis, Southwest J. Pure Appl. Math. 1 (2000) 44–54.
[79] F. Hosseinzadeh Lotfi, G.R. Jahanshahloo, M.R. Mozzaffari, J. Gerami, Finding DEA-efficient hyperplanes using MOLP efficient faces, J. Comput. Appl. Math. 235 (2011) 1227–1231.
[80] F. Hosseinzadeh Lotfi, G.R. Jahanshahloo, M. Soltanifar, A. Ebrahimnejad, S.M. Manosourzadeh, Relationship between MOLP and DEA based on output oriented CCR dual model, Exp. Syst. Appl. 37 (2010) 4331–4336.
[81] F. Hosseinzadeh Lotfi, A.A. Noora, G.R. Jahanshahloo, J. Gerami, M.R. Moffari, Characterization of efficient points of the production possibility set under variable returns to scale in DEA, Iran. J. Oper. Res. 2 (2010) 50–61.
[82] F. Hosseinzadeh Lotfi, A.A. Noora, G.R. Jahanshahloo, J. Jablonsky, M.R. Mozaffari, J. Gerami, An MOLP based procedure for finding efficient units in DEA models, Cent. Eur. J. Oper. Res. 17 (2009) 1–11.
[83] C.L. Hwang, A. S. M. d. Masud, Multiple objective decision making: methods and applications, Springer-Verlag, 1979.
[84] A. Jafarian-Moghaddam, K. Ghoseir, Multi-objective data envelopment analysis model in fuzzy dynamic environment with missing values, Int. J. Adv. Manufact. Technol. 61 (2012) 771–785.
[85] A.R. Jafarinan-Moghaddam, K. Ghoseiri, Fuzzy dynamic multi-objective data envelopment analysis model, Expert Syst. Appl. 38 (2011) 850–855.
[86] G.R. Jahanshahloo, A.A. Foroughi, Efficiency analysis, generating an efficient extreme point for an MOLP, and some comparisons, Appl. Math. Comput. 162 (2005) 991–1005.

[87] G.R. Jahanshahloo, A. Hadi-Vencheh, A.A. Foroughi, R. Kazemi Matin, Inputs/outputs estimation in DEA when some factors are undesirable, Appl. Math. Comput. 156 (2004) 19–32.
[88] G.R. Jahanshahloo, F. Hosseinzadeh Lotfi, Y. Jafari, R. Maddahi, Selecting symmetric weights as a secondary goal in DEA cross-efficiency evaluation, Appl. Math. Model. 35 (2011) 544–549.
[89] G.R. Jahanshahloo, F. Hosseinzadeh Lotfi, M. Khanmohammadi, M. Kazemimanesh, V. Rezaie, Ranking of units by positive ideal DMU with common weights, Expert Syst. Appl. 37 (2010) 7483–7488.
[90] G.R. Jahanshahloo, F. Hosseinzadeh Lotfi, F. Rezai Balf, H. Zhiani Rezai, Discriminant analysis of interval data using Monte Carlo method in assessment of overlap, Appl. Math. Comput. 191 (2007) 521–532.
[91] G.R. Jahanshahloo, F. Hosseinzadeh Lotfi, M. Rostamy-Malkhalifeh, A. Darigh, Value efficiency analysis for incorporating preference information in data envelopment analysis with interval data, J. Basic Appl. Sci. Res. 3 (2013) 689–694.
[92] G.R. Jahanshahloo, F. Hosseinzadeh Lotfi, N. Shoja, G. Tohidi, A method for solving 0−1 multiple objective linear programming problem using DEA, J. Oper. Res. Soc. Jpn. 46 (2003) 189–202.
[93] G.R. Jahanshahloo, F. Hosseinzadeh Lotfi, N. Shoja, G. Tohidi, A method for generating all efficient solutions of 0−1 multi-objective linear programming problem, Appl. Math. Comput. 169 (2005) 874–886.
[94] G.R. Jahanshahloo, F. Hosseinzadeh Lotfi, N. Shoja, G. Tohidi, S. Razavyan, The outputs estimation of a DMU according to improvement of its efficiency, Appl. Math. Comput. 147 (2004) 409–413.
[95] G.R. Jahanshahloo, F. Hosseinzadeh Lotfi, N. Shoja, G. Tohidi, S. Razavyan, Input estimation and identification of extra inputs in inverse DEA models, Appl. Math. Comput. 156 (2004) 427–437.
[96] G.R. Jahanshahloo, F. Hosseinzadeh Lotfi, N. Shoja, G. Tohidi, S. Razavyan, Sensitivity of efficiency classifications in the inverse DEA models, Appl. Math. Comput. 169 (2005) 905–916.
[97] G.R. Jahanshahloo, F. Hosseinzadeh Lotfi, H. Zhiani Rezai, F. Rezai Balf, Using Monte Carlo method for ranking efficient DMUs, Appl. Math. Comput. 162 (2005) 371–379.
[98] G.R. Jahanshahloo, A. Memariani, F. Hosseinzadeh Lotfi, H.Z. Rezai, A note on some of DEA models and finding efficiency and complete ranking using common set of weights, Appl. Math. Comput. 166 (2005) 265–281.
[99] G.R. Jahanshahloo, M. Soleimani-damaneh, M. Reshadi, On the Pareto (dynamically) efficient paths, Int. J. Comput. Math. 83 (2006) 629–633.
[100] G.R. Jahanshahloo, M. Soleimani-damaneh, S. Ghobadi, Inverse DEA under inter-temporal dependence using multiple-objective programming, Eur. J. Oper. Res. 240 (2015) 447–456.
[101] G.R. Jahanshahloo, M. Zohrehbandian, A. Alinezhad, S. Abbasian Naghneh, H. Abbasian, R. Kiani Mavi, Finding common weights based on the DM's preference information, J. Oper. Res. Soc. 62 (2011) 1796–1800.
[102] T. Joro, P. Korhonen, J. Wallenius, Structural comparison of data envelopment analysis and multiple objective linear programming, Manage. Sci. 44 (1998) 962–970.
[103] T. Joro, P. Korhonen, S. Zionts, An interactive approach to improve estimates of value efficiency in data envelopment analysis, Eur. J. Oper. Res. 149 (2003) 688–699.
[104] C. Kao, Malmquist productivity index based on common-weights DEA: the case of Taiwan forests after reorganization, Omega 38 (2010) 484–491.
[105] C. Kao, H.T. Hung, Data envelopment analysis with common weights: the compromise solution approach, J. Oper. Res. Soc. 56 (2005) 1196–1203.
[106] E.E. Karsak, S.S. Ahishka, Practical common weight multi-criteria decision-making approach with an improved discriminating power for technology selection, Int. J. Prod. Res. 43 (2005) 1437–1554.
[107] E. Keshavarz, M. Toloo, Efficiency status of a feasible solution in the multi-objective integer linear programming problems: a DEA methodology, Appl. Math. Model. 39 (2014) 3236–3247.

[108] E. Keshavarz, M. Toloo, Finding efficient assignments: an innovative DEA approach, Measurement 58 (2014) 448–458.

[109] R. Kiani Mavi, R. Farzipoor Saen, M. Goh, Joint analysis of eco-efficiency and eco-innovation with common weights in two-stage network DEA: a big data approach, Technol. Forecast. Soc. Change 144 (2019) 553–562.

[110] T.C. Koopmans, Activity Analysis of Production and Allocation, Wiley, New York, 1951.

[111] J.S.H. Korenbluth, Analysing policy effectiveness using cone restricted DEA, Eur. J. Oper. Res. 42 (1991) 1097–1104.

[112] P. Korhonen, VIG-a visual interactive support system for multiple criteria decision making, Belgian J. Oper. Res. Stat. Comput. Sci. 27 (1987) 3–15.

[113] P. Korhonen, S. Stenfors, M. Syrjanen, Multiple objective approach as an alternative to radial projection in DEA, J. Product. Anal. 20 (2003) 305–321.

[114] P. Korhonen, J. Wallenius, A Pareto race, Nav. Res. Logist. 35 (1988) 15–62.

[115] P.J. Korhonen, M.J. Syrjanen, On the interpretation of value efficiency, J. Product. Anal. 24 (2005) 197–201.

[116] R. Lahdelma, P. Salminen, SMAA-2: stochastic multicriteria acceptability analysis for group decision making, Oper. Res. 49 (2001) 444–454.

[117] K.F. Lam, In the determination of weight sets to compute cross-efficiency ratios in DEA, J. Oper. Res. Soc. 61 (2010) 134–143.

[118] S. Lertworasirikul, P. Charnsethikul, S.C. Fang, Inverse data envelopment analysis model to preserve relative efficiency values: the case of variable returns to scale, Comput. Ind. Eng. 61 (2011) 1017–1023.

[119] X.B. Li, G.R. Reeves, A multiple criteria approach to data envelopment analysis, Eur. J. Oper. Res. 115 (1999) 507–517.

[120] L. Liang, J. Wu, W.D. Cook, J. Zhu, Alternative secondary goals in DEA cross efficiency evaluation, Int. J. Prod. Econ. 113 (2008) 1025–1030.

[121] S. Lim, An aggressive formulation of cross-efficiency in DEA, J. Kor. Oper. Res. Manage. Soc. 33 (2008) 85–102. (in Korean)

[122] R. Lin, Z. Chen, W. Xiong, An iterative method for determining weights in cross efficiency evaluation, Comput. Ind. Eng. 101 (2016) 91–102.

[123] F.H.F. Liu, H.H. Peng, Ranking of units on the DEA frontier with common weight, Comput. Oper. Res. 35 (2008) 1624–1637.

[124] F.H.F. Liu, H.H. Peng, A systematic procedure to obtain a preferable and robust ranking of units, Comput. Oper. Res. 36 (2009) 1012–1025.

[125] X. Liu, J. Chu, P. Yin, J. Sun, DEA cross-efficiency evaluation considering undesirable output and ranking priority: a case study of eco-efficiency analysis of coal-fired power plants, J. Clean. Prod. 142 (2017) 877–885.

[126] S. Lozano, N. Soltani, A. Dehnokhalaji, A compromise programming approach for target setting in DEA, Ann. Oper. Res. 288 (2020) 363–390.

[127] S. Lozano, G. Villa, Multiobjective target setting in data envelopment analysis using AHP, Comput. Oper. Res. 36 (2009) 549–564.

[128] N. Malekmohamadi, F. Hosseinzadeh Lotfi, A.B. Jaafar, Target setting target setting in data envelopment analysis using MOLP, Appl. Math. Model. 35 (2011) 328–338.

[129] H. Nakayama, Y. Sawaragi, Satisfying trade-off method for multiobjective programming. an interactive decision analysis, in: M. Grauer, A.P. Wierzbicki (Eds.), Lecture Notes in Economics and Mathematical Systems, vol. 229, Springer, Laxenburg, 1984, pp. 113–122.

[130] I. Nykowski, Z. Zolkiewski, A compromise produce for the multiple objective linear fractional programming problem, Eur. J. Oper. Res. 19 (1985) 91–97.

[131] H. Omrani, Common weights data envelopment analysis with uncertain data: a robust optimization approach, Comput. Ind. Eng. 66 (2013) 1163–1170.
[132] M. Oral, G.R. Amin, A. Oukil, Cross-efficiency in DEA: a maximum resonated appreciative model, Measurement 63 (2015) 159–167.
[133] H.H. Orkcu, H. Bal, Goal programming approaches for data envelopment analysis cross efficiency evaluation, Appl. Math. Comput. 218 (2011) 346–356.
[134] H.H. Orkcu, H. Bal, A new approach to cross efficiency in data envelopment analysis and performance evolution of Turkey cities, Gazi Univ. J. Sci. 25 (2012) 107–117.
[135] H.H. Orkcu, V. Soner Ozsoy, M. Orkcu, H. Bal, A neutral cross efficiency approach for basic two stage production systems, Expert Syst. Appl. 125 (2019) 333–344.
[136] J.T. Pastor, J.L. Ruiz, I. Sirvent, An enhanced DEA Russell graph efficiency measure, J. Oper. Res. Soc. 115 (1999) 596–607.
[137] A. Payan, F. Hosseinzadeh Lotfi, A.A. Noora, M. Khodabakhshi, A modified common set of weights method to complete ranking DMUs, Int. J. Math. Models Methods Appl. Sci. 5 (2011) 1143–1153.
[138] Q. Frota, J. Netol, L. Angulo-Meza, J. Quariguasi Frota Netol, L. Angulo-Meza, Alternative targets for data envelopment analysis through multi-objective linear programming: Rio de Janeiro odontological public health system case study, J. Oper. Res. Soc. 58 (2007) 865–873.
[139] S. Ramezani-Tarkhorani, M. Khodabakhshi, S. Mehrabian, F. Nuri-Bahmani, Ranking decision-making units using common weights in DEA, Appl. Math. Model. 38 (2014) 3890–3896.
[140] N. Ramon, J.L. Ruiz, I. Sirvent, A multiplier bound approach to assess relative efficiency in DEA without slacks, Eur. J. Oper. Res. 203 (2010) 261–269.
[141] N. Ramon, J.L. Ruiz, I. Sirvent, On the choice of weights profiles in cross-efficiency evaluations, Eur. J. Oper. Res. 207 (2010) 1564–1572.
[142] N. Ramon, J.L. Ruiz, I. Sirvent, Common sets of weights as summaries of DEA profiles of weights: with an application to the ranking of professional tennis players, Expert Syst. Appl. 39 (2012) 4882–4889.
[143] S.H. Razavi Hajiaghaa, H. Amoozad Mahdiraji, M. Tavana, S.S. Hashemi, A novel common set of weights method for multi-period efficiency measurement using mean-variance criteria, Measurement 129 (2018) 569–581.
[144] D.L. Retzlaff-Roberts, Relating discriminant analysis and data envelopment analysis to one another, Comput. Oper. Res. 23 (1996) 311–322.
[145] Y. Roll, B. Golany, Alternative methods of treating factor weights in DEA, Omega 21 (1993) 99–103.
[146] M. Salahi, N. Torabi, A. Amiri, An optimistic robust optimization approach to common set of weights in DEA, Measurement 93 (2016) 67–73.
[147] S. Sayin, An algorithm based on facial decomposition for finding the efficient set in multiple objective linear programming, Oper. Res. Lett. 19 (1996) 87–94.
[148] M. Seiford, J. Zhu, Modeling undesirable factors in efficiency evaluation, Eur. J. Oper. Res. 142 (2002) 16–20.
[149] T.R. Sexton, R.H. Silkman, R. Hogan, Data envelopment analysis: critique and extension, in: R.H. Silkman (Ed.), Measuring Efficiency: An Assessment of Data Envelopment Analysis Jossey-Bass: San Francisco. 73-105, 1986.
[150] S.J. Sharahi, K. Khalili-Damghani, Fuzzy type-II de-novo programming for resource allocation and target setting in network data envelopment analysis: a natural gas supply chain, Expert Syst. Appl. 117 (2019) 312–329.
[151] Z. Sinuany-Stern, L. Friedman, DEA and the discriminant analysis of ratios for ranking units, Eur. J. Oper. Res. 111 (1998) 470–478.
[152] Z. Sinuany-Stern, A. Mehrez, A. Barboy, Academic departments efficiency in DEA, Comput. Oper. Res. 21 (1994) 543–556.

[153] R.E. Steuer, An interactive multiple objective linear programming procedure, TIMS Stud. Manage. Sci. 6 (1977) 225–239.
[154] R.E. Stuere, Multiple Criteria Optimization: Theory, Computation, and Application, Wiley, New York, 1986.
[155] R.E. Steuer, E.U. Choo, An interactive weighted Tchebycheff procedure for multiple objective programming, Math. Program. 26 (1983) 326–344.
[156] T.J. Stewart, Relationships between data envelopment analysis and multi criteria decision analysis, J. Oper. Res. Soc. 47 (1996) 654–665.
[157] T.J. Stewart., Goal directed benchmarking for organizational efficiency, Omega 38 (2010) 534–539.
[158] T. Sueyoshi, DEA-discriminant analysis in the view of goal programming, Eur. J. Oper. Res. 115 (1999) 564–582.
[159] T. Sueyoshi, Extended DEA-discriminant analysis, Eur. J. Oper. Res. 131 (2001) 324–351.
[160] T. Sueyoshi, Financial ratio analysis of electric power industry, Asia-Pac. J. Oper. Res. 22 (2004) 349–376.
[161] T. Sueyoshi, Mixed integer programming approach of extended-discriminant analysis, Eur. J. Oper. Res. 152 (2004) 45–55.
[162] T. Sueyoshi, DEA-discriminant analysis: methodological comparison among eight discriminant analysis approaches, Eur. J. Oper. Res. 169 (2006) 247–272.
[163] T. Sueyoshi, S.N. Hwang, A use of nonparametric test for DEA-DA: a methodological comparison, Asia-Pac. J. Oper. Res. 21 (2004) 179–195.
[164] T. Sueyoshi, Y. Kirihara, Efficiency measurement and strategic classification of Japanese banking institutions, Int. J. Syst. Sci. 29 (1998) 1249–1263.
[165] J. Sun, J. Wu, D. Guo, Performance ranking of units considering ideal and anti-ideal DMU with common weights, Appl. Math. Model. 37 (2013) 6301–6310.
[166] M. Tavassoli, R. Farzipoor Saen, Predicting group membership of sustainable suppliers via data envelopment analysis and discriminant analysis, Sustain. Prod. Consum. 18 (2019) 41–52.
[167] E. Thanassoulis, R.G. Dyson, Estimating preferred target input-output levels using data envelopment analysis, Eur. J. Oper. Res. 56 (1992) 80–97.
[168] R.G. Thompson, L.N. Langemeier, C.T. Lee, E. Lee, R.M. Thrall, The role of multipliers bounds in efficiency analysis with application to Kansas farming, J. Econ. 46 (1990) 93–108.
[169] R.G. Thompson, J.R.F.D. Singleton, R.M. Thrall, B.A. Smith, Comparative site evaluations for locating a high-energy physics lab in Texas, Interface 16 (1986) 35–49.
[170] M. Toloo, On finding the most BCC-efficient DMU: a new integrated MIP-DEA model, Appl. Math. Model. 36 (2012) 5515–5520.
[171] M. Toloo, The most efficient unit without explicit inputs: an extended MILP-DEA model, Measurement 46 (2013) 3628–3634.
[172] M. Toloo, Alternative minimax model for finding the most efficient unit in data envelopment analysis, Comput. Ind. Eng. 81 (2014) 186–194.
[173] M. Toloo, An epsilon-free approach for finding the most efficient unit in DEA, Appl. Math. Model. 38 (2014) 3182–3192.
[174] M. Toloo, Selecting and full ranking suppliers with imprecise data: a new DEA method, Int. J. Adv. Manufact. Technol. 74 (2014) 1141–1148.
[175] M. Toloo, The role of non-Archimedean epsilon in finding the most efficient unit: with an application of professional tennis players, Appl. Math. Model. 38 (2014) 5334–5346.
[176] M. Toloo, T. Ertay, The most cost efficient automotive vendor with price uncertainty: a new DEA approach, Measurement 52 (2014) 135–144.
[177] M. Toloo, A. Kresta, Finding the best asset financing alternative: a DEA-WEO approach, Measurement 55 (2014) 288–294.

[178] M. Toloo, S. Nalchigar, A new integrated DEA model for finding most BCC-efficient DMU, Appl. Math. Model. 33 (2009) 597–604.
[179] M. Toloo, S. Nalchigar, A new DEA method for supplier selection in presence of both cardinal and ordinal data, Expert Syst. Appl. 38 (2011) 14726–14731.
[180] K. Tone, A slacks-based measure of efficiency in data envelopment analysis, Eur. J. Oper. Res. 130 (2001) 498–509.
[181] W.K. Wang, W.M. Lu, P.Y. Liu, A fuzzy multi-objective two-stage DEA model for evaluating the performance of US bank holding companies, Expert Syst. Appl. 41 (2014) 4290–4297.
[182] Y.M. Wang, K.S. Chin, A neutral DEA model for cross-efficiency evaluation and its extension, Expert Syst. Appl. 37 (2010) 3666–3675.
[183] Y.M. Wang, K.S. Chin, Some alternative models for DEA cross-efficiency evaluation, Int. J. Prod. Econ. 128 (2010) 332–338.
[184] Y.M. Wang, K.S. Chin, J. P., Weight determination in the cross-efficiency evaluation, Comput. Ind. Eng. 61 (2011) 497–502.
[185] Y.M. Wang, K.S. Chin, S. Wang, DEA models for minimizing weight disparity in cross-efficiency evaluation, J. Oper. Res. Soc. 63 (2012) 1079–1088.
[186] Y.M. Wang, P. Jiang, Alternative mixed integer linear programming models for identifying the most efficient decision making unit in data envelopment analysis, Comput. Ind. Eng. 62 (2012) 546–553.
[187] S. Washio, S. Yamada, Evaluation method based on ranking in data envelopment analysis, Expert Syst. Appl. 40 (2013) 257–262.
[188] M. Wegener, G.R. Amin, Minimizing greenhouse gas emissions using inverse DEA with an application in oil and gas, Expert Syst. Appl. 122 (2019) 369–375.
[189] Q.L. Wei, G. Yu, Analyzing the properties of K-cone in a generalized Data Envelopment Analysis model, CCS Research Report 700. Center for Cybernetic Studies, The University of Texas at Austin, 1993.
[190] Q.L. Wei, G. Yu, J.S. Lu, Necessary and sufficient conditions for return-to-scale properties in generalized data envelopment analysis models. CCS Research Report 708. Center for Cybernetic Studies, The University of Texas at Austin, 1993.
[191] Q.L. Wei, J. Zhang, X. Zhang, An inverse DEA model for inputs/outputs estimate, Eur. J. Oper. Res. 121 (2000) 151–163.
[192] A. Wierzbicki, The use of reference objectives in multiobjective optimization, in: G. Fandel, T. Gal (Eds.), Multiple Objective Decision Making. Theory and Application, Springer-Verlag, New York, 1980.
[193] B.Y.H. Wong, M. Luquec, J.B. Yang, Using interactive multiobjective methods to solve DEA problems with value judgements, Comput. Oper. Res. 36 (2009) 623–636.
[194] J. Wu, J. Chu, J. Sun, Q. Zhu, L. Liang, Extended secondary goal models for weights selection in DEA cross-efficiency evaluation, Comput. Ind. Eng. 93 (2016) 143–151.
[195] J. Wu, L. Liang, Y. Zha, F. Yang, Determination of cross-efficiency under the principle of rank priority in cross evaluation, Expert Syst. Appl. 36 (2009) 4826–4829.
[196] J. Wu, J. Sun, L. Liang, Cross efficiency evaluation method based on weight-balanced data envelopment analysis model, Comput. Ind. Eng. 63 (2012) 513–519.
[197] H. Yan, Q.L. Wei, G. Hao, DEA models for resource reallocation and production input/output estimation, Eur. J. Oper. Res. 136 (2002) 19–31.
[198] B. Yang, Y. Zhang, H. Zhang, R. Zhang, B. Xu, Factor-specific Malmquist productivity index based on common weights DEA, Oper. Res. 16 (2016) 51–70.
[199] F. Yang, S. Ang, Q. Xia, C. Yang, Ranking DMUs by using interval DEA cross efficiency matrix with acceptability analysis, Eur. J. Oper. Res. 223 (2012) 483–488.
[200] J.B. Yang, Gradient projection and local region search for multi-objective optimisation, Eur. J. Oper. Res. 112 (1999) 432–459.

[201] J.B. Yang, D. Li, Normal vector identification and interactive trade-off analysis using minimax formulation in multi-objective optimization, IEEE Trans. Syst. Man Cybern. 32 (2002) 305–319.
[202] J.B. Yang, P. Sen, Preference modelling by estimating local utility functions for multi-objective optimization, Eur. J. Oper. Res. 95 (1996) 115–138.
[203] J.B. Yang, B.Y.H. Wong, D.L. Xu, T.J. Stewart, Integrating DEA-oriented performance assessment and target setting using interactive MOLP methods, Eur. J. Oper. Res. 195 (2009) 205–222.
[204] J.B. Yang, D.L. Xu, S. Yang, Integrated efficiency and trade-off analyses using a DEA-oriented interactive minimax reference point approach, Comput. Oper. Res. 39 (2012) 1062–1073.
[205] Y.Q. You, T. Jie, Y. Xin, Erratum to "finding the most efficient DMUs in DEA: an improved integrated model" [Comput. Ind. Eng. 52 (2007) 71-77], Comput. Ind. Eng. 66 (2013) 1178–1179.
[206] J.W. Yougbare, J. Teghem, Using DEA to obtain efficient solutions for multi-objective 0-1 linear programming and DEA efficiency, Eur. J. Oper. Res. 183 (2007) 608–617.
[207] G. Yu, Q.L. Wei, P. Brockett, A generalized data envelopment analysis model: a unification and extension of existing methods for efficiency analysis of decision making units, Ann. Oper. Res. 66 (1996) 47–89.
[208] P.L. Yu, M. Zeleny, The set of all nondominated solutions in linear cases and multicriteria simplex method, J. Math. Anal. Appl. 49 (1975) 430–468.
[209] P. Zamani, F. Hosseinzadeh Lotfi, Using MOLP based procedures to solve DEA problems, Int. J. Data Envelop. Anal. 1 (2013) 15.
[210] X. Zhang, J. Cui, A project evaluation system in the state economic information system of China. 1996. Presented at IFORS'96 Conference. Vancouver.
[211] J. Zhang, Z. Liu, Calculating some inverse linear programming problems, J. Comput. Appl. Math. 72 (1996) 261–273.
[212] P. Zhou, B.W. Ang, K.L. Poh, A mathematical programming approach to constructing composite indicators, Ecol. Econ. 62 (2007) 291–297.
[213] J. Zhu, Data envelopment analysis with preference structure, J. Oper. Res. Soc. 47 (1996) 136–150.
[214] H.J. Zimmermann, Fuzzy programming and linear programming with several objective function, Fuzzy Sets Syst. 1 (1978) 45–55.
[215] S. Zionts, J. Wallenius, An interactive programming method for solving the multiple criteria problem, Manage. Sci. 22 (1976) 625–663.
[216] M. Zohrebandian, Using Ziants–Wallenius method to improve estimate of value efficiency in DEA, Appl. Math. Model. 35 (2011) 3769–3776.
[217] M. Zohrehbandian, A. Makui, A. Alinezhad, A compromise solution approach for finding common weights in DEA: an improvement to Kao and Hung's approach, J. Oper. Res. Soc. 61 (2010) 604–610.

Improved crow search algorithm based on arithmetic crossover—a novel metaheuristic technique for solving engineering optimization problems

S.N. Kumar[a], A. Lenin Fred[b], L. R. Jonisha Miriam[b], Parasuraman Padmanabhan[c], Balázs Gulyás[c], Ajay Kumar H[b] and Nisha Dayana[d]

[a]Amal Jyothi College of Engineering, Kerala, India, [b]Mar Ephraem College of Engineering and Technology, Elavuvilai, Tamil Nadu, India, [c]Lee Kong Chian School of Medicine, Nanyang Technological University, Singapore, [d]Dhanraj Baid Jain College, Chennai, Tamil Nadu, India.

4.1 Introduction

Optimization techniques play a vital role in solving complex engineering problems. Crows are the world's most knowledgeable animals [1] that belong to the genus of birds. Using the cognizance obtained from the behavior of crows, crow optimization (CO) algorithm was developed to arrive at the optimal solution in optimization problems. In the real-time scenario, the estimation of optimal or suboptimal solution is a complex task for the engineering optimizations problems that are nonlinear and are solved using metaheuristic algorithms [2–4]. The metaheuristic algorithms generate best solution for most of the global optimization problems [5]. The nature-inspired optimizations are metaheuristic algorithms that are efficient and best for handling complex engineering optimization problems [6]. The research in the optimization for solving complex engineering problems has also increased rapidly in the last few decades. A wide number of metaheuristic algorithms involved in solving the engineering problems are available in the literature [7–9]. The different types of nature-inspired metaheuristic algorithm were developed with only a few characteristics inspired from nature and many more characteristics are yet to be explored.

The different types of the metaheuristic algorithms used for solving engineering problems are described by Almufti [10]. Still there are many metaheuristic optimization algorithms whose application to the engineering problems depends on the performance, evolutionary operators, parameter tuning, method of finding an optimal solution, etc. [11].

A vector-evaluated Artificial Bee Colony algorithm for discrete variables was implemented for minimizing the total cost and weight of the composites, which is a design optimization for composite structures. It produced better results compared to other nature-inspired optimization [12]. For the economic as well as emission load dispatch problem, the modified bacterial foraging optimization (BFO) algorithm was applied in the fuzzy decision-making approach. The study was conducted in various central load dispatch centers and it generated robust results [13]. The chaotic Bat algorithm was introduced in [14] and evaluated in the benchmark problems with various chaotic maps. The increased global search capacity of chaotic bat algorithm generates robust results when compared with the classical bat algorithm.

The pitch adjustment operation in harmony search optimization algorithm was employed in cuckoo search algorithm to enhance the search and speed of convergence. The benchmark datasets were used for evaluation and comparing with other optimization algorithms [15]. In [16], Firefly algorithm was proposed for the optimization of computer-aided process planning turning machining parameters for cutting tool selection and minimizing the unit production cost [16]. The traits of the Lion were utilized in the formulation of an algorithm for optimizing the area and minimizing the dead space in the very large-scale integration (VLSI) floor planning [17]. A hybrid discrete multiobjective gray wolf optimizer was used for the casting production-scheduling problem; robust results were produced while comparing with the traditional optimization techniques [18]. A Selective Refining Harmony Search algorithm, which incorporates the population refinement step into the original Harmonic Search (HS) algorithm was proposed for solving engineering problems. [19]. Social Spider Optimization method was developed for the job shop scheduling problem and generates efficient results in contrast with other optimization processes [20].

For solving optimization problems, Grasshopper optimization approach was employed in [21]. The comparative analysis of grasshopper optimization algorithm (GOA) with other seven optimization approaches for unimodal, multimodal, composite, and congress on evolutionary computation (CEC) benchmark functions in terms of average and standard deviation shows that the proposed GOA yields best results [21]. The amalgamation of Cuckoo search and Levy flight achieves optimal solution, when compared with various algorithms and this method was applied to 13 structural optimization problems [22]. Mine blast approach (MBA) was applied to 16 constrained benchmarks problems. It uses accuracy and function evaluation as performance metrics. Results reveal that, the proposed method provides outstanding results and it has a small number of function evaluations [23]. In [24], ant lion optimizer optimizes the shape of two ship propellers and it solves three-bar truss, cantilever beam, and gear train design problems. Moreover, the attributes are computed from mathematical functions.

The multilevel thresholding named Kapur's entropy based crow search optimization is demonstrated in [25] which utilizes histogram and was validated using peak signal-to-noise ratio (PSNR), structural similarity index (SSIM) and feature similarity index for image (FSIM) metrics. The automatic detection of lung cancer was achieved by Probabilistic Neural Network with Chaotic Crow-Search Algorithm. The main steps involved are preprocessing, feature extraction, and feature selection [26]. Crow search optimization based intuitionistic fuzzy clustering approach with neighborhood attraction was applied to mini-mammographic image analysis society (MIAS) database for the detection of breast cancer to provide better accuracy [27]. In [28], improved crow search optimization was implemented by using a spiral search technique to update the weights. Results were tested with 23 benchmark functions and 4 design problems, and this method achieves better accuracy and faster convergence. The Otsu multilevel thresholding in [29] utilizes crow search optimization for optimal threshold values selection and yields better results, when it was tested on five benchmark functions and compared with other evolutionary algorithms such as Particle Swarm Optimization (PSO) and firefly algorithm (FA) [29]. A detailed study was carried out in [30] for the modification, hybridization, and application of crow search optimization.

Materials and methods comprises optimization algorithms concept proposed in this chapter, and in results and discussion, the improved crow search optimization was validated on 13 benchmark functions, 2 design problems, and incorporated in the multiclass support vector machine (SVM) for the classification of brain tumors.

4.2 Materials and methods

This chapter focuses on the improved crow search optimization based on the arithmetic crossover concept of genetic algorithm (GA). This section comprises crow search optimization, arithmetic crossover concept of GA and the improved crow search optimization based on arithmetic crossover. The arithmetic crossover concept was incorporated in the position update of crow search optimization.

4.2.1 Crow search optimization

Crows are regarded as the most intelligent bird and it imitates the manners of human beings. The main notion behind the crow is that it conceals their excess food in a secret place. Also, even after a few months the crow remembers the secret place of food. By incorporating this, a novel optimization methodology named crow search algorithm (CSA) was implemented.

Let us deliberate the clusters of crows as population and each population comprises a specific crow. The crow's position is determined by

$$g_q^i = [g_{q1}^i, g_{q2}^i, \ldots \ldots \ldots, g_{qm}^i] \tag{4.1}$$

where i indicates the iteration and m represents the dimensionality of the problem.

The group of crows and its memory is denoted as

$$G = \begin{bmatrix} g_{11}^j & g_{12}^j & \cdots & g_{1m}^j \\ g_{21}^j & g_{22}^j & \cdots & g_{2m}^j \\ \vdots & \vdots & \cdots & \vdots \\ g_{n1}^j & g_{n2}^j & \cdots & g_{nm}^j \end{bmatrix} \tag{4.2}$$

$$E = \begin{bmatrix} e_{11}^j & e_{12}^j & \cdots & e_{1m}^j \\ e_{21}^j & e_{22}^j & \cdots & e_{2m}^j \\ \vdots & \vdots & \cdots & \vdots \\ e_{n1}^j & e_{n2}^j & \cdots & e_{nm}^j \end{bmatrix} \tag{4.3}$$

To standardize each solution the fitness function $f\left(g_q^{(i+1)}\right)$ has to be estimated. Depending on the awareness probability, the new position of the crow varies by either of the two conditions. Deliberate one crow as q and the other crow as l.

Condition 1. Let us assume that the crow q is coming after crow l. However, the crow l is not aware of that the crow q is pursuing. Simultaneously the crow q finds its new position as well as the unique place of food.

$$g_q^{(i+1)} = g_q^i + r_q * fl_q^{(i)} * (e_l^{(i)} - g_q^{(i)}) \quad (4.4)$$

where r_q is the random number lying between 0 and 1, $fl_q^{(i)}$ symbolizes flight length.

Condition 2. The crow l is coming after the crow q. However, the crow q is not aware of that the crow l is pursuing. Simultaneously the crow l progress arbitrary to halfwit the crow q.

$$g_q^{(i+1)} = \begin{cases} g_q^{(i)} + r_q * fl_q^{(i)} * (e_l^{(i)} - g_q^{(i)}), & r_l \geq AP_q^i \\ \text{Moves to a random position,} & \text{Otherwise} \end{cases} \quad (4.5)$$

where AP_q^i symbolizes awareness probability.

By calculating the objective function, the update of memory is determined using the formula.

$$e_q^{(i+1)} = \begin{cases} g_q^{(i+1)}, & \text{if } f(g_q^{(i+1)}) \text{ is better than } f(e_q^{(i)}) \\ e_q^i, & \text{Otherwise} \end{cases} \quad (4.6)$$

4.2.2 Arithmetic crossover based on genetic algorithm

The heuristic evolutionary computation named GA is evolved to solve both constrained and unconstrained optimization problems. It works on the principle of natural selection to achieve an optimal solution. In natural selection, the best characters from individuals are transferred to offspring. GA is initialized with a set of solutions known as population and it is chosen randomly. The solution to a problem is said to be an individual. The encoding of a solution is termed as chromosome and the fragment of it is called a gene. Fitness function yields the quality of solution or optimal solution. When the optimal solution is not attained, fitness function proceeds with the heart of the GA named breeding. The methodologies in breeding are selection, crossover, and mutation. Selection is also termed as reproduction in which two individuals are selected from the population for crossing. Crossover is a method of combining two individuals to generate offspring. When a crossover is carried out, mutation may occur. Hence, both crossover and mutation produce new offspring.

The crow position update is carried out using the arithmetic crossover operation of GA. In the arithmetic crossover, consider two randomly selected individuals α_1 and α_2 to generate two offspring, which are a linear combination of their parents, denoted as

$$\alpha_1^{New} = r\alpha_1 + (1-r)\alpha_2 \quad (4.7)$$

$$\alpha_2^{New} = (1-r)\alpha_1 + r\alpha_2 \quad (4.8)$$

where α symbolizes the individuals, α^{New} symbolizes the offspring and r symbolizes the random number.

4.2.3 Hybrid CO algorithm

The improved crow search optimization based on the arithmetic crossover is termed as the hybrid CO. In the arithmetic crossover operation of GA, two chromosomes selected for crossover C_1 and C_2 may produce two offspring that are a linear combination of their parents. The Flowchart of Hybrid CO Algorithm is shown in Fig. 4.1. The arithmetic crossover concept of the GA was incorporated in the crow position updation and the expressions are as follows:

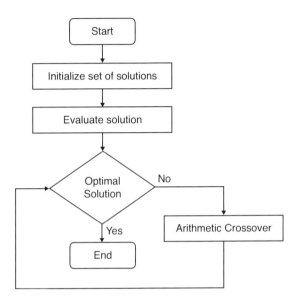

FIGURE 4.1

Flowchart of Hybrid Crow Optimization Algorithm.

Step 1: The crow l is not aware of that the crow q is coming. In this case, the new position of crow q is obtained as

$$g^{q.iter+1} = r \times g^{q.iter} + (1-r) \times \left(m^{l.iter} - g^{q.iter}\right) \quad (4.9)$$

where m represents the lth crows hiding place, g represents the $q$thcrows current position and r is the random number.

Step 2: The crow l is aware of that the crow q is pursuing, hence crow l will fool crow q by moving to another spot of the search space in a random manner.

$$g^{q.iter+1} = r \times m^{q.iter} + (1-r) \times (L + (U-L) \times rand) \quad (4.10)$$

where L is search space lower limit and U is search space upper limit.

4.3 Results and discussion

The crow search optimization was found to be efficient in solving the engineering problems and also validated in terms of benchmark functions [6]. The crow search optimization yields efficient results when compared with the PSO, GA, HS algorithms [6]. The hybrid crow search optimization was validated in terms of the benchmark functions shown in Table 4.1. The benchmark functions are evaluated for different values of iterations (50, 100,150, and 200). The convergence rate values for crow search and hybrid crow search optimization for iteration value 200 are depicted in this table. The convergence rate of hybrid CO was found to be good when compared with crow search

Table 4.1 Benchmark functions used for the validation of optimization techniques.

Benchmark function	Formula
Rosenbrock function	$f(x, y) = (a - x)^2 + b(y - x^2)^2$
Sum of different powers function	$f(x) = \sum_{i=1}^{d} \|x_i\|^{i+1}$
Sum squares function	$r_k(n) = \|\{(a_1, a_2, \ldots a_k) \in Z^k : n = a_1^2 + a_2^2 + \cdots + a_k^2\}\|$
Sphere function	$f(x) = \frac{1}{899}\left(\sum_{i=1}^{6} x_i^2 2^i - 1745\right)$
Schweefel function	$f(x) = 418.9829d - \sum_{i=1}^{d} x_i \sin(\sqrt{\|x_i\|})$
Rastrigin function	$f(x) = 10d - \sum_{i=1}^{d}\left[x_i^2 - 10\cos(2\pi x_i)\right]$
Griewank function	$f(x) = \sum_{i=1}^{d} \frac{x_i}{4000} - \prod_{i=1}^{d} \cos\left(\frac{x_i}{\sqrt{i}}\right) + 1$
Ackley function	$f(x) = -a \exp\left(-b\sqrt{\frac{1}{d}\sum_{i=1}^{d} x_i^2}\right) - \exp\left(\frac{1}{d}\sum_{i=1}^{d} \cos(cx_i)\right) + a + \exp(1)$

Table 4.2 Performance of improved crow search optimization for benchmark functions I and II.

Iterations	Algorithms	$Min_{fitness}$	$Max_{fitness}$	$Avg_{fitness}$	$Std_{fitness}$
Objective function: Rosenbrock function					
50	Crow Optimization	1.308367e+02	9.514827e+02	3.467484e+02	1.623264e+02
	Hybrid Crow Optimization	1.889723e+01	1.897691e+01	1.893867e+01	1.832538e-02
100	Crow Optimization	6.607216e+01	3.007670e+02	1.306503e+02	5.145978e+01
	Hybrid Crow Optimization	1.886461e+01	1.895217e+01	1.892618e+01	1.953378e-02
150	Crow Optimization	3.100475e+01	1.308928e+02	7.005889e+01	2.355379e+01
	Hybrid Crow Optimization	1.885933e+01	1.894508e+01	1.891316e+01	1.809183e-02
200	Crow Optimization	2.637029e+01	1.217840e+02	4.950789e+01	1.727670e+01
	Hybrid Crow Optimization	1.885223e+01	1.893649e+01	1.890804e+01	1.795604e-02
Objective function: Sum of different powers function					
50	Crow Optimization	1.767649e+00	6.013826e+04	2.151886e+03	8.540907e+03
	Hybrid Crow Optimization	3.510160e-16	1.475424e-09	3.867608e-11	2.084296e-10
100	Crow Optimization	7.577129e-01	6.965294e+02	5.190284e+01	1.047474e+02
	Hybrid Crow Optimization	6.101781e-20	9.135727e-14	3.336166e-15	1.316430e-14
150	Crow Optimization	9.907866e-02	1.078222e+02	1.021934e+01	1.832053e+01
	Hybrid Crow Optimization	9.780369e-23	1.050909e-15	2.988649e-17	1.506644e-16
200	Crow Optimization	3.083433e-03	1.659123e+01	1.483021e+00	3.060307e+00
	Hybrid Crow Optimization	4.768502e-25	1.697936e-18	6.542472e-20	2.464084e-19

optimization. Tables 4.2–4.5 depict the maximum, minimum, average, and standard deviation of fitness values, respectively. The parameters used in the validation of optimization algorithms are $Min_{fitness}$ (minimum value of fitness function), $Max_{fitness}$ (maximum value of fitness function), $Avg_{fitness}$ (average value of fitness function), $Std_{fitness}$ (standard deviation value of fitness function). The parameters of the

Table 4.3 Performance of improved crow search optimization for benchmark functions III and IV.

Iterations	Algorithms	$Min_{fitness}$	$Max_{fitness}$	$Avg_{fitness}$	$Std_{fitness}$
Objective function: Sum squares function					
50	Crow Optimization	3.693672e+01	2.066779e+02	1.305435e+02	3.980279e+01
	Hybrid Crow Optimization	1.965657e-11	1.249901e-05	1.054315e-06	2.132667e-06
100	Crow Optimization	1.585424e+01	9.518087e+01	4.605588e+01	1.779393e+01
	Hybrid Crow Optimization	4.265173e-14	1.421464e-09	1.074724e-10	2.118655e-10
150	Crow Optimization	9.659870e+00	7.282447e+01	2.331272e+01	1.022290e+01
	Hybrid Crow Optimization	6.654707e-17	1.184189e-11	9.952327e-13	2.311735e-12
200	Crow Optimization	3.997993e+00	3.056998e+01	1.136525e+01	5.514796e+00
	Hybrid Crow Optimization	5.427745e-20	2.958538e-13	1.566090e-14	4.350869e-14
Objective function: Sphere function					
50	Crow Optimization	4.804130e+02	2.637603e+03	1.382754e+03	4.939709e+02
	Hybrid Crow Optimization	1.428563e-10	6.507153e-05	6.149010e-06	1.284611e-05
100	Crow Optimization	1.812698e+02	9.870226e+02	5.102805e+02	1.505138e+02
	Hybrid Crow Optimization	4.788857e-12	4.724575e-08	2.724900e-09	7.336188e-09
150	Crow Optimization	6.857410e+01	4.222252e+02	2.002540e+02	7.839481e+01
	Hybrid Crow Optimization	6.756627e-15	1.321018e-10	9.298150e-12	2.071080e-11
200	Crow Optimization	1.933117e+01	1.670223e+02	7.767440e+01	3.434234e+01
	Hybrid Crow Optimization	1.228309e-16	7.735249e-13	1.262408e-13	1.896705e-13

Table 4.4 Performance of improved crow search optimization for benchmark functions V and VI.

Iterations	Algorithms	$Min_{fitness}$	$Max_{fitness}$	$Avg_{fitness}$	$Std_{fitness}$
Objective function: Schwef function					
50	Crow Optimization	8.305298e+03	8.317432e+03	8.311383e+03	2.789642e+00
	Hybrid Crow Optimization	8.315340e+03	8.329147e+03	8.322115e+03	3.070559e+00
100	Crow Optimization	8.302609e+03	8.312032e+03	8.305759e+03	2.226435e+00
	Hybrid Crow Optimization	8.310903e+03	8.325239e+03	8.318214e+03	2.710184e+00
150	Crow Optimization	8.301616e+03	8.306780e+03	8.303170e+03	1.054122e+00
	Hybrid Crow Optimization	8.310211e+03	8.320511e+03	8.316221e+03	2.071048e+00
200	Crow Optimization	8.301089e+03	8.304530e+03	8.301911e+03	6.307231e-01
	Hybrid Crow Optimization	8.309853e+03	8.319850e+03	8.314594e+03	2.236917e+00
Objective function: Rastrigin function					
50	Crow Optimization	1.066382e+01	5.971210e+01	2.983921e+01	1.014526e+01
	Hybrid Crow Optimization	1.097362e-10	1.593402e-06	1.806746e-07	2.894474e-07
100	Crow Optimization	3.850562e+00	3.446557e+01	1.200397e+01	6.205222e+00
	Hybrid Crow Optimization	1.421085e-13	6.097878e-10	6.031030e-11	1.204763e-10
150	Crow Optimization	1.184313e+00	1.012021e+01	4.886730e+00	2.077438e+00
	Hybrid Crow Optimization	0	1.887202e-11	7.167955e-13	2.724332e-12
200	Crow Optimization	5.483633e-01	9.818894e+00	2.084439e+00	1.438114e+00
	Hybrid Crow Optimization	0	2.842171e-14	5.684342e-16	4.019437e-15

Table 4.5 Performance of improved crow search optimization for benchmark functions VII and VIII.

Iterations	Algorithms	$Min_{fitness}$	$Max_{fitness}$	$Avg_{fitness}$	$Std_{fitness}$
Objective function: Griewank function					
50	Crow Optimization	2.545049e-01	9.041427e-01	6.563461e-01	1.355122e-01
	Hybrid Crow Optimization	8.278489e-12	6.449963e-0	9.417428e-09	1.713020e-08
100	Crow Optimization	1.248844e-01	7.327561e-01	3.709557e-01	1.378181e-01
	Hybrid Crow Optimization	5.551115e-16	1.250985e-10	4.281815e-12	1.771596e-11
150	Crow Optimization	7.907898e-02	5.002895e-01	2.115311e-01	9.418317e-02
	Hybrid Crow Optimization	0	1.086908e-13	5.853096e-15	1.635974e-14
200	Crow Optimization	3.070635e-02	2.520312e-01	1.163677e-01	5.116532e-02
	Hybrid Crow Optimization	0	1.554312e-15	7.549517e-17	2.571526e-16
Objective function: Ackley function					
50	Crow Optimization	2.598945e+00	3.892524e+00	3.299941e+00	2.896106e-01
	Hybrid Crow Optimization	2.221956e-06	7.485124e-04	1.090239e-04	1.277280e-04
100	Crow Optimization	1.801073e+00	3.684971e+00	2.640500e+00	3.984669e-01
	Hybrid Crow Optimization	2.991771e-08	4.996794e-06	1.464290e-06	1.254436e-06
150	Crow Optimization	1.154453e+00	3.169327e+00	2.244080e+00	5.181379e-01
	Hybrid Crow Optimization	3.011665e-09	4.035642e-07	1.215791e-07	1.129147e-07
200	Crow Optimization	4.633185e-01	2.930164e+00	1.990906e+00	5.173816e-01
	Hybrid Crow Optimization	8.057643e-11	1.150647e-07	1.359041e-08	1.705117e-08

hybrid crow search optimization are as follows: population size = 20, flight length = 2, awareness probability = 0.1. The optimization technique is executed for 50 times. The dimension of the objective function is 20 and the number of iterations is changed from 50,100, 150, and 200. The fitness values plot for Benchmark function, Rosenbrock function, sum of different powers function, sum squares function, sphere function, Schwefel function, Rastrigin function, Griewank function, and Ackley function are displayed in Figs. 4.2–4.5.

The improved crow search optimization based on the arithmetic crossover was also evaluated on the following multiobjective combinatorial optimization problems; three-bar truss design and cantilever beam design problems. Ray and Saini suggested the design optimization problem of three-bar truss to choose a position with the minimized weight of bars [31]. The three-bar truss problem consists of three restrictive functions and two design parameters. The three-bar truss design is depicted in Fig. 4.6.

Objective functions

$$\min f(y) = \left(2\sqrt{2}y_1 + y_2\right) \times L \quad (4.11)$$

Restrictive function

$$g_1 = \frac{\sqrt{2}y_1 + y_2}{\sqrt{2}y_1^2 + 2y_1 y_2} P - \sigma \leq 0 \quad (4.12)$$

$$g_2 = \frac{y_2}{\sqrt{2}y_1^2 + 2y_1 y_2} P - \sigma \leq 0 \quad (4.13)$$

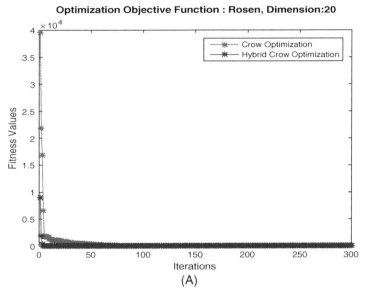
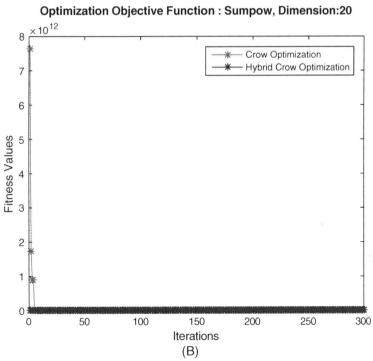

FIGURE 4.2

Fitness values plot for objective functions I and II.

80 Chapter 4 Improved crow search algorithm based on arithmetic crossover

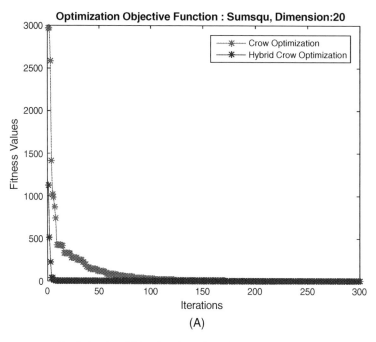

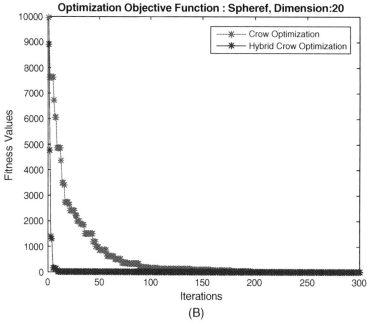

FIGURE 4.3

Fitness values plot for objective functions III and IV.

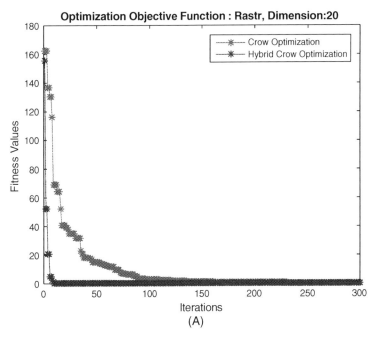

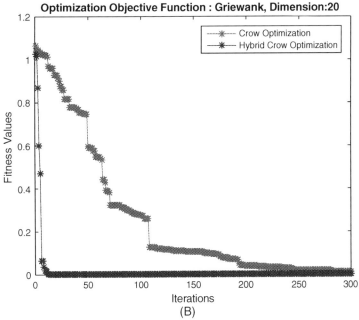

FIGURE 4.4

Fitness values plot for objective functions V and VI.

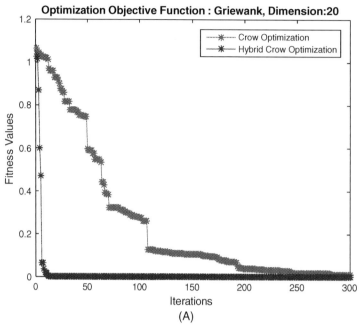

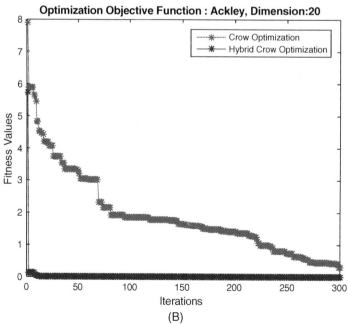

FIGURE 4.5

Fitness values plot for objective functions VII and VIII.

4.3 Results and discussion

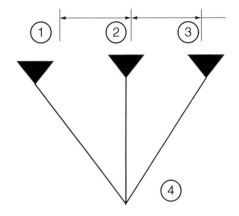

FIGURE 4.6

Three-bar truss design problem.

Table 4.6 Comparative analysis of ICO results with classical approaches for three-bar truss design problem.

Algorithm	y_1	y_2	Optimal weight (lb)
ICO	0.78432	0.41558	263.8184
GA	0.78439	0.42051	263.9096
ALO [33]	0.788662816000317	0.408283133832901	263.8958434
DEDS [34]	0.78867513	0.40824828	263.8958434
PSO-DE [35]	0.7886751	0.4082482	263.8958433
MBA [36]	0.7885650	0.4085597	263.8958522
Ray and Saini [37]	0.795	0.395	264.3
Tsa [38]	0.788	0.408	263.68
CS [39]	0.78867	0.40902	263.9716

$$g_3 = \frac{1}{y_1 + \sqrt{2}y_2} P - \sigma \leq 0 \qquad (4.14)$$

where

Variable ranges $0 \leq y_1, y_2 \leq 1$.

$L = 100$ cm; $P = 2$ KN/cm^2; $\sigma = 2$ KN/cm^2.

The objective of the problem is to minimize the weight of truss, which depends on stress, deflection, and buckling constraints. The suggested crow search optimization algorithm with 100 search agents and 500 iterations was engaged in this three-bar truss problem. The comparative analysis of improved crow search optimization (ICO) results with classical approaches for three-bar truss design problem is depicted in Table 4.6. For comparative analysis, classical optimization approaches are taken into account and the GA was found to be efficient in the design optimization of pin-jointed structures [32].

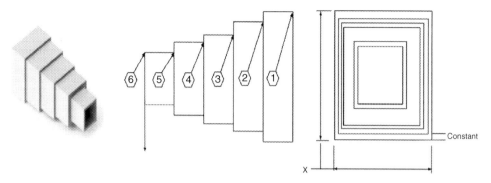

FIGURE 4.7

Cantilever beam design problem.

Table 4.7 Comparative analysis of ICO results with classical approaches for cantilever beam design problem.

Algorithm	y_1	y_2	y_3	y_4	y_5	Optimal weight
ICO	6.0086	5.3009	4.4988	3.4787	2.1553	1.33578
GA	6.06105	5.30368	4.49446	3.33352	2.11016	1.340761
ALO [33]	6.01812	5.31142	4.48836	3.49751	2.158329	1.33995
MMA [41]	6.0100	5.3000	4.4900	3.4900	2.1500	1.3400
GCA_I [41]	6.0100	5.3000	4.4900	3.4900	2.1500	1.3400
GCA_II [41]	6.0100	5.3000	4.4900	3.4900	2.1500	1.3400
CS [39]	6.0089	5.3049	4.5023	3.5077	2.1504	1.33999
SOS [42]	6.01878	5.30344	4.49587	3.49896	2.15564	1.33996

The cantilever beam design problem, a well-known structural design problem, is depicted in Fig. 4.7 [40]. This problem is formulated as follows:

Objective functions

$$\min f(y) = 0.6224(y_1 + y_2 + y_3 + y_4 + y_5) \tag{4.15}$$

Restrictive function

$$g(y) = \frac{61}{y_1^3} + \frac{27}{y_2^3} + \frac{19}{y_3^3} + \frac{7}{y_4^3} + \frac{1}{y_5^3} - 1 \leq 0 \tag{4.16}$$

where

Variable ranges $0.01 \leq y_1, y_2, y_3, y_4, y_5 \leq 100$.

The cantilever beam is build using five, hollow, square-section, box girders and the lengths of those girders are the design parameters of the problem. The suggested crow search optimization algorithm with 100 search agents and 500 iterations was engaged on this Cantilever beam design problem. The comparative analysis of ICO results with classical approaches for cantilever beam design problem is depicted in Table 4.7.

Table 4.8 Confusion matrix of SA–SVM (testing stage).

	Glioma	Meningioma	Pituitary	Precision
Glioma	177	34	74	62.10
Meningioma	33	84	25	59.15
Pituitary	75	24	87	46.77
Recall	62.10	59.15	46.77	

Table 4.9 Confusion matrix of GA–SVM (testing stage).

	Glioma	Meningioma	Pituitary	Precision
Glioma	209	34	42	73.33
Meningioma	21	88	33	61.97
Pituitary	55	20	111	59.67
Recall	73.33	61.97	59.67	

Table 4.10 Confusion matrix of CO–SVM (testing stage).

	Glioma	Meningioma	Pituitary	Precision
Glioma	236	22	27	82.80
Meningioma	20	113	9	79.57
Pituitary	29	7	150	80.64
Recall	82.80	79.57	80.64	

The real-time application of hybrid crow search optimization was performed in the classification of MR brain tumor images using the multiclass SVM algorithm. The parameter tuning is crucial in classical multiclass SVM algorithm; hence, for the tuning of parameters, various optimization techniques are employed. The hybrid CO, when coupled with the multiclass SVM algorithm, generates efficient results when compared with the multiclass SVM coupled with the simulated annealing, GA, and crow search optimization. The SVM parameters considered were Kernel type, degree (d), gamma (g), cost (c), coefficient (r) and weight (wi). The classification was made on the images taken from Fig.share public database (http://dx.doi.org/10.6084/m9.Fig.share.1512427). The database comprises 3064 T1-weighted contrast enhanced MR images with three types of brain tumor from 233 patients: meningioma (708 slices), glioma (1426 slices), and pituitary tumor (930 slices).

From Table 4.8, it is clear from the Confusion Matrix, simulated annealing (SA)–SVM approach generates a classification accuracy of 56.77% during the testing stage.

It is clear from the Confusion Matrix (Table 4.9), that the GA–SVM approach generated a classification accuracy of 66.55% during the testing stage.

It is clear from the results of the Confusion Matrix (Table 4.10), the CO–SVM approach generates a classification accuracy of 81.40% during the testing stage. The improved crow search optimization algorithm results are depicted in Figs. 4.8 and 4.9.

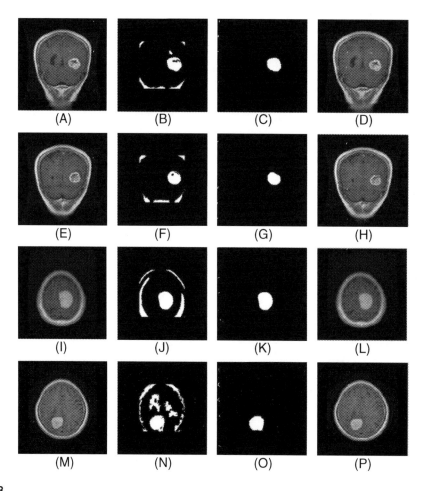

FIGURE 4.8

Improved Crow optimization algorithm output for the data sets (D1-D4).

Table 4.11 Confusion matrix of ICO–SVM (testing stage).				
	Glioma	Meningioma	Pituitary	Precision
Glioma	245	12	28	85.96
Meningioma	17	118	7	83.09
Pituitary	23	12	151	81.18
Recall	85.96	83.09	81.18	

The improved CO when coupled with the SVM algorithm was found to yield satisfactory results and highest accuracy of 83.85% in the testing phase when compared with the crow algorithm with a testing efficiency of 81.4%, GA with a testing accuracy of 66.55%, and simulated annealing with a testing accuracy of 56.77%. (Table 4.11).

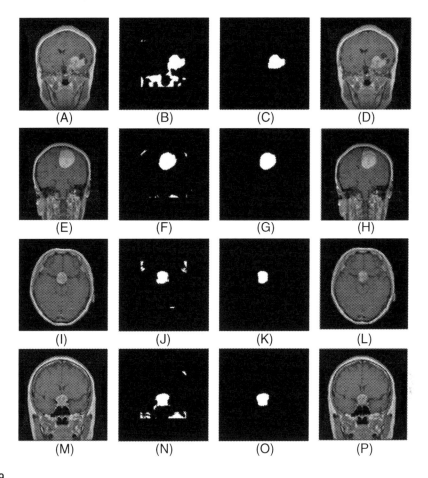

FIGURE 4.9

Improved Crow optimization algorithm output for the data sets (D5-D8).

4.4 Conclusion

This chapter proposes hybrid CO algorithm for solving engineering problems. The hybrid CO is an improved version of CO formulated by the incorporation of arithmetic crossover concept of the GA.

The following inferences are made from the results and discussion. The first experimental analysis of hybrid CSA was carried out on eight benchmark functions. The algorithm was executed for various iteration values of 50,100,150, and 200. Satisfactory results were produced when compared with the classical crow search optimization. The crow search optimization was found to be robust for standard benchmark functions when compared with the classical PSO and GA. The second experimental analysis was carried out on structural design problems: three-truss bar and cantilever beam design problems. The hybrid CSA results were found to be proficient in the comparative analysis with other classical optimization approaches. The third experimental analysis of hybrid crow search optimization was carried on the classification of MR brain tumor stages, the optimization algorithm was employed for the optimal tuning of parameters of the multiclass SVM algorithm. The hybrid crow search optimization coupled with the multiclass SVM algorithm generates efficient results when compared with the SVM coupled with the simulated annealing, GA, and crow search optimization.

The future works will be the exploring of hybrid crow search optimization application on medical image segmentation and compression, development of chaotic improved crow search optimization, and its validation on benchmark data sets, structural design problems, and real-time applications. The less number of parameter tuning and fast convergence rate makes it an attractive solution for solving real-world complex problems.

Acknowledgments

The authors would like to acknowledge the support provided by Nanyang Technological University under NTU Ref: RCA-17/334 for providing the medical images and supporting us in the preparation of the manuscript. Parasuraman Padmanabhan and Balazs Gulyas also acknowledge the support from Lee Kong Chian School of Medicine and Data Science and AI Research (DSAIR) center of NTU (Project Number ADH-11/2017-DSAIR) and the support from the Cognitive NeuroImaging Centre (CONIC) at NTU. The author S.N Kumar would like to acknowledge the support from the Schmitt Centre for Biomedical Instrumentation (SCBMI) of Amal Jyothi College of Engineering.

References

[1] P. Rincon. Science/Naturel crows and jays top bird IQ scale. BBC News. (2005), http://news.bbc.co.uk/2/hi/science/nature/4286965.stm.
[2] H. Prior, A. Schwarz, O. Güntürkün, Mirror-induced behavior in the magpie (Pica pica): evidence of self-recognition, PLoS Biol. 6 (8) (2008) e202.
[3] A. Askarzadeh, A novel metaheuristic method for solving constrained engineering optimization problems: crow search algorithm, Comput. Struct. 169 (2016) 1–2.
[4] C. Blum, A. Roli, Metaheuristics in combinatorial optimization: overview and conceptual comparison, ACM Comput. Surv. 35 (2003) 268–308.
[5] X.S. Yang, Metaheuristic optimization, Scholarpedia 6 (2011) 11472.
[6] A. Askarzadeh, A novel metaheuristic method for solving constrained engineering optimization problems: crow search algorithm, Comput. Struct. 169 (2016) 1–2.
[7] D. Karaboga, An Idea Based on Honey Bee Swarm for Numerical Optimization. Technical report-tr06, Erciyes University, Engineering faculty, Computer Engineering Department, 2005 Oct.
[8] S. Das, A. Biswas, S. Dasgupta, A. Abraham, Bacterial foraging optimization algorithm: theoretical foundations, analysis, and applications, in: A. Abraham, A.-E. Hassanien, P. Siarry, A. Engelbrecht (Eds.), Foundations of Computational Intelligence, 3, Springer, Berlin, Heidelberg, 2009, pp. 23–55.
[9] X.S. Yang, Bat algorithm for multi-objective optimisation, Int. J. Bioinspired Comput. 3 (5) (2011) 267–274.
[10] S.M. Almufti, Historical survey on metaheuristics algorithms, Int. J. Sci. World 7 (1) (2019) 1.
[11] T. Dokeroglu, E. Sevinc, T. Kucukyilmaz, A. Cosar, A survey on new generation metaheuristic algorithms, Comput. Ind. Eng. 137 (2019) 106040.
[12] S.N. Omkar, J. Senthilnath, R. Khandelwal, G.N. Naik, S. Gopalakrishnan, Artificial Bee Colony (ABC) for multi-objective design optimization of composite structures, Appl. Soft. Comput. 11 (1) (2011) 489–499.

[13] P.K. Hota, A.K. Barisal, R. Chakrabarti, Economic emission load dispatch through fuzzy based bacterial foraging algorithm, Int. J. Electric. Power Energy Syst. 32 (7) (2010) 794–803.
[14] A.H. Gandomi, X.S. Yang, Chaotic bat algorithm, J. Comput. Sci. 5 (2) (2014) 224–232.
[15] G.G. Wang, A.H. Gandomi, X. Zhao, H.C. Chu, Hybridizing harmony search algorithm with cuckoo search for global numerical optimization, Soft Comput. 20 (1) (2016) 273–285.
[16] A.F. Zubair, M.S. Mansor, Embedding firefly algorithm in optimization of CAPP turning machining parameters for cutting tool selections, Comput. Ind. Eng. 135 (2019) 317–325.
[17] L.L. Laudis, N. Ramadass, A lion's pride inspired algorithm for VLSI floorplanning, J. Circuits Syst. Comput. 29 (01) (2020) 2050003.
[18] H. Qin, P. Fan, H. Tang, P. Huang, B. Fang, S. Pan, An effective hybrid discrete grey wolf optimizer for the casting production scheduling problem with multi-objective and multi-constraint, Comput. Ind. Eng. 128 (2019) 458–476.
[19] M. Shabani, S.A. Mirroshandel, H. Asheri, Selective refining harmony search: a new optimization algorithm, Exp. Syst. Appl. 81 (2017) 423–443.
[20] S. Kavitha, P. Venkumar, N. Rajini, P. Pitchipoo, An efficient social spider optimization for flexible job shop scheduling problem, J. Adv. Manufact. Syst. 17 (02) (2018) 181–196.
[21] S. Saremi, S. Mirjalili, A. Lewis, Grasshopper optimisation algorithm: theory and application, Adv. Eng. Softw. 105 (2017) 30–47.
[22] A.H. Gandomi, X.S. Yang, A.H. Alavi, Cuckoo search algorithm: a metaheuristic approach to solve structural optimization problems, Eng. Comput. 29 (1) (2013) 17–35.
[23] A. Sadollah, A. Bahreininejad, H. Eskandar, M. Hamdi, Mine blast algorithm: a new population based algorithm for solving constrained engineering optimization problems, Appl. Soft. Comput. 13 (5) (2013) 2592–2612.
[24] S. Mirjalili, The ant lion optimizer, Adv. Eng. Softw. 83 (2015) 80–98.
[25] P. Upadhyay, J.K. Chhabra, Kapur's entropy based optimal multilevel image segmentation using crow search algorithm, Appl. Soft. Comput. 27 (2019) 105522.
[26] S.R. Sannasi Chakravarthy, H. Rajaguru, Lung cancer detection using probabilistic neural network with modified crow-search algorithm, Asian Pac. J. Cancer Prev. 20 (7) (2019) 2159–2166.
[27] S. Parvathavarthini, N. Karthikeyani Visalakshi, S. Shanthi, Breast cancer detection using crow search optimization based Intuitionistic fuzzy clustering with neighborhood attraction, Asian Pac. J. Cancer Prev. 20 (1) (2019) 157–165.
[28] X. Han, Q. Xu, L. Yue, Y. Dong, G. Xie, X. Xu, An improved crow search algorithm based on spiral search mechanism for solving numerical and engineering optimization problems, IEEE Access 8 (2020) 92363–92382.
[29] F. Shahabi, F. Poorahangaryan, S.A. Edalatpanah, H. Beheshti, A multilevel image thresholding approach based on crow search algorithm and Otsu method, International J. Comput. Intell. Appl. 19 (02) (2020) 2050015.
[30] Y. Meraihi, A.B. Gabis, A. Ramdane-Cherif, D. Acheli, A comprehensive survey of crow search algorithm and its applications, Artific. Intell. Rev. 28 (2020) 1–48.
[31] T. Ray, P. Saini, Engineering design optimization using a swarm with an intelligent information sharing among individuals, Eng. Optimiz. 33 (6) (2001) 735–748.
[32] T. Talaslioglu, A new genetic algorithm methodology for design optimization of truss structures: bipopulation-based genetic algorithm with enhanced interval search, Model. Simulat. Eng. 2009 (2009) 615162.
[33] S. Mirjalili, The ant lion optimizer, Adv. Eng. Softw. 83 (2015) 80–98.
[34] M. Zhang, W. Luo, X. Wang, Differential evolution with dynamic stochastic selection for constrained optimization, Inf. Sci. 178 (2008) 3043–3074.

[35] H. Liu, Z. Cai, Y. Wang, Hybridizing particle swarm optimization with differential evolution for constrained numerical and engineering optimization, Appl. Soft. Comput. 10 (2010) 629–640.

[36] A. Sadollah, A. Bahreininejad, H. Eskandar, M. Hamdi, Mine blast algorithm: a new population based algorithm for solving constrained engineering optimization problems, Appl. Soft Comput. 13 (5) (2013) 2592–2612.

[37] T. Ray, P. Saini, Engineering design optimization using a swarm with an intelligent information sharing among individuals, Eng. Optim. 33 (2001) 735–748.

[38] J.-F. Tsai, Global optimization of nonlinear fractional programming problems in engineering design, Eng. Optim. 37 (2005) 399–409.

[39] A.H. Gandomi, X.S. Yang, A.H. Alavi, Cuckoo search algorithm: a metaheuristic approach to solve structural optimization problems, Eng. Comput. 29 (1) (2013) 17–35.

[40] S. Saremi, S. Mirjalili, A. Lewis, Grasshopper optimisation algorithm: theory and application, Adv. Eng. Softw. 105 (2017) 30–47.

[41] H. Chickermane, H. Gea, Structural optimization using a new local approximation method, Int. J. Numer. Methods Eng. 39 (1996) 829–846.

[42] M.Y. Cheng, D. Prayogo, Symbiotic organisms search: a new metaheuristic optimization algorithm, Comput. Struct. 139 (2014) 98–112.

CHAPTER 5

MOGROM: Multiobjective Golden Ratio Optimization Algorithm

A.F. Nematollahi[a], A. Rahiminejad[b] and B. Vahidi[a]
[a]*Electrical Engineering Department, Amirkabir University of Technology, Tehran, Iran,* [b]*Department of Electrical and Computer Science, Esfarayen University of Technology, Esfarayen, North Khorasan, Iran*

5.1 Introduction

During the last decades, fast improvement of technology, particularly computers, propels the engineers to deploy computers for dealing with complex problems related to the design and operation of sophisticated systems. There is no doubt that computer development has had a great impact on solving complex problems; however, experts' knowledge is still vitally required [1]. On the other hand, different experts with different knowledge levels may have various opinions. This may make the desired answer unreliable. Hence, extensive studies have been performed to program the systems by computers without prior knowledge of the problem in addition to automated optimum design, while the results are reliable and cost effective. To this end, researchers have developed and deployed metaheuristic algorithms where no detail is needed for designing problems, and the system can be optimally architected by having a general overview of the issues [2,3].

Generally speaking, the optimization methods can be classified into two main classes including the numerical (traditional) and population-based methods. Using the former, the exact optimum point would be obtained where the gradient/derivation is zero. The methods falling into this group are hard to implement especially for those problems with a huge number of variables and/or complicated objective functions. The second group, that is, population-based methods, does not need derivation and details of the problem and is easy to implement [4–6].

Real-world optimization problems should be optimized, considering different objectives simultaneously. They are known as Multiobjective Optimization Problems (MOPs) and considered as Multicriteria Decision-Making problems. Sometimes the objectives are in contrast to each other and that would be hard to optimize all the objectives. Thus, a tradeoff needs to be made between different objectives. Solving these problems results in a set of solutions that do not take any precedence over each other. At the final stage, the experts will select the best solution among the obtained set, called Pareto, based on the system requirements [7,8].

The researchers have introduced numerous Multiobjective Optimization Methods (MOMs) during recent years. Nevertheless, based on the Nonfree Lunch theory, there is no unique method that can deal with all types of optimization problems and gain the best results [9]. This is the motivation for many researchers to keep looking for a method with excellent performance in solving a larger portion of problems. In this regard, this chapter presents a novel multiobjective optimization algorithm inspired by a very recently introduced optimization method known as the Golden Ratio Optimization

Method (GROM). Twenty-two benchmark test functions are employed to evaluate the performance of the proposed approach through different criteria. The obtained results are also compared to those of five well-known algorithms, including NSGA_II [10], MOPSO [11–13], MOGWO [14,15], MOALO [16,17], and MOLAPO [18,19], and the effectiveness of the introduced method is demonstrated.

The remainder of the chapter is organized as follows: In Section 5.1.1, the Multi-Objective problem will be defined, and in Section 5.1.2, the literature will be reviewed. The proposed multiobjective method will be illustrated in Section 5.2. In Section 5.3, the results will be obtained, compared, and investigated. Lastly, Section 5.4 concludes the chapter.

5.1.1 Definition of multiobjective problems (MOPs)

An MOP can be formulated as follows:

$$\text{Minimize } F(x) = [f_1(x) f_2(x) \ldots F_m(x)] \quad \text{S.T.} \quad x \in \Omega \tag{5.1}$$

where $F(x)$ consists of a set of objective functions, x is the vector of decision variables, and Ω denotes the search space. In a real-world optimization problem, changes in some variables might improve an objective and affect another negatively. Simply put, it is almost impossible to find a solution that optimizes all the objective functions. In these types of problems, instead of one single solution, a set of solutions, known as Pareto-Optimal Front (POF), is obtained. This will help the operator in selecting the best solution considering the operation point of the system and requirement.

In the concept of POF, there are two main definitions [8,20–22] stated by Edgeworth and Pareto [22] as follows:

Definition 1. Dominance

A solution $\vec{u} = (u_1, u_2, \ldots, u_n)$ dominates the solution $\vec{v} = (v_1, v_2, \ldots, v_n)$ if and only if $F(u) < F(v)$. In other words, $[f_1(u) f_2(u) \cdots f_m(u)] < [f_1(v) f_2(v) \cdots f_m(v)]$. It should also be noticed that if $m-1$ objective functions of solution u are equal to solution v and only one objective of u is lower than v, solution u still dominates the solution v. Mathematically, it can be formulated as follows [16]:

$$\forall i \in \{1, 2, \ldots, m\} : f_i(u) \leq f_i(v) \land \exists i \in \{1, 2, \ldots, m\} : f_i(u) < f_i(v) \tag{5.2}$$

Definition 2. Nondominated

If neither u dominates v nor v dominates u, these solutions are nondominated.

Definition 3. Pareto optimal

Solution u will be called Pareto optimal if and only if no solution in the population dominates the solution u. Mathematically [20]:

$$PF = \{u \in \Omega | \nexists v \in \Omega, F(v) < F(u)\} \tag{5.3}$$

Definition 4. Pareto-optimal set (POS)

A number of solutions fall into Pareto optimal, called POS, as follows:

$$P_s := \{\vec{u}, \vec{v} \in \Omega | \nexists F(\vec{v}) < F(\vec{u})\} \tag{5.4}$$

Definition 5. POF

The corresponding objective functions of Pareto solutions are called POF as follows:

$$P_f := \{F(\vec{u}) | \vec{u} \in P_s\} \quad (5.5)$$

5.1.2 Literature review

The methods introduced to solve MOPs fall into two categories: Priori [23] and Posteriori [24]. The former converts MOPs into Single-objective Optimization Problems (SOPs) using different ways [25,26] (e.g., weighting coefficient [27]) and obtain an individual solution as the optimal result.

The main drawbacks of this group are that they neglect the Pareto solutions, and the final best solution depends hugely on the weighted coefficients and objective priorities. Moreover, as the objective priorities or weighted coefficients are altered, the whole optimization problem needs to be solved once again to achieve the new results. It should also be mentioned that for those problems that the dimensions of different objectives are not the same, they must be normalized [8].

The latter group, that is, Posteriori, solves MOPs in a multiobjective form and obtains a set of solutions as the best final results [28,29]. The methods in this group can achieve a huge number of best solutions in just one trial. From the above, it can be concluded that the posterior approaches are more beneficial as (1) they solve the MOPs in the form of multiobjective, not SOP, (2) a wide range of possible solutions are obtained, and (3) the operators can select their desired solution from the obtained Pareto based on their needs.

The precise definition of MOMs is given in References [8,20,22,30]. As MOPs are inherently population-based, population-based optimization algorithms have better performance in comparison to numerical methods. These methods were utilized for solving MOPs in 1984, for the very first time [31]. Then, Coello [28,32] indicated that population-based evolutionary approaches are capable of solving MOPs. Since then, numerous evolutionary algorithms have been introduced and deployed to solve MOPs [8,33–40].

Evolutionary approaches follow the same process for solving SOPs and MOPs. The only difference is that in SOPs, the optimization method's output will be only one individual solution, known as the global optimum solution. In these types of problems, one can easily compare the solutions and determine the best and the worst ones. In MOPs, on the other side, the comparison cannot be performed likewise [18]. Instead of one single best solution, a set of solutions, called optimal Pareto, is obtained as the output of the multiobjective evolutionary algorithm (MOEA). These solutions do not have any preference over each other, and they are nondominated.

The quality of the obtained Pareto depends on several factors. The first and most important factor is that the obtained Pareto should be close to the True POF (TPOF). Second, the solutions, constituting the Pareto, should be uniformly distributed [41,42]. Third, the more the solutions obtained, the better the search space is searched. In summary, the closer the Pareto is to the actual one, and the more the solutions are uniformly distributed, the better the quality of the method [8,43–45].

5.1.3 Background and related work

As already mentioned, for any single-objective evolutionary method, there is a multiobjective form. Although the quality of the results obtained by different methods is not the same, according to the No

Free Lunch (NFL) theory, there is no single algorithm that is capable of solving all the problems in the best way [9]. Hence, optimization algorithms are proposed by different researchers on a daily basis.

Among different MOEAs, the genetic algorithm is one of the most well-known algorithms [46]. The genetic algorithm, along with the spatial information system, provides a powerful combination for solving large-scale and spatial hybrid optimization problems [41]. Niche Pareto Genetic Algorithm (NPGA) is one of the multiobjective forms of the genetic algorithm that is able to solve larger problems in a shorter time compared to the previous versions [42]. Strength Pareto Evolutionary Algorithm (SPEA) used the concept of nondominated to solve MOPs in 1999 [47]. This method is capable of obtaining the best solution with the help of multicriteria evolutionary algorithms. Nondominated Sorting Genetic Algorithm (NSGA) is one of the commonly used and robust algorithms available for solving MOPs, first proposed by [10]. This method's main advantages are fast convergence, elitism, no need for an external population, and relative ease of implementation.

During the recent years, there have been many efforts done on MOEA. The most important articles in this area can be found in NSGA_I [30], NSGA-II [10], MOPSO [12,48,49], MODE [43,50,51], MOTLBO [52], MOSFL [53–55], MOABC [36,56], MOALO [16], MOWGO [14], and MOLAPO [18].

This chapter proposes Multiobjective GROM (MOGROM), the multiobjective version of the GROM, which is recently presented in [57]. The performance of the introduced algorithm is evaluated by means of 22 benchmark multiobjective functions. Then the obtained results are compared with the results of five well-known methods, including MOPSO [11], MOLAPO [18], NSGAII [10], MOALO [16], and MOGWO [14]. The evaluations reveal the effectiveness, robustness, and excellent performance of the proposed method.

5.2 GROM and MOGROM

It is essential for optimization algorithms to have the ability to solve both Sops and MOPs. As the solving procedure of these two kinds of problems is quite different, it is necessary to introduce the multiobjective version of already presented single-objective optimization algorithms and, then, evaluate their performance in comparison to the existing methods.

GROM has been introduced in 2019 in [57]. The method is inspired by the golden ratio, which is a mathematical ratio and can be found widely in natural behavior. This ratio, which is also known as Golden Average, Golden ratio or ϕ, is as follows:

$$\phi = \frac{1 + \sqrt{5}}{2} \approx 1.6180339887 \tag{5.6}$$

Regarding the Fibonacci series, there is an impressive ratio between consecutive numbers as 1.61, which is exactly the golden ratio. This ratio can be found in DNA spirals, human ear spiral, the snail, the spiral structure of galaxies, the leaves of trees, lines and engravings on peacock feathers, and sunflower spirals, to name but not limited. It is noteworthy to mention that the masterpieces or draws that contain the golden ratio are more eye-catching.

To solve an optimization problem by means of a metaheuristic method, initially, a set of solutions is defined randomly in the search space. These solutions are changed based on a set of mathematical formulations to achieve the global best solution. In GROM, the solutions are updated in different steps. Initially, the algorithm tries to enhance the quality of the worst solution in the population. To this end, the mean and the worst solutions are extracted from society based on their fitness values. If the mean solution has better quality than the worst ones, the latter is discarded and replaced by the former. This step, where the worst solution is improved, can speed up the algorithm convergence.

Then, the solutions are directed toward the best solution in society. To determine the step's length of each solution movement toward the best global one, the golden ratio is employed. To avoid getting stuck in local optimum points, a random term is also factored in. The following formulations illustrate the solution's updating procedure:

$$F_{best} > F_{medium} > F_{worst}$$
$$\vec{X_t} = \vec{X_{medium}} - \vec{X_{worst}} \tag{5.7}$$

$$F_t = GF \times \frac{\varphi^T - (1-\varphi)^T}{\sqrt{5}}, \quad GF = 1.618$$
$$T = \frac{t}{t_{max}} \tag{5.8}$$
$$X_{new} = (1 - F_t)X_{best} + rand \times F_t \times X_t$$

In the third step, an individual solution is selected and updated by interacting with another member. Moreover, this interacting is also affected by the other members of society. Put in other words, for any sample solution i, another solution j is selected randomly. The solution that is constructed by the mean value of all solutions is also considered. Then, among these three solutions, the best and the worst ones are determined. The selected solution i, then, is moved toward the best one and in the opposite direction of the worst one. The step's length, here, is also based on the golden ratio. The mathematical interpretation of this step is formulated as follows:

$$Xnew = Xold + rand \times \left(\frac{1}{GF}\right) \times (X_{best} - X_{worst})$$
$$\frac{1}{GF} = 0.618 \tag{5.9}$$

In each step, if the new solution has better quality than the old one, they are replaced; otherwise, the updated one is discarded.

5.2.1 MOGROM

To create the multiobjective version of GROM, nearly the same approach must be followed. In the single-objective version, the best and the worst solutions are needed for updating all the individuals in the population. However, in multiobjective form, it is not as easy as the single form to find these individuals. The proposed multiobjective version of GROM, known as MOGROM, is illustrated in the following.

The first step, like any other metaheuristic method, is generating randomly a set of solutions in the search space. Then, these solutions are compared to each other based on their fitness values. A nondominated sorting approach is utilized in the proposed MOEA to sort the solutions from the best to the worst. In this approach, each solution is compared to the other members of society. Based on how frequent the solution is dominated by the others, it is placed in a class known as *front*. Those solutions that are not dominated by any other solution in the whole population are positioned in the first front. This approach is repeated for the remaining solutions to obtain the other fronts. Then, the solutions of front 1 are stored in a matrix called *Repository*. Due to the limited space of *Repository*, if the number of solutions in the first front is higher than the *Repository* capacity, a certain number of solutions will be stored. The nominated solutions for storing into the *Repository* are selected based on crowding distance. This criterion represents the distance of a solution to its neighbors. It is crystal clear that the space around a solution with high crowding distance is not well searched. Thus, the solutions with higher crowding distance are stored in the *Repository* to have an appropriate diversity in the search space. Given that the GROM needs the best and the worst solutions to update all the individuals, the solutions from the *Repository* with the highest and lowest crowding distance are considered as the best and the worst solutions, respectively. This helps the algorithm to update the solutions toward the areas that are not searched very well and avoid congesting the solutions in one area.

In summary, first of all, the population is created by defining the solutions randomly in the predefined search area. Then, the objective functions for any solution are calculated, and the whole solutions with their corresponding fitness values are stored in the matrix *Population*. The solutions are placed on different fronts using the nondominated sorting approach. The solutions were placed in the first front, then sorted in a descending order considering crowding distances. If the number of these solutions is less than the *Repository* matrix capacity, they are all stored; otherwise, they are stored from the first solution until the *Repository* matrix is filled.

Now, the solutions are updated based on the GROM algorithm. The average solution (x_{av}), is obtained, and the corresponding objective values are calculated. If this solution dominates the worst solution in *Repository*, it is replaced; if they are nondominated to each other, the replacement is performed considering a probability value of 50%; otherwise, no action is needed.

For any solution i, a random solution j, is selected and the comparison is implemented for three solutions, that is, solution i, j, and the average. The two different cases may occur as follows:

Case I: If one of the solutions dominates the others, it is considered as the best one. Among the two other solutions, if one of the solutions dominates the other one, it is considered as the average, and the last one is the worst. If these solutions are nondominated, the average and the worst are determined randomly by the probability of 50%.

Case II: For any other cases, the best, the average, and the worst are selected randomly by the probability of 50%.

By defining the best, the worst, and the average solutions among the three aforementioned solutions, solution i, is updated as follows:

$$\vec{X_t} = \overrightarrow{X_{medium}} - \overrightarrow{X_{worst}} \tag{5.10}$$

$$F_t = GF \times \frac{\varphi^T - (1-\varphi)^T}{\sqrt{5}}, \quad GF = 1.618$$

$$T = \frac{t}{t_{max}}$$

$$X_{new} = (1 - F_t)X_{best} + rand \times F_t \times X_t$$

Then, the objectives corresponding to the new updated solutions are calculated and compared to the previous one. If the new solution dominates the old one, the former is kept and the latter is discarded, if they are nondominated to each other, the replacement is performed randomly considering 50% probability, otherwise, no action is needed.

When all the solutions are updated, these solutions with those stored in the *Repository* are sorted again to update the *Repository* matrix. Now, the solutions must be updated based on the new best and the worst solutions. These solutions are the ones from the *Repository* with the highest and the lowest crowding distances.

Then, each solution is updated as follows:

$$Xnew = Xold + rand \times \left(\frac{1}{GF}\right) \times (X_{best} - X_{worst}) \quad (5.11)$$

$$\frac{1}{GF} = 0.618$$

The same replacement procedure is performed again. It should be mentioned that the boundaries of variables are checked all the time. If the variables violate their boundaries, they will be corrected. If there is any constraint related to objective functions, a penalty factor will be considered.

This procedure is repeated as long as the convergence criterion is not achieved. Like any other evolutionary algorithm, the convergence criterion is the Maximum Cycle Number, which is a predefined value. As the algorithm is repeated a certain number of times, the optimization process is completed. The stepwise procedure of the proposed MOEA is explained in the following, and the flowchart is depicted in Fig. 5.1.

Step 1. A set of solutions are generated randomly in the search space.
Step 2. The objective function corresponding to each solution is calculated, the first front is obtained, and the *Repository* matrix is formed.
Step 3. The average solution (x_{av}) and its corresponding objective functions (F_{av}) are calculated.
Step 4. The worst solution is updated based on F_{av}.
Step 5. For any solution i, a solution j is select randomly.
Step 6. The objective function of solution i, j, and x_{av} are compared and the best, the worst, and the average are determined.
Step 7. The new position of each solution is updated based on Eq. (5.10)
Step 8. If the objective function of the new solution "i" dominates the old solution "i," it is replaced. If they are nondominated one of the solutions is randomly selected.
Step 9. The *Repository* is updated.
Step 10. All solutions update based on Eq. (5.11)
Step 11. If the objective function of the new solution "i" dominates the old solution "i," it is replaced. If they are nondominated one of the solutions randomly selected.
Step 12. The *Repository* is update.
Step 13. If convergence criterion is satisfied, print the *Repository* as the best achieved objective functions, else go to Step 3.

5.3 Simulation results, investigation, and analysis

To evaluate the performance of the proposed method, it is applied to different benchmark test functions and the results are compared from statistic and quality points of view. The algorithms used for comparisons include MOGWO, MOLAPO, NSGA_II, MOPSO, and MOALO. For static analysis,

Chapter 5 MOGROM: Multiobjective Golden Ratio Optimization Algorithm

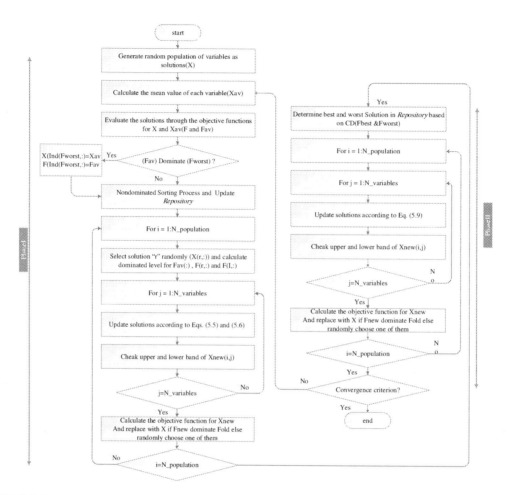

FIGURE 5.1

Flowchart of MOGROM.

30 trials are run for each algorithm, and the statistical measures such as mean value and standard deviation are compared.

Twenty-two benchmark test functions, extracted from [58], are used to evaluate the performance of the proposed algorithm. These test functions are simple mathematical formulation that can be easily implemented. The front shape is known and they are adequate to compare the results with TPOF. As these test functions have different front shapes, the proposed algorithm can be evaluated in solving different types of problems.

To compare the results with the TPOF, four criteria are employed. The first one is Generational Distance (GD), showing the distance between the obtained POF and TPOF. The less the distance, the better the performance. The second criterion is Inverted Generational Distance (IGD), which measures the convergence and the diversity of the obtained results. The lower value of this criterion guarantees

5.3 Simulation results, investigation, and analysis

the better performance of an algorithm. The third criterion measures the distribution of the obtained solutions in comparison to TPOF. The higher value shows that the obtained POF is well covered. The fourth criterion is Spacing, which compares the obtained POF distribution compared to TPOF. The lower value states that the obtained solutions are equally distributed through the first front.

The 22 test function can be categorized into 5 groups based on the number of fronts, type of fronts, and the number of objective functions. Table 5.1 illustrates functions in detail, and the five groups are explained in the following.

1. The first group contains the test functions with two objective functions. These functions include one overall and one local front, and the fronts are continued. The test functions are differentiated based on linear or nonlinear fronts, and also the existence of contact between the fronts.
2. The second group encompasses test functions with three objectives. In these test functions, there is more than one local front. Using these test cases, the performance of the proposed method in solving problems with several local fronts will be evaluated.
3. The third group includes four test functions, three of which have two objectives and the other one has three. This group of test functions includes different discrete fronts. Hence, it is worthy of evaluating algorithm performance in solving these kinds of problems.
4. The fourth group contains two test functions having fronts that change both in a continuous and stepwise manner.
5. The fifth class evaluates the proposed method in dealing with test functions that have three objectives.

5.3.1 First class

This class contains nine test functions with the same formulation but different variables. Their global and local front's shapes are completely different as a consequence of different variables. In this class, the fronts are convex or linear. In some cases, there is contact between the fronts. The formulations and the variables of these benchmark problems are explained in Eq. (5.12). The results of solving this class using the proposed algorithm are depicted in Figs. 5.2–5.4. In the figures, the True global and local fronts, as well as those obtained by the proposed MOGROM, are presented.

The numerical results are tabulated in Table 5.2. The comparisons between the results obtained by the proposed method and the other five methods using the aforementioned criteria reveal that the POF achieved by the proposed MOEA is very close to TPOF. In terms of computational time, MOGROM is the second best after MOALO, and the others are far worse than these two algorithms. From the result quality viewpoint, the proposed method is always among the two best methods for all the test functions and all the criteria. For IGD and spacing criteria, the introduced algorithm outperforms the others in all nine test functions.

$$f_1(x) = x_1$$
$$f_2(x) = H(x_2) \times \{G(x) + S(x_1)\} + \omega$$
$$H(x) = \frac{1}{\sqrt{2\Pi}} e^{-0.5\left(\frac{x-1.5}{0.5}\right)^2} + \frac{2}{\sqrt{2\Pi}} e^{-0.5\left(\frac{x-0.5}{\alpha}\right)^2}$$
$$G(x) = \sum_{i=3}^{N} 50x_i^2$$
$$S(x) = -x_1^\beta$$
$$\alpha > 0$$

(5.12)

Table 5.1 Test functions detail.

Test function	Objective function number	Variable of function	Variable limit	Number of fronts	Pareto shape
F1	2	2	[Min,Max]=[0,2]	2	Global front: linear Local front: linear
F2	2	2	[Min,Max]=[0,2]	2	Global front: Concave Local front: Concave
F3	2	2	[Min,Max]=[0,2]	2	Global front: Convex Local front: Convex
F4	2	2	[Min,Max]=[0,2]	2	Global front: convex Local front: linear
F5	2	2	[Min,Max]=[0,2]	2	Global front: linear Local front: convex
F6	2	2	[Min,Max]=[0,2]	2	Global front: linear Local front: concave
F7	2	2	[Min,Max]=[0,2]	2	Global front: concave Local front: linear
F8	2	2	[Min,Max]=[0,2]	2	Global front: concave Local front: convex
F9	2	2	[Min,Max]=[0,2]	2	Global front: convex Local front: concave
F10	2	2	[Min,Max]=[0,1]	5	Global front: Convex Local fronts: Convex
F11	2	2	[Min,Max]=[0,1]	5	Global front: linear Local fronts: linear
F12	2	2	[Min,Max]=[0,1]	5	Global front: Concave Local front: Concave
F13	2	2	[Min,Max]=[0,1]	6	Discrete fronts
F14	2	2	[Min,Max]=[0,1]	6	Discrete fronts
F15	2	2	[Min,Max]=[0,1]	6	Discrete fronts
F16	3	2	[Min,Max]=[0,1]	Plane	Discrete plane
F17	2	2	[Min,Max]=[0,1]	1	Step wise front
F18	2	2	[Min,Max]=[0,1]	1	Step wise front
F19	3	2	[Min,Max]=[0,1]	Plane	Plane
F20	3	2	[Min,Max]=[0,1]	Plane	Plane
F21	3	2	[Min,Max]=[0,1]	Plane	Plane
F22	3	2	[Min,Max]=[0,1]	Plane	Plane

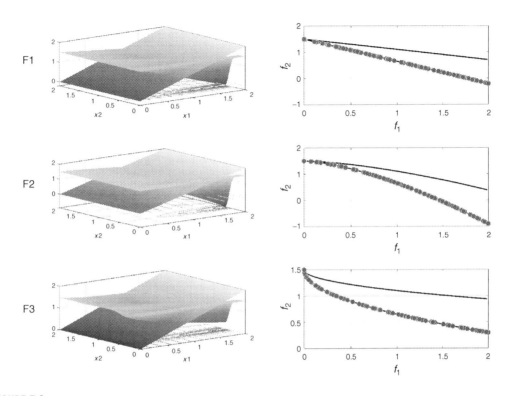

FIGURE 5.2

Search Space, Pareto Front, and Obtained Pareto Front by MOGROM for the Class 1 test functions (F1–3).

	F1	F2	F3	F4	F5	F6	F7	F8	F9
α	0.1	0.1	0.1	0.1	0.1	0.1	0.1	0.1	0.1
ω	1.5	1.5	1.5	1.5	1.5	1.5	1.5	1.5	1.5
β_1	1	1.5	0.5	0.5	1	1	1.5	1.5	0.5
β_2	1	1.5	0.5	1	0.5	1.5	1	0.5	1.5

5.3.2 Second class

The test functions of this class have more local fronts in comparison to those in the previous class. This class includes three test functions, and the mathematical formulation and the variables are illustrated in Eq. (5.13). The difference between these three test functions is related to the front's shape (e.g., linear, convex, or concave). The fronts obtained for these test functions by means of MOGROM are depicted

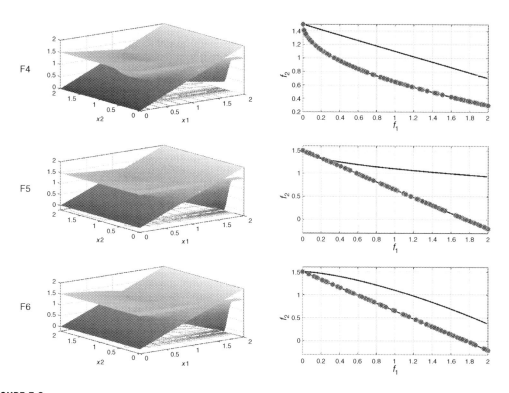

FIGURE 5.3

Search Space, Pareto Front, and Obtained Pareto Front by MOGROM for the Class 1 test functions (F4–6).

in Fig. 5.5. It is shown that the proposed MOGROM obtained the TPOF among all the local fronts and distributed the solutions very well through the best front.

$$
\begin{aligned}
& f_1(x) = x_1 \\
& f_2(x) = H(x_2) \times \{G(x) + S(x_1)\} + \omega \\
& H(x) = \frac{e^{-x^2} \cos(\lambda \times 2\pi x) - x}{\gamma} + 0.5 \\
& G(x) = \sum_{i=3}^{N} 50 x_i^2 \\
& S(x) = -x_1^{\beta} \\
& \gamma > 1.3 \\
& \lambda > 1
\end{aligned}
\tag{5.13}
$$

The criteria for different methods are gathered in Table 5.3. The method outperforms the others for test functions 10 and 12. For test function F11, the proposed algorithm has worse answers than Multiobjective Particle Swarm Optimization (MPSO) for the spread criterion. In terms of computational time, it has excellent performance.

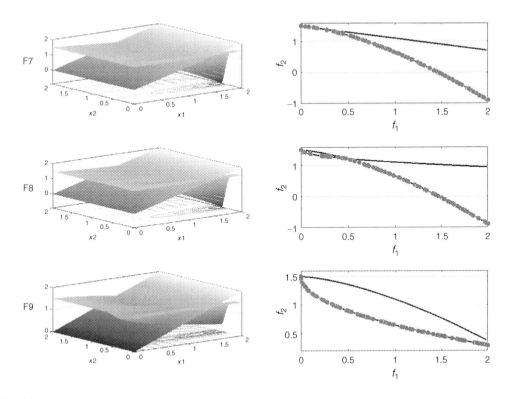

FIGURE 5.4

Search Space, Pareto Front, and Obtained Pareto Front by MOGROM for the Class 1 test functions (F7–9).

5.3.3 Third class

This class encompasses four test functions. The first three functions have two objectives and contain discrete and several local fronts. The fourth one contains three objectives and discrete fronts. The formulation and the variable are illustrated in Eqs. (5.14) and (5.15). The front shapes and the results related to this class of functions are depicted in Figs. 5.6 and 5.7. The results evaluation (Table 5.4) demonstrates that the proposed method has the best performance and is the second-best one from a computational time perspective.

$$\begin{aligned}
&f_1(x) = x_1 \\
&f_2(x) = H(V_2) \times \{G(x) + S(x_1)\} + \omega \\
&H(x) = \frac{e^{-x^2}\cos(\lambda \times 2\pi x) - x}{\gamma} + 0.5 \\
&G(x) = \sum_{i=3}^{N} 50 x_i^2 \\
&S(x) = -x_1^\beta \\
&\gamma > 1.3 \\
&\lambda > 1
\end{aligned} \quad (5.14)$$

Table 5.2 Obtained comparison criterion in solving class 1 test functions by different method.

		GD criteria		Metric of spread		Metric of spacing		IGD criteria		Computation time(s)
		Ave.	Std.	Ave.	Std.	Ave.	Std.	Ave.	Std.	
F1	MOGROM	5.68E-04	4.95E-06	7.17E-01	6.41E-02	6.73E-03	1.09E-03	1.51E-03	1.79E-04	2.37E+01
	MOLAPO[18]	5.54E-04	2.51E-05	5.91E-01	5.36E-02	1.11E-02	1.09E-03	8.54E-03	3.40E-04	2.42E+01
	MOPSO[18]	9.28E-04	4.27E-05	7.39E-01	8.79E-02	1.80E-02	1.71E-03	1.39E-02	1.49E-03	6.73E+01
	NSGA_II[18]	7.50E-04	1.72E-05	3.90E-01	2.96E-02	1.20E-02	6.92E-04	8.59E-03	4.80E-04	2.75E+02
	MOALO[18]	8.47E-04	1.46E-04	4.78E-01	9.04E-02	2.20E-02	6.44E-03	4.77E-02	8.49E-03	2.35E+01
	MOGWO[18]	1.93E-03	9.62E-04	5.66E-01	7.26E-02	3.20E-02	9.71E-03	2.55E-02	5.67E-03	9.68E+01
F2	MOGROM	5.80E-04	5.30E-05	6.74E-01	2.03E-03	1.19E-02	1.54E-03	7.62E-03	7.86E-04	2.15E+01
	MOLAPO[18]	5.86E-04	4.42E-05	5.79E-01	4.55E-02	1.12E-02	1.32E-03	8.98E-03	8.30E-04	2.31E+01
	MOPSO[18]	9.95E-04	3.77E-05	6.94E-01	4.32E-02	1.88E-02	1.44E-03	1.45E-02	1.30E-03	6.99E+01
	NSGA_II[18]	9.00E-04	4.78E-05	4.03E-01	4.14E-02	1.60E-02	8.20E-04	1.03E-02	3.10E-04	2.65E+02
	MOALO[18]	1.68E-03	6.40E-04	5.64E-01	4.65E-02	3.71E-02	1.98E-02	5.22E-02	5.41E-03	2.08E+01
	MOGWO[18]	1.10E-03	6.21E-05	4.25E-01	2.17E-02	3.63E-02	5.43E-03	2.71E-02	5.09E-03	1.07E+02
F3	MOGROM	3.34E-04	7.92E-05	5.33E-01	2.77E-02	3.76E-04	2.64E-04	4.63E-04	1.29E-04	2.19E+01
	MOLAPO[18]	6.81E-04	9.67E-05	6.22E-01	3.14E-02	9.45E-03	7.63E-04	8.03E-03	1.06E-03	2.13E+01
	MOPSO[18]	9.55E-04	8.45E-05	7.35E-01	2.84E-02	1.49E-02	1.89E-03	1.16E-02	5.02E-04	6.41E+01
	NSGA_II[18]	1.00E-03	5.55E-05	4.14E-01	5.25E-02	1.19E-02	7.80E-04	8.40E-03	4.80E-04	2.31E+02
	MOALO[18]	1.14E-03	3.54E-04	5.71E-01	1.17E-01	3.25E-02	1.58E-02	4.07E-02	1.21E-02	2.24E+01
	MOGWO[18]	7.82E-04	7.51E-05	5.57E-01	1.30E-01	1.81E-02	3.83E-03	2.25E-02	6.67E-03	1.00E+02
F4	MOGROM	6.03E-04	1.97E-05	6.91E-01	3.28E-03	5.35E-03	4.63E-04	7.44E-03	3.18E-04	2.71E+01
	MOLAPO[18]	6.94E-04	4.97E-05	5.82E-01	7.62E-02	9.75E-03	9.70E-04	7.77E-03	3.20E-04	2.61E+01
	MOPSO[18]	9.80E-04	1.13E-04	7.58E-01	3.97E-02	1.50E-02	5.00E-04	1.11E-02	1.02E-03	5.26E+01
	NSGA_II[18]	1.06E-03	4.41E-05	4.09E-01	3.31E-02	1.21E-02	1.30E-03	8.05E-03	3.50E-04	2.43E+02
	MOALO[18]	1.26E-03	5.96E-04	4.67E-01	7.91E-02	2.37E-02	4.41E-03	5.41E-02	1.43E-02	2.45E+01
	MOGWO[18]	8.33E-04	1.20E-04	6.50E-01	3.98E-02	2.28E-02	4.16E-03	1.68E-02	1.31E-03	1.02E+02

(continued on next page)

Table 5.2 (continued)

		GD criteria		Metric of spread		Metric of spacing		IGD criteria		Computation time(s)
		Ave.	Std.	Ave.	Std.	Ave.	Std.	Ave.	Std.	
F5	MOGROM	5.67E-04	2.40E-05	6.73E-01	3.52E-02	1.12E-02	4.83E-04	8.96E-03	6.55E-04	2.30E+01
	MOLAPO[18]	9.26E-04	4.31E-05	5.86E-01	3.97E-02	1.19E-02	1.40E-03	1.06E-02	8.00E-04	1.47E+01
	MOPSO[18]	1.24E-03	1.27E-04	6.95E-01	7.31E-02	1.84E-02	2.43E-03	1.45E-02	5.80E-04	6.31E+01
	NSGA_II[18]	1.12E-03	1.05E-05	3.97E-01	6.42E-02	1.34E-02	1.10E-03	1.00E-02	5.10E-04	1.92E+02
	MOALO[18]	1.53E-03	5.60E-04	4.30E-01	8.28E-02	2.13E-02	8.79E-03	5.57E-02	4.04E-02	1.88E+01
	MOGWO[18]	1.37E-03	3.59E-04	5.21E-01	7.75E-02	2.93E-02	6.14E-03	1.97E-02	2.46E-03	1.01E+02
F6	MOGROM	8.54E-05	1.40E-05	6.91E-01	1.91E-02	8.02E-03	2.35E-04	8.13E-03	6.38E-05	2.40E+01
	MOLAPO[18]	5.74E-04	2.24E-05	5.76E-01	4.49E-02	1.10E-02	8.40E-04	8.66E-03	2.00E-04	2.52E+01
	MOPSO[18]	8.80E-04	1.10E-04	6.57E-01	6.10E-02	1.61E-02	1.20E-03	1.30E-02	2.01E-03	6.01E+01
	NSGA_II[18]	7.40E-04	2.71E-05	4.42E-01	7.16E-02	1.40E-02	1.41E-03	8.79E-03	3.30E-04	2.12E+02
	MOALO[18]	1.19E-03	4.52E-04	5.80E-01	5.15E-02	2.79E-02	9.10E-03	3.92E-02	6.94E-03	2.36E+01
	MOGWO[18]	9.16E-04	8.84E-05	5.28E-01	8.35E-02	2.73E-02	9.37E-03	2.58E-02	7.26E-03	9.81E+01
F7	MOGROM	5.65E-04	3.18E-05	6.97E-01	3.70E-02	2.28E-03	1.60E-04	3.26E-03	5.25E-04	8.60E+00
	MOLAPO[18]	6.46E-04	2.85E-05	5.81E-01	5.41E-02	1.32E-02	8.60E-04	1.01E-02	5.00E-04	2.72E+01
	MOPSO[18]	1.02E-03	5.39E-05	6.19E-01	1.04E-01	1.90E-02	2.30E-03	1.41E-02	1.40E-03	7.46E+01
	NSGA_II[18]	9.59E-04	4.99E-05	3.74E-01	5.50E-02	1.50E-02	1.76E-03	1.04E-02	8.85E-04	2.03E+02
	MOALO[18]	1.11E-03	2.16E-04	6.07E-01	1.20E-01	3.75E-02	8.55E-03	5.51E-02	9.38E-03	2.21E+01
	MOGWO[18]	9.94E-04	6.94E-05	5.51E-01	3.09E-02	3.17E-02	5.75E-04	2.47E-02	9.80E-04	9.98E+01

(continued on next page)

Table 5.2 (continued)

		GD criteria		Metric of spread		Metric of spacing		IGD criteria		Computation time(s)
		Ave.	Std.	Ave.	Std.	Ave.	Std.	Ave.	Std.	
F8	MOGROM	2.40E-03	1.71E-05	6.99E-01	6.39E-03	3.41E-03	3.86E-04	1.67E-02	2.54E-04	2.05E+01
	MOLAPO[18]	2.36E-03	5.62E-05	6.01E-01	1.46E-02	1.39E-02	1.23E-03	2.17E-02	4.30E-04	2.38E+01
	MOPSO[18]	2.84E-03	1.50E-04	5.93E-01	6.24E-02	1.89E-02	1.30E-03	2.44E-02	1.31E-03	6.11E+01
	NSGA_II[18]	2.90E-03	1.93E-04	4.08E-01	2.60E-02	1.57E-02	1.20E-03	2.24E-02	8.00E-04	2.22E+02
	MOALO[18]	3.60E-03	1.14E-03	5.40E-01	1.05E-01	4.23E-02	1.21E-02	5.35E-02	9.79E-03	1.88E+01
	MOGWO[18]	2.41E-03	5.81E-04	9.37E-01	2.39E-01	3.13E-02	6.72E-03	3.18E-02	1.27E-02	9.84E+01
F9	MOGROM	6.57E-04	3.75E-05	6.05E-01	2.20E-02	3.33E-03	9.65E-04	6.95E-03	1.33E-04	2.01E+01
	MOLAPO[18]	6.33E-04	5.13E-05	5.89E-01	1.92E-02	9.33E-03	1.02E-03	7.38E-03	2.80E-04	2.83E+01
	MOPSO[18]	1.00E-03	4.64E-05	6.73E-01	2.26E-02	1.38E-02	1.10E-03	1.08E-02	5.10E-04	5.84E+01
	NSGA_II[18]	1.10E-03	1.20E-04	4.37E-01	4.93E-02	1.18E-02	7.98E-04	8.14E-03	4.10E-04	1.93E+02
	MOALO[18]	1.16E-03	8.70E-04	4.27E-01	1.27E-01	2.29E-02	1.17E-02	6.01E-02	1.98E-02	2.24E+01
	MOGWO[18]	9.14E-04	1.60E-04	5.98E-01	4.37E-02	2.16E-02	2.08E-03	2.15E-02	5.66E-03	9.91E+01

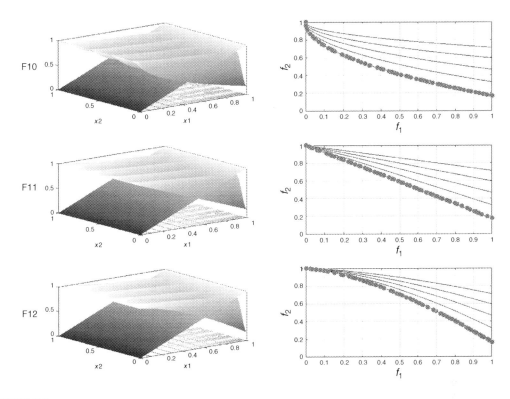

FIGURE 5.5

Search Space, Pareto Front, and Obtained Pareto Front by MOGROM for the Class 2 test functions (F10–12).

	F15	F14	F13	
	8	4	2	ζ
	6	6	6	λ
	3	3	3	γ
	0.5	0.5	0.5	ω

$$F16 = \begin{cases} f_1(x) = x_1 \\ f_2(x) = x_2 \\ f_3(x) = G(x) \times \left(\left(\left\{ 1 - \sqrt{\dfrac{x_1}{G(x)}} - \dfrac{x_1}{G(x)} \sin(\zeta \times 2\pi x_1) \right\} + H(x_1) \right) \right) \times \\ \quad \left((\{1 - \sqrt{\dfrac{x_2}{G(x)}} - \dfrac{x_2}{G(x)} \sin(\zeta \times 2\pi x_2)\} + H(x_3)) \right) \\ H(x) = \dfrac{e^{-2x^2} \sin(\lambda \times 2\pi(x + \dfrac{\pi}{4\lambda})) - x}{\gamma} + 0.5 \\ G(x) = 1 + 10 \dfrac{\sum_{i=2}^{N} x_i}{N} \\ 1 \leq \lambda \\ 1 \leq \gamma \end{cases} \quad (5.15)$$

Table 5.3 Obtained comparison criterion in solving class 2 test functions by different methods.

		GD criteria		Metric of spread		Metric of spacing		IGD criteria		Time (s)
		Ave.	Std.	Ave.	Std.	Ave.	Std.	Ave.	Std.	
F10	MOGROM	3.34E-05	2.65E-06	8.24E-01	6.10E-02	5.25E-03	7.58E-05	3.19E-03	1.07E-04	2.13E+01
	MOLAPO[18]	3.52E-04	3.93E-05	7.90E-01	6.26E-02	6.21E-03	4.67E-04	4.48E-03	1.83E-04	2.25E+01
	MOPSO[18]	6.10E-04	6.02E-05	7.61E-01	6.49E-02	8.20E-03	1.12E-03	6.33E-03	8.62E-04	6.29E+01
	NSGA_II[18]	7.49E-04	8.50E-05	4.23E-01	2.16E-02	6.71E-03	2.44E-04	4.72E-03	3.46E-04	1.62E+02
	MOALO[18]	1.09E-03	5.30E-04	7.30E-01	1.19E-01	2.41E-02	2.11E-02	6.26E-02	4.21E-02	2.88E+01
	MOGWO[18]	4.50E-04	1.00E-04	6.36E-01	1.58E-01	1.57E-02	4.35E-03	1.22E-02	4.08E-03	1.76E+02
F11	MOGROM	9.27E-05	1.39E-06	6.89E-01	1.85E-02	3.92E-03	1.02E-04	3.19E-03	1.25E-04	2.07E+01
	MOLAPO[18]	2.20E-04	6.87E-06	6.94E-01	4.76E-02	5.94E-03	4.93E-04	4.24E-03	3.80E-04	2.53E+01
	MOPSO[18]	4.74E-04	9.24E-05	7.91E-01	5.62E-02	9.02E-03	7.10E-04	7.00E-03	8.42E-04	6.90E+01
	NSGA_II[18]	3.66E-04	2.19E-05	4.75E-01	5.99E-02	7.10E-03	3.32E-04	4.32E-03	2.30E-04	1.83E+02
	MOALO[18]	4.17E-04	8.94E-05	5.42E-01	1.05E-01	1.55E-02	7.51E-03	2.33E-02	1.93E-02	2.87E+01
	MOGWO[18]	3.85E-04	1.95E-05	7.67E-01	2.60E-02	1.07E-02	1.36E-03	7.56E-03	6.80E-04	1.71E+02
F12	MOGROM	1.75E-04	7.92E-06	8.36E-01	5.66E-03	1.63E-03	2.85E-04	8.25E-04	2.37E-04	2.27E+01
	MOLAPO[18]	2.03E-04	8.47E-06	7.74E-01	5.40E-02	5.70E-03	2.50E-04	4.33E-03	2.30E-04	3.37E+01
	MOPSO[18]	3.85E-04	1.60E-05	7.23E-01	7.29E-02	7.72E-03	5.80E-04	6.00E-03	4.80E-04	7.24E+01
	NSGA_II[18]	4.01E-04	1.15E-05	4.15E-01	5.27E-02	6.43E-03	8.91E-04	4.33E-03	3.22E-04	2.14E+02
	MOALO[18]	8.83E-04	6.76E-04	6.37E-01	6.49E-02	2.46E-02	1.53E-02	2.51E-02	1.67E-02	3.01E+01
	MOGWO[18]	3.60E-04	3.87E-05	6.55E-01	1.69E-01	1.42E-02	2.24E-03	1.21E-02	2.89E-03	1.77E+02

5.3 Simulation results, investigation, and analysis 109

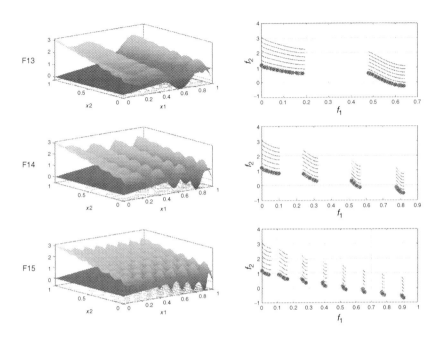

FIGURE 5.6

Search Space, Pareto Front, and Obtained Pareto Front by MOGROM for the Class 3 test functions (F13–15).

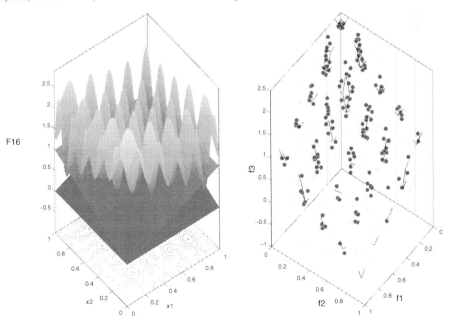

FIGURE 5.7

Search Space, Pareto Front, and Obtained Pareto Front by MOGROM for the Class 3 test functions (F16).

Table 5.4 Obtained comparison criterion in solving class 3 test functions by a different method.

		GD criteria		Metric of spread		Metric of spacing		IGD criteria		Time(s)
		Ave.	Std.	Ave.	Std.	Ave.	Std.	Ave.	Std.	
F13	MOGROM	1.05E-04	7.26E-06	8.69E-01	6.87E-03	4.37E-03	2.79E-06	1.83E-01	5.01E-04	2.35E+01
	MOLAPO[18]	8.44E-04	7.43E-05	8.25E-01	5.15E-02	5.30E-03	4.82E-04	1.89E-01	4.90E-04	2.16E+01
	MOPSO[18]	1.53E-03	4.71E-05	8.68E-01	3.86E-02	8.80E-03	6.32E-04	1.92E-01	2.64E-03	5.96E+01
	NSGA_II[18]	1.65E-03	6.48E-05	6.54E-01	3.67E-02	7.50E-03	4.91E-04	1.88E-01	3.20E-04	1.10E+02
	MOALO[18]	1.19E-03	1.80E-04	7.02E-01	9.27E-02	3.82E-02	2.61E-02	2.41E-01	5.33E-02	2.01E+01
	MOGWO[18]	2.71E-03	4.22E-04	8.53E-01	5.66E-02	1.39E-02	4.70E-03	2.16E-01	5.68E-03	6.31E+01
F14	MOGROM	1.47E-04	3.61E-05	9.66E-01	5.32E-02	3.75E-03	5.98E-04	4.28E-02	1.88E-04	2.39E+01
	MOLAPO[18]	1.39E-03	7.23E-05	8.98E-01	5.32E-02	6.96E-03	5.25E-04	1.88E-01	5.70E-04	2.14E+01
	MOPSO[18]	2.20E-03	2.13E-04	8.52E-01	6.10E-02	9.95E-03	5.96E-04	1.90E-01	1.10E-03	5.46E+01
	NSGA_II[18]	2.43E-03	1.11E-04	6.06E-01	7.67E-02	8.90E-03	1.27E-03	1.87E-01	3.20E-04	1.01E+02
	MOALO[18]	1.88E-03	3.46E-04	7.21E-01	1.32E-01	4.51E-02	3.45E-02	2.66E-01	5.65E-02	1.95E+01
	MOGWO[18]	1.13E-04	3.35E-05	7.62E-01	3.24E-02	1.25E-02	6.74E-04	2.16E-01	6.00E-03	7.01E+01
F15	MOGROM	2.99E-04	1.33E-04	9.17E-01	2.56E-02	1.43E-03	1.06E-03	1.58E-01	3.95E-04	2.00E+01
	MOLAPO[18]	1.76E-03	1.22E-04	8.51E-01	3.73E-02	7.89E-03	9.97E-04	2.04E-01	9.00E-04	2.49E+01
	MOPSO[18]	2.90E-03	2.75E-04	8.80E-01	6.91E-02	1.20E-02	1.00E-03	2.07E-01	2.11E-03	5.39E+01
	NSGA_II[18]	3.04E-03	1.45E-04	6.87E-01	6.36E-02	1.04E-02	8.50E-04	2.03E-01	1.90E-04	1.04E+02
	MOALO[18]	2.83E-03	3.86E-04	8.18E-01	6.95E-02	3.87E-02	1.21E-02	2.20E-01	1.26E-02	1.82E+01
	MOGWO[18]	3.29E-03	3.10E-04	7.68E-01	7.69E-02	1.88E-02	6.59E-03	2.28E-01	4.38E-03	7.85E+01
F16	MOGROM	7.33E-03	1.70E-05	6.77E-01	3.59E-03	1.27E-02	1.14E-03	1.68E-01	4.40E-04	2.87E+01
	MOLAPO[18]	1.46E-02	1.80E-04	5.91E-01	2.33E-02	5.54E-02	3.80E-03	3.23E-01	7.09E-03	2.54E+01
	MOPSO[18]	3.54E-02	1.03E-03	6.59E-01	2.74E-02	7.59E-02	1.08E-02	3.07E-01	4.89E-03	5.78E+01
	NSGA_II[18]	6.14E-02	7.39E-03	6.43E-01	3.81E-02	8.48E-02	7.89E-03	2.95E-01	3.56E-03	1.07E+02
	MOALO[18]	4.49E-02	3.15E-03	5.71E-01	1.42E-01	1.16E-01	7.53E-03	3.21E-01	1.13E-02	2.31E+01
	MOGWO[18]	3.69E-02	1.40E-03	6.35E-01	5.60E-02	9.18E-02	0.00E+00	2.99E-01	7.84E-03	6.94E+01

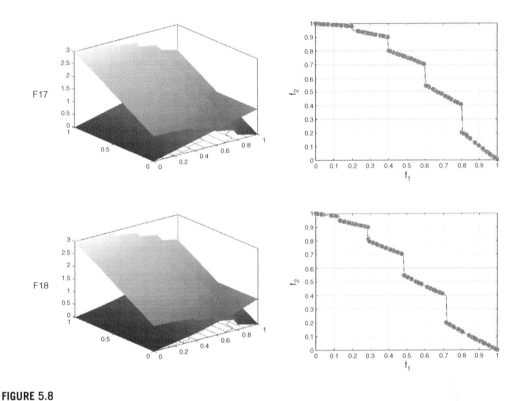

FIGURE 5.8

Search Space, Pareto Front, and Obtained Pareto Front by MOGROM for the Class 4 test functions (F17–18).

5.3.4 Fourth class

Using the test functions of this class, the proposed method is tested in solving the problems with stepwise fronts. This class includes two test functions and the formulations are explained in Eqs. (5.16) and (5.17). The fronts obtained by the proposed algorithm well cover the true fronts (Fig. 5.8). The obtained results of MOGROM are compared with those of five other methods in Table 5.5. The results reveal that the proposed method outperforms the other methods in Spacing, GD, and IGD criteria. In just F17 in criteria Spread, the best results belong to MOPSO followed by the proposed method with a negligible difference. In terms of computational time, the proposed method is the fastest in F17 and the second fastest after MOALO in F18.

$$F17 = \begin{cases} f_1(x) = x_1 \\ f_2(x) = g(x) - c(x_1) \\ g(x) = 1 + 10\dfrac{\sum_{i=2}^{N} x_i}{N} \\ C(x_1) = \begin{cases} 0 & 0 \le x_1 \le 0.2 \\ 0.25 & 0.2 \le x_1 \le 0.4 \\ 0.5 & 0.4 \le x_1 \le 0.6 \\ 0.75 & 0.6 \le x_1 \le 0.8 \\ 1 & 0.8 \le x_1 \le 1 \end{cases} \end{cases} \quad (5.16)$$

Table 5.5 Obtained comparison criterion in solving class 4 test functions by different methods.

		GD criteria		Metric of spread		Metric of spacing		IGD criteria		Time(s)
		Ave.	Std.	Ave.	Std.	Ave.	Std.	Ave.	Std.	
F17	MOGROM	7.12E-02	7.61E-04	7.25E-01	2.24E-02	5.96E-03	1.78E-04	5.56E-04	6.76E-04	2.13E+01
	MOLAPO[18]	6.03E-02	1.24E-03	6.96E-01	5.31E-02	5.30E-03	3.86E-04	4.00E-03	6.93E-04	2.31E+01
	MOPSO[18]	4.68E-04	2.28E-04	7.75E-01	5.36E-02	7.96E-03	1.04E-03	5.90E-03	3.77E-04	6.24E+01
	NSGA_II[18]	3.70E-04	1.93E-05	4.46E-01	5.20E-02	6.08E-03	2.93E-04	4.01E-03	1.81E-04	2.07E+02
	MOALO[18]	1.37E-03	8.53E-04	6.26E-01	1.17E-01	1.24E-02	4.35E-03	1.99E-02	3.61E-03	2.71E+01
	MOGWO[18]	3.30E-04	6.12E-06	6.93E-01	1.36E-01	8.86E-03	1.62E-03	9.61E-03	5.13E-04	7.31E+01
F18	MOGROM	1.30E-04	7.07E-05	8.19E-01	4.14E-03	6.36E-03	3.38E-04	2.90E-03	1.77E-05	3.47E+01
	MOLAPO[18]	4.30E-04	1.40E-04	6.79E-01	2.51E-02	5.40E-03	2.83E-04	4.04E-03	2.37E-04	3.64E+01
	MOPSO[18]	6.10E-04	2.51E-04	7.09E-01	4.70E-02	7.00E-03	1.46E-03	5.90E-03	4.20E-04	5.71E+01
	NSGA_II[18]	4.30E-04	2.49E-05	4.33E-01	3.55E-02	6.34E-03	4.33E-04	4.15E-03	3.60E-04	2.12E+02
	MOALO[18]	8.91E-01	1.32E-02	6.14E-01	9.60E-03	2.78E-02	2.55E-02	9.39E+00	4.83E-01	3.15E+01
	MOGWO[18]	3.98E-04	2.58E-05	5.45E-01	7.17E-02	9.18E-03	1.24E-03	6.69E-03	9.53E-04	7.59E+01

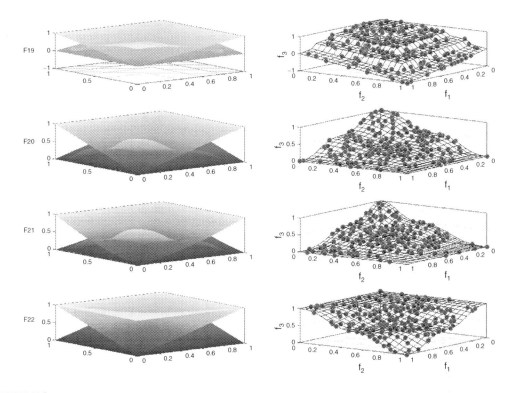

FIGURE 5.9

Search Space, Pareto Front, and Obtained Pareto Front by MOGROM for the Class 5 test functions (F19–22).

$$F18 = \begin{cases} f_1(x) = \dfrac{e^{x_1} - 1}{e - 1} \\ f_2(x) = g(x) - C(x_1) \\ g(x) = 1 + 10\dfrac{\sum_{i=2}^{N} x_i}{N} \\ C(x_1) = \begin{cases} 0 & 0 \leq x_1 \leq 0.2 \\ 0.25 & 0.2 \leq x_1 \leq 0.4 \\ 0.5 & 0.4 \leq x_1 \leq 0.6 \\ 0.75 & 0.6 \leq x_1 \leq 0.8 \\ 1 & 0.8 \leq x_1 \leq 1 \end{cases} \end{cases} \quad (5.17)$$

5.3.5 Fifth class

This class investigates how good MOGROM is in solving MOPs with three objectives. There are four test functions in this class, and the formulations and the variables are illustrated in Eqs. (5.18), (5.19), (5.20), and (5.21). The fronts achieved by the proposed method are presented in Fig. 5.9. As one can see, the obtained front covers the surface of the global front very well, and the solutions are distributed along it nearly equally. The comparisons can be found in Table 5.6. For these test functions, the proposed method has the best and faster performance.

$$F19 = \begin{cases} f_1(x) = x_1 \\ f_2(x) = x_2 \\ yf_3(x) = g(x) - C(x_1, x_2) \\ g(x) = 1 + 10\frac{\sum_{i=2}^{N} x_i}{N} \\ C(x_1, x_2) = \begin{cases} 0 & (0 \le x_1 \le 0.2)V(0 \le x_2 \le 0.2) \\ 0.25 & (0.2 \le x_1 \le 0.4)V(0.2 \le x_2 \le 0.4) \\ 0.5 & 0.4 \le x_1 \le 0.6V(0.4 \le x_2 \le 0.6) \\ 0.75 & 0.6 \le x_1 \le 0.8V(0.6 \le x_2 \le 1) \\ 1 & 0.8 \le x_1 \le 1V(0.8 \le x_2 \le 1) \end{cases} \end{cases} \quad (5.18)$$

$$F20 = \begin{cases} f_1(x) = x_1 \\ f_2(x) = x_2 \\ f_3(x) = g(x)\left[\frac{\sin(4\pi x_1) - 15x_1}{15} + 1\right]\left[\frac{\sin(4\pi x_2) - 15x_2}{15} + 1\right] \\ g(x) = 1 + 10\frac{\sum_{i=2}^{N} x_i}{N} \end{cases} \quad (5.19)$$

$$F21 = \begin{cases} f_1(x) = \frac{e^{x_1} - 1}{e - 1} \\ f_2(x) = \frac{e^{x_2} - 1}{e - 1} \\ f_3(x) = g(x)\left[\frac{\sin(4\pi x_1) - 15x_1}{15} + 1\right]\left[\frac{\sin(4\pi x_2) - 15x_2}{15} + 1\right] \\ g(x) = 1 + 10\frac{\sum_{i=2}^{N} x_i}{N} \end{cases} \quad (5.20)$$

$$F22 = \begin{cases} f_1(x) = x_1 \\ f_2(x) = x_2 \\ f_3(x) = g(x)\left[\frac{\sin(4\pi x_1 x_2) - 15x_1 x_2}{15} + 1\right] \\ g(x) = 1 + 10\frac{\sum_{i=2}^{N} x_i}{N} \end{cases} \quad (5.21)$$

Table 5.6 Obtained comparison criterion in solving class 5 test functions by different method.

		GD criteria		Metric of spread		Metric of spacing		IGD criteria		Time(s)
		Ave.	Std.	Ave.	Std.	Ave.	Std.	Ave.	Std.	
F19	MOGROM	1.42E-03	7.66E-05	7.76E-01	1.71E-03	9.64E-03	1.28E-03	4.54E-02	1.50E-03	3.65E+01
	MOLAPO[18]	3.21E-03	7.46E-05	4.03E-01	3.18E-02	4.57E-02	1.58E-03	4.97E-02	1.85E-03	4.71E+01
	MOPSO[18]	5.10E-03	2.89E-04	4.25E-01	2.29E-02	5.80E-02	6.27E-03	6.60E-02	1.35E-03	1.01E+02
	NSGA_II[18]	1.70E-01	1.26E-02	5.60E-01	2.95E-02	1.02E-01	1.74E-02	1.33E-01	3.06E-02	3.86E+02
	MOALO[18]	5.87E-03	4.60E-04	4.45E-01	4.17E-02	8.86E-02	9.52E-03	1.60E-01	1.77E-02	5.43E+01
	MOGWO[18]	6.74E-03	1.70E-04	5.23E-01	5.23E-02	4.25E-02	2.36E-03	2.58E-01	5.16E-02	1.03E+02
F20	MOGROM	2.09E-03	4.07E-05	4.44E-01	5.01E-03	3.08E-02	2.39E-04	2.12E-02	2.36E-03	1.99E+01
	MOLAPO[18]	3.06E-03	1.10E-04	3.70E-01	2.12E-02	3.59E-02	1.04E-03	4.61E-02	3.02E-03	5.12E+01
	MOPSO[18]	4.42E-03	2.05E-04	4.25E-01	5.68E-02	4.73E-02	7.98E-03	6.17E-02	3.02E-03	1.07E+02
	NSGA_II[18]	4.97E-02	2.15E-03	4.64E-01	9.30E-03	5.02E-02	3.19E-03	6.05E-02	8.21E-03	3.84E+02
	MOALO[18]	4.43E-03	2.99E-04	4.42E-01	5.34E-02	7.11E-02	1.22E-02	1.27E-01	9.78E-03	1.27E+01
	MOGWO[18]	4.67E-03	9.50E-05	3.23E-01	1.59E-02	3.59E-02	4.35E-03	1.76E-01	7.90E-03	1.17E+02
F21	MOGROM	3.02E-04	3.30E-05	3.55E-01	1.83E-03	1.91E-02	1.02E-03	1.70E-02	6.80E-04	3.28E+01
	MOLAPO[18]	2.98E-03	2.84E-05	3.66E-01	3.00E-03	3.50E-02	1.82E-03	4.45E-02	1.07E-03	5.13E+01
	MOPSO[18]	4.30E-03	1.70E-04	3.87E-01	4.12E-02	4.22E-02	4.94E-03	5.92E-02	3.73E-03	1.05E+02
	NSGA_II[18]	4.70E-02	1.45E-03	4.94E-01	1.57E-02	4.67E-02	2.41E-03	5.53E-02	9.85E-03	3.81E+02
	MOALO[18]	4.22E-03	5.40E-04	4.93E-01	4.34E-02	7.42E-02	9.62E-03	1.13E-01	7.60E-03	6.01E+01
	MOGWO[18]	4.80E-03	1.49E-04	4.65E-01	3.13E-02	3.53E-02	1.73E-03	1.88E-01	2.03E-02	1.07E+02
F22	MOGROM	3.08E-03	8.34E-05	5.87E-01	1.50E-02	3.46E-02	1.03E-03	2.06E-02	4.14E-03	4.35E+01
	MOLAPO[18]	3.01E-03	8.39E-05	3.57E-01	3.16E-02	3.41E-02	1.54E-03	4.70E-02	3.53E-03	4.89E+01
	MOPSO[18]	4.23E-03	8.66E-05	4.05E-01	4.04E-02	4.51E-02	3.91E-03	5.86E-02	3.21E-03	9.81E+01
	NSGA_II[18]	1.67E-01	1.46E-02	5.61E-01	4.24E-02	7.82E-02	1.00E-02	1.72E-01	3.82E-02	3.72E+02
	MOALO[18]	4.31E-03	3.10E-04	4.78E-01	6.22E-02	7.50E-02	1.90E-02	1.19E-01	1.62E-02	5.95E+01
	MOGWO[18]	4.84E-03	4.17E-05	4.96E-01	5.06E-02	3.45E-02	4.94E-03	2.21E-01	1.80E-02	1.04E+02

5.4 Conclusion

A new MOEA inspired by a recently introduced optimization method known as GROM is presented in this chapter. The procedure for solutions' updating in MOGROM is the same as the single version, except for two other mechanisms that are added. The first mechanism is the nondominated sorting of solutions to find the best front for storing into the *Repository* matrix. The second one is crowding distance, which helps us in sorting the solutions in the *Repository* matrix. Using the *Repository* matrix the best solutions during the optimization procedure are guaranteed. The crowding distance, on the other side, improves the convergence behavior and searching for the whole search space.

For the sake of the evaluation, the proposed method is applied to 22 test functions from 5 different classes. These classes are different regarding the front shape, the number of local fronts, and also the number of objectives. The results obtained from the MOGROM are contrasted to those of five well-known multiobjective optimization algorithms, including NSGA-II, MOLAPO, MOPSO, MOGWO, and MOALO. The comparison's criteria are General Distance (GD), IGD, metric of spread, and metric of Spacing. The comparisons demonstrate the effectiveness, robustness, and performance of the proposed method. Finally, from this study it can be inferred that the GROM is an outstanding optimization algorithm able to solve both the single- and MOPs.

References

[1] S. Mirjalili, Moth-flame optimization algorithm: a novel nature-inspired heuristic paradigm, Knowledge-Based Syst. 89 (2015) 228–249.
[2] J. Arora, Introduction to Optimum Design, Academic Press, MA, 2004.
[3] P. K. Miettinen, Preface By-Neittaanmaki, Evolutionary Algorithms in Engineering and Computer Science: Recent Advances in Genetic Algorithms, Evolution Strategies, Evolutionary Programming, GE, John Wiley & Sons, Inc., NJ, 1999.
[4] A.D. Belegundu, J.S. Arora, A study of mathematical programming methods for structural optimization. Part I: Theory, Int. J. Numer. Methods Eng. 21 (1985) 1583–1599.
[5] H. Chickermane, H.C. Gea, Structural optimization using a new local approximation method, Int. J. Numer. Methods Eng. 39 (1996) 829–846.
[6] D. Karaboga, B. Basturk, A powerful and efficient algorithm for numerical function optimization: artificial bee colony (ABC) algorithm, J. Glob. Optim. 39 (2007) 459–471.
[7] X. Hu, R. Eberhart, Solving constrained non-linear optimization problems with particle swarm optimization, in: Proc. Sixth World Multiconference Syst. Cybern. Informatics, Citeseer, 2002, pp. 203–206.
[8] K. Deb, Multi-objective Optimization Using Evolutionary Algorithms, John Wiley & Sons, NJ, 2001.
[9] D.H. Wolpert, W.G. Macready, No free lunch theorems for optimization, IEEE Trans. Evol. Comput. 1 (1997) 67–82.
[10] K. Deb, S. Agrawal, A. Pratap, T. Meyarivan, A fast elitist non-dominated sorting genetic algorithm for multi-objective optimization: NSGA-II, in: Proc. Int. Conf. Parallel Probl. Solving From Nat., Springer, 2000, pp. 849–858.
[11] S. Mostaghim, J. Teich, Strategies for finding good local guides in multi-objective particle swarm optimization (MOPSO), in: Proc. Swarm Intell. Symp. 2003. SIS'03. Proc. 2003 IEEE, IEEE, 2003, pp. 26–33.
[12] A.J. Nebro, J.J. Durillo, J. Garcia-Nieto, C.A.C. Coello, F. Luna, E. Alba, Smpso: A new PSO-based metaheuristic for multi-objective optimization, in: Proc. Comput. Intell. Miulti-Criteria Decis. 2009. mcdm'09. IEEE Symp., IEEE, 2009, pp. 66–73.

[13] X. Hu, R. Eberhart, Multi-objective optimization using dynamic neighborhood particle swarm optimization, In: Proc. Evol. Comput. 2002, in: CEC'02. Proc. 2002 Congr., IEEE, 2002, pp. 1677–1681.

[14] S. Mirjalili, S. Saremi, S.M. Mirjalili, L. dos S. Coelho, Multi-objective grey wolf optimizer: a novel algorithm for multi-criterion optimization, Expert Syst. Appl. 47 (2016) 106–119.

[15] S. Gupta, K. Deep, A novel Random Walk Grey Wolf Optimizer, Swarm Evol. Comput. 44 (2019) 101–112.

[16] S. Mirjalili, P. Jangir, S. Saremi, Multi-objective ant lion optimizer: a multi-objective optimization algorithm for solving engineering problems, Appl. Intell. 46 (2017) 79–95.

[17] S. Mirjalili, The ant lion optimizer, Adv. Eng. Softw. 83 (2015) 80–98.

[18] A.F. Nematollahi, A. Rahiminejad, B. Vahidi, A novel multi-objective optimization algorithm based on Lightning Attachment Procedure Optimization algorithm, Appl. Soft. Comput. 75 (2019) 404–427.

[19] A.F. Nematollahi, A. Rahiminejad, B. Vahidi, A novel physical based meta-heuristic optimization method known as Lightning Attachment Procedure Optimization, Appl. Soft. Comput. (2017).

[20] P. Ngatchou, A. Zarei, A. El-Sharkawi, Pareto multi objective optimization, in: in: Proc. Intell. Syst. Appl. to Power Syst. 2005. Proc. 13th Int. Conf., IEEE, 2005, pp. 84–91.

[21] K. Deb, J. Sundar, Reference point based multi-objective optimization using evolutionary algorithms, in: Proc. 8th Annu. Conf. Genet. Evol. Comput., ACM, 2006, pp. 635–642.

[22] W. Stadler, A survey of multi-criteria optimization or the vector maximum problem, part I: 1776–1960, J. Optim. Theory Appl. 29 (1979) 1–52.

[23] J. Branke, T. Kaußler, H. Schmeck, Guidance in evolutionary multi-objective optimization, Adv. Eng. Softw. 32 (2001) 499–507.

[24] R.T. Marler, J.S. Arora, Survey of multi-objective optimization methods for engineering, Struct. Multidiscip. Optim. 26 (2004) 369–395.

[25] R.E. Bellman, L.A. Zadeh, Decision-making in a fuzzy environment, Manage. Sci. 17 (1970) 141 B-.

[26] K. Tomsovic, M.Y. Chow, Tutorial on Fuzzy Logic Application in Power Systems, IEEE Press, NY, 1999.

[27] W. Zhang, Y. Liu, Multi-objective reactive power and voltage control based on fuzzy optimization strategy and fuzzy adaptive particle swarm, Int. J. Electr. Power Energy Syst. 30 (2008) 525–532.

[28] C.A.C. Coello, A comprehensive survey of evolutionary-based multi-objective optimization techniques, Knowl. Inf. Syst. 1 (1999) 129–156.

[29] C.A.C. Coello, E.M. Montes, Constraint-handling in genetic algorithms through the use of dominance-based tournament selection, Adv. Eng. Inform. 16 (2002) 193–203.

[30] N. Srinivas, K. Deb, Muiltiobjective optimization using nondominated sorting in genetic algorithms, Evol. Comput. 2 (1994) 221–248.

[31] C.A.C. Coello, G.B. Lamont, D.A. Van Veldhuizen, Evolutionary Algorithms for Solving Multi-objective Problems, Springer, Cham, 2007.

[32] C.A.C. Coello, Use of a self-adaptive penalty approach for engineering optimization problems, Comput. Ind. 41 (2000) 113–127.

[33] V.B. Semwal, J. Singha, P.K. Sharma, A. Chauhan, B. Behera, An optimized feature selection technique based on incremental feature analysis for bio-metric gait data classification, Multimed. Tools Appl. (2016) 1–19.

[34] A. Rahimi-Vahed, A.H. Mirzaei, A hybrid multi-objective shuffled frog-leaping algorithm for a mixed-model assembly line sequencing problem, Comput. Ind. Eng. 53 (2007) 642–666.

[35] S.N. Omkar, J. Senthilnath, R. Khandelwal, G.N. Naik, S. Gopalkrishnan, Artificial Bee Colony (ABC) for multi-objective design optimization of composite structures, Appl. Soft Comput. 11 (2011) 489–499.

[36] A. Saad, S.A. Khan, A. Mahmood, A multi-objective evolutionary artificial bee colony algorithm for optimizing network topology design, Swarm Evol. Comput. 38 (2018) 187–201.

[37] N.K. Madavan, Multi-objective optimization using a Pareto differential evolution approach, in: Proc. Evol. Comput. 2002. CEC'02. Proc. 2002 Congr., IEEE, 2002, pp. 1145–1150.

[38] T. Ray, K.M. Liew, A swarm metaphor for multi-objective design optimization, Eng. Optim 34 (2002) 141–153.
[39] S. Salcedo-Sanz, A. Pastor-Sánchez, D. Gallo-Marazuela, A. Portilla-Figueras, A novel coral reefs optimization algorithm for multi-objective problems, in: Proc. Int. Conf. Intell. Data Eng. Autom. Learn., Springer, 2013, pp. 326–333.
[40] S. Yazdani, H. Nezamabadi-pour, S. Kamyab, A gravitational search algorithm for multimodal optimization, Swarm Evol. Comput. 14 (2014) 1–14.
[41] C.M. Fonseca, P.J. Fleming, Genetic algorithms for multi-objective optimization: formulation discussion and generalization, Proc. ICGA (1993) 416–423.
[42] J. Horn, N. Nafpliotis, D.E. Goldberg, A niched Pareto genetic algorithm for multi-objective optimization, in: Proc. Evol. Comput. 1994. IEEE World Congr. Comput. Intell. Proc. First IEEE Conf., IEEE, 1994, pp. 82–87.
[43] B.V Babu, M.M.L. Jehan, Differential evolution for multi-objective optimization, in: Proc. Evol. Comput. 2003. CEC'03. 2003 Congr., IEEE, 2003, pp. 2696–2703.
[44] E. Zitzler, K. Deb, L. Thiele, Comparison of multi-objective evolutionary algorithms: empirical results, Evol. Comput. 8 (2000) 173–195.
[45] E. Zitzler, L. Thiele, Multi-objective optimization using evolutionary algorithms—a comparative case study, in: Proc. Int. Conf. Parallel Probl. Solving from Nat, Springer, 1998, pp. 292–301.
[46] D.E. Goldberg, Genetic algorithms in search, optimization, and machine learning, Addison-Wesley, MA, 1989.
[47] E. Zitzler, L. Thiele, Multi-objective evolutionary algorithms: a comparative case study and the strength Pareto approach, IEEE Trans. Evol. Comput. 3 (1999) 257–271.
[48] M.R. Sierra, C.A.C. Coello, Improving PSO-based multi-objective optimization using crowding, mutation and e-dominance, in: C.A. Coello, A. Hernández Aguirre, E. Zitzler (Eds.), Evolutionary Multi-Criterion Optimization. EMO 2005. Lecture Notes in Computer Science, vol 3410, Springer, Berlin, Heidelberg, 2005, pp. 505–519.
[49] E.J. Alvarez-Benitez, R.M. Everson, J.E. Fieldsend, A MOPSO algorithm based exclusively on pareto dominance concepts, *International Conference on Evolutionary Multi-Criterion Optimization*, Springer, Berlin, Heidelberg, 2005.
[50] H.A. Abbass, R. Sarker, C. Newton, PDE: a Pareto-frontier differential evolution approach for multi-objective optimization problems, in: Proc. Evol. Comput. 2001. Proc. 2001 Congr., IEEE, 2001, pp. 971–978.
[51] K. Price, R.M. Storn, J.A. Lampinen, Differential Evolution: a Practical Approach to Global Optimization, Springer Science & Business Media, Berlin/Heidelberg, Germany, 2006.
[52] F. Zou, L. Wang, X. Hei, D. Chen, B. Wang, Multi-objective optimization using teaching-learning-based optimization algorithm, Eng. Appl. Artif. Intell. 26 (2013) 1291–1300.
[53] A. Rahimi-Vahed, M. Dangchi, H. Rafiei, E. Salimi, A novel hybrid multi-objective shuffled frog-leaping algorithm for a bi-criteria permutation flow shop scheduling problem, Int. J. Adv. Manuf. Technol. 41 (2009) 1227–1239.
[54] T. Niknam, M. rasoul Narimani, M. Jabbari, A.R. Malekpour, A modified shuffle frog leaping algorithm for multi-objective optimal power flow, Energy 36 (2011) 6420–6432.
[55] J. Li, Q. Pan, S. Xie, An effective shuffled frog-leaping algorithm for multi-objective flexible job shop scheduling problems, Appl. Math. Comput. 218 (2012) 9353–9371.
[56] R. Akbari, R. Hedayatzadeh, K. Ziarati, B. Hassanizadeh, A multi-objective artificial bee colony algorithm, Swarm Evol. Comput. 2 (2012) 39–52.
[57] A.F. Nematollahi, A. Rahiminejad, B. Vahidi, A novel meta-heuristic optimization method based on golden ratio in nature, Soft Comput (2019) 1–35.
[58] S. Mirjalili, A. Lewis, Novel frameworks for creating robust multi-objective benchmark problems, Inf. Sci. (Ny). 300 (2015) 158–192.

CHAPTER 6

Multiobjective charged system search for optimum location of bank branch

Siamak Talatahari[a], Abolfazl Ranjbar[a,b], Mohammad Toloueia[a] and Iman Rahimi[c]

[a]*Department of Civil Engineering, University of Tabriz, Tabriz, Iran,* [b]*Department of GIS Engineering, Faculty of Surveying Engineering, Tehran University, Tehran, Iran,* [c]*University of Technology Sydney, Sydney, Australia*

6.1 Introduction

Bank branches location–allocation problem belongs to non-deterministic polynomial-time hardness (NP-hard) problems [1–3] which can be possibly solved only in exponential time by the increase in the number of banks and customers [4]; especially when the location model includes various datasets, several objectives, and constraints [5]. As a consequent, we need to use heuristic methods to solve this type of problems [6,7]. Also, as the majority of data and analyses applied in the location–allocation problems are spatial; GIScience's abilities should be employed beside optimization methods [8].

Nowadays, to perform particular financial tasks, bank customers often need to be present at their bank. For the sake of its customers, a bank should increase its branches in the city to attract more customers in the competition with other banks. However, establishing new branches is too expensive and banks prefer to carry out an optimal location finding procedure. Such procedures should consider many criteria and objectives including spatial data of customers, new and existing bank branches as well as level of attraction of banks. Customers often select a bank that is closer to them, has better services or financial records, and also consider other human or physical factors (based on Huff law, [9]). Hence, planning to increase the number of customers for a new branch of a bank considering spatial criteria and various other objectives appears necessary. This chapter proposes solution for optimum location modeling of bank branches under competitive conditions, which is categorized as a Multiobjective Optimization problem.

Recently, many of metaheuristics optimization algorithms are used in multiobjective facility location problems. For example, Beheshtifar and Alimohammadi [5] applied Multiobjective Evolutionary Algorithm (MOEA) for modeling site suitability for health-care facilities and Neema and Ohgai [10] presented a Genetic Algorithm (GA) based method model to obtain optimum locations for urban parks and open spaces and they proved that this algorithm is effective. Li and Yeh [3] integrated Geospatial Information System and GA for searching the optimal locations and compared Simulate Annealing (SA) with GA. In another study, Rahmati et al. [11] presented multiobjective multiserver facility allocation problem with Multiobjective Harmony Search (MOHS) and compared it with two popular algorithms Nondominated Sorting Genetic Algorithm II (NSGA-II) and Nondominated Ranking Genetic Algorithm and according to their result MOHS has better performance than two mentioned algorithms in terms of computational time. In another study, Xiao et al. [12] applied MOEAs to optimize the shape and location of sites by considering the cost surface. Li et al. [13] used the Ant Colony Optimization to solve the site selection problem by minimization of the total costs. Masoomi et al. [14] presented Multiobjective Particle Swarm Optimization (MOPSO) algorithm to find the optimum arrangement of urban land uses in parcel level and their result shows that MOPSO is effective. Duh and Brown [15] develop a knowledge-informed Pareto SA approach to tackle specifically multiobjective allocation problems that consider spatial patterns as objectives, and then proved that is effective.

Location selection problem in banking is an important issue for the commercial success in a competitive environment [16]. However, the studies on finding optimal Location of Bank Branches under Competitive Conditions (LBBCC) using spatial data with multiple purposes are limited. Like many practical problems, the LBBCC have more than one objective. From marketing point of view, the location of a new branch should be as far away as possible from branches of the same bank and sum of distance between all customers and every branch should be minimized (customer focus). In reality planning these objectives conflicts each other. Therefore, the LBBCC is categorized as a multi-objective problem. This chapter presents a multiobjective spatial optimization model to solve the above-mentioned problem. To fulfill this aim, we present a Multiobjective Charged System Search (MOCSS) optimization algorithm that indicated good performance display over multiobjective benchmark and engineering function [17], for determining the optimum LBBCC using spatial data of Tabriz.

The rest of the chapter is organized as follows: Section 6.2 reviews the Pareto Front concept and performance metrics. The utilized algorithms are presented in Section 6.3. Analytic Hierarchy Process (AHP) is presented in Section 6.4. Section 6.5 describes the problem of LBBCC in details and then provides a model formulation. Then the results of the selected algorithms for solving the LBBCC problem using spatial data problems are presented in Section 6.6 . Finally, some relevant issues are discussed and conclusions are drawn in Section 6.7. Meanwhile, we will also discuss the unique features of the proposed algorithm as well as topics for further studies.

6.2 Multiobjective backgrounds

In this section, some required multiobjective backgrounds are defined as Pareto Front and performance metrics in the following subsections.

6.2.1 Dominance and Pareto Front

Consider an Multi-Objective (MO) problem with k objectives,
    Find **x** that minimize

$$F(\mathbf{x}) = f_1(\mathbf{x}), f_2(\mathbf{x}), \ldots, f_k(\mathbf{x}) \tag{6.1}$$

subject to

$$g_i(\mathbf{x}) \leq 0, i = 1, 2, \ldots, h \tag{6.2}$$

where $\mathbf{x} = (x_1, x_2, \ldots, x_{nVar})$ is the vector of solution that minimizes objective function(s), $F(\mathbf{x})$, while satisfying the constraint(s), $g_i(\mathbf{x}) \leq 0$. The numbers of design parameters, objective functions, and constraint are $nVar$, k, and h, respectively.

The concept of dominance is generally used to compare two solutions, $\mathbf{x}_i$ and $\mathbf{x}_j$. The solution $\mathbf{x}_i$ is said to dominate solution $\mathbf{x}_j$ if $f_k(\mathbf{x}_i)$ is no worse than $f_k(\mathbf{x}_j)$ for all the objectives and it is better for at least one of them [18,19,32]:

$$f_l(\mathbf{x}_i) \leq f_l(\mathbf{x}_j), \ \forall l = 1, 2, \ldots, k \quad \text{and} \quad f_m(\mathbf{x}_i) < f_m(\mathbf{x}_j), \ \exists m \in \{1, 2, \ldots, k\} \tag{6.3}$$

As a result, a set of solutions is said to be a Pareto Front if no solution can dominate any solution in this set [10]. For more details about Pareto-optimal solutions, one can refer to Coello et al. [20] and Deb [21].

6.2.2 Performance metrics

A general approach for evaluating MO algorithms is the quantitative comparison of the performance of different algorithms [22] or how the solution points are distributed uniformly on the Pareto Front for considered functions. In this work, we use Spacing and Coverage of Two Set (CS) performance metrics for comparing the proposed approach.

6.2.2.1 Spacing (S)

The Spacing (S) metric numerically describes the spread of the vectors in PF_{known} [18,23]. This metric does not require the researcher to know PF_{true}. Note that this becomes important in the deception problems where all Pareto Front solutions are equally spaced. Equations (6.4) and (6.5) define this metric.

$$S = \left[\frac{1}{n-1} \sum_{i=1}^{n} (d_i - \bar{d})^2 \right]^{\frac{1}{2}} \quad \text{where} \quad \bar{d} = \frac{\sum_{i=1}^{n} d_i}{n} \tag{6.4}$$

$$d_i = \min_j \left\{ \sum_{l=1}^{k} |f_l(\mathbf{x}_i) - f_l(\mathbf{x}_j)| \right\}, \quad i, j = 1, 2, \ldots, n, \ i \neq j \tag{6.5}$$

where n is the number of solutions in PF_{known}.

6.2.2.2 Coverage of Two Sets (CS)

To compare the dominance relationship between two populations resulting from two different MO algorithms, Zitzler et al. [22] proposed the CS that is measured to show how the final population of one algorithm dominates the final population of another algorithm, [24]. The CS value can be calculated as follows:

$$CS(\mathbf{x}', \mathbf{x}'') = \frac{|\{a'' \in \mathbf{x}''; \exists a' \in \mathbf{x}' : a' \leq a''\}|}{|\mathbf{x}''|} \tag{6.6}$$

where $\mathbf{x}'$ and $\mathbf{x}''$ are two sets of solutions resulting from different algorithms. $a' \leq a''$ means that a' dominate a'' if and only if $a' < a''$ or $a' = a''$. Function CS is defined as the mapping of the order pair $(\mathbf{x}', \mathbf{x}'')$ to the interval $[0, 1]$. In general, if all solutions in $\mathbf{x}'$ dominate all solutions in $\mathbf{x}''$, then $CS(\mathbf{x}', \mathbf{x}'') = 1$. Also, $CS(\mathbf{x}', \mathbf{x}'') = 0$ implies that none of the solutions in $\mathbf{x}''$ are dominated by $\mathbf{x}'$. Note that both $CS(\mathbf{x}', \mathbf{x}'')$ and $CS(\mathbf{x}'', \mathbf{x}')$ need to be considered independently as they have the distinct meanings, as $CS(\mathbf{x}', \mathbf{x}'')$ is not necessarily equal to $1 - CS(\mathbf{x}'', \mathbf{x}')$. The advantage of this Pareto-compliant metric is that it is easy to calculate and provides a relative comparison based on dominance numbers between two algorithms [25].

6.3 Utilized methods

This section reviews the utilized methods briefly.

6.3.1 NSGA-II algorithm

Srinivas and Deb [26] introduced NSGA to deal with MO problems. In this algorithm, Goldberg's nondomination criterion is used to determine solution ranks, where fitness sharing is used to control diversity of solutions in the search space. Deb et al. [19] have proposed an improved version of the NSGA algorithm, called NSGA-II that includes a second-order sorting criterion called crowding distance (CD), which is faster and more reliable than NSGA. It uses this CD in its selection operator to keep a diverse front by making sure each member stays a CD apart. This keeps the population diverse and helps the algorithm to explore the fitness landscape.

In an evolution cycle of the NSGA-II, a mating pool is first created and filled using binary tournament selection. Then, crossover and mutation operators are applied to the members of the mating pool. Next, the old set of solutions and newly created solutions are merged to create a larger population. This new population is sorted based on two criterions: (1) rank and (2) CD. Finally, certain number of individuals in the sorted population is selected and others are deleted. These steps are repeated until a stopping condition is met. After NSGA-II terminates, nondominated solutions of the final population are the approximate Pareto Front of the problem.

6.3.2 MOPSO algorithm

MOPSO is an extension of Particle Swarm Optimization (PSO) proposed by Coello et al. [18] to deal with MO problems. MOPSO stores the nondominated solutions found so far in an external archive of solutions, namely "repository." The members of repositories are not dominating each other and they provide an approximation of real Pareto Front of the optimization problem. In MOPSO, each particle randomly selects a solution in the repository as its leader, instead of a unique global best for all particles. MOPSO uses the concept of Pareto dominance to determine the flight direction of a particle and it maintains previously found nondominated vectors in a global repository that is later used by other particles to guide their own flight.

6.3.3 MOCSS algorithm

MOCSS is the based on the standard Charged System Search (CSS) proposed by [27] The CSS is a population-based algorithm that considers a number of charged particles (CPs). It is based on electrostatic and Newtonian mechanics laws. More details of these algorithms can be found in [27]. The initial position of CPs is obtained by Equation (6.7) in the search space

$$x_{i,j} = x_{i,min} + rand(x_{i,max} - x_{i,min}), \quad i = 1, 2, 3, \ldots, n \quad (6.7)$$

where $x_{i,j}$ determines the initial value of the ith variable for the jth CP; $x_{i,min}$ and $x_{i,max}$ are the minimum and the maximum allowable values for the ith variable; $rand$ is a random number in the interval [0,1]; and n is the number of variables. The magnitude of the charge is calculated by the quality of solutions as follows:

$$q_i = \frac{fit(i) - fit(worst)}{fit(best) - fit(worst)}, \quad i = 1, 2, \ldots, m \qquad (6.8)$$

where *fit(best)* and *fit(worst)* are the best and the worst fitness of all particles, so far; *fit(i)* represents the objective function value or the fitness of the solution i; and m is the total number of CPs. The separation distance r_{ij} between two CPs is also defined as follows:

$$r_{ij} = \frac{||X_i - X_j||}{\left\| \frac{(X_i + X_j)}{2} - X_{best} \right\| + \varepsilon} \qquad (6.9)$$

where X_i and X_j are the positions of the ith and jth CPs, X_{best} is the position of the best current CP, and ε is a small positive number to avoid singularities. Also, the main formula of CSS uses Newton's laws (with some modifications) for calculating the new position and velocity of each CP as follows:

$$X_{j,new} = 0.5 \, rand_{j1} \cdot \left(1 + \frac{iter}{iter_{max}}\right) \sum_{i, i \neq j} \left(\frac{q_i}{a^3} r_{ij} \cdot i_1 + \frac{q_i}{r_{ij}^2} \cdot i_2 \right) P_{ij}(X_i - X_j)$$

$$+ 0.5 \, rand_{j2} \cdot \left(1 - \frac{iter}{iter_{max}}\right) V_{j,old} + X_{j,old} \qquad (6.10)$$

$$V_{j,new} = X_{j,new} - X_{j,old} \qquad (6.11)$$

where *iter* is the current iteration number and *iter*$_{max}$ is the maximum number of iterations.

The MOCSS algorithm [17] uses Nondominated Sorting (NS) and high diversity of Pareto Front concepts [19]. NS sorts the solutions on the base of nondominate and then forms the levels of Pareto Fronts. Now for selecting the numbers of best solution, first the solutions are chosen that have the highest Pareto Front rank and if needed the other solutions are selected from the next Pareto Front such that CD condition is satisfied. This means that solutions are selected with further more distance at mentioned Pareto Front according to Fig. 6.1. Also, to prevent the early convergence, a mutation function is used.

The following pseudocode summarized the MOCSS algorithm:

Level 1: Initialization

Step 1 Initialize specific parameters of the optimization problem.
Step 2 Initialize the initial positions of CPs and their associated velocities.
Step 3 Evaluate each of the CPs.
Step 4 Determine the Nondominated solutions.

Level 2: Search

Step 1 Determine the probability of moving and calculate the attracting force vector for each CP.
Step 2 Select leader.
Step 3 Move each CP to the new position and find the velocities.
Step 4 Perform mutation part.
Step 5 Rank CPs according to the NS approach.

Level 3: Terminating criterion controlling
Repeat search level steps until a terminating criterion is satisfied.

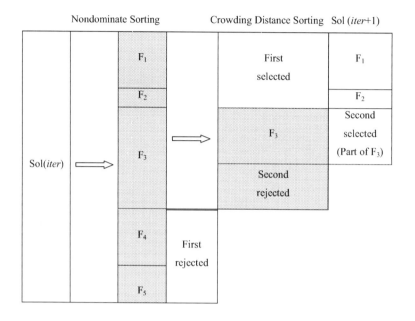

FIGURE 6.1
Flow diagram that shows the way in which the NS works. Sol(*iter*) is the solutions at iteration *iter*. F_1 are the best solution. F_2 are the second-best solutions and so on.

6.4 Analytic Hierarchy Process

The AHP which is a powerful tool in applying Multicriteria Decision Making was introduced and developed by Saaty in 1980. In comparison with other multiple-criteria decision-making (MCDM) methods, the AHP method has widely been used in MCDM and has been applied successfully in many practical decision-making problems [28]. In AHP the decision maker starts by laying out the overall hierarchy of the decision. This hierarchy reveals the factors to be considered as well as the various alternatives in the decision. Here, both qualitative and quantitative criteria can be compared using a number of pairwise comparisons, which result in the determination of factor weights. Finally, the alternative with the highest total weighted score is selected as the best cone [28].

- *Gravity model*

According to the gravity model, the probability that a customer at *i* refers at a bank *j* is given by [9]:

$$p_{ij} = \frac{a_j}{d_{ij}^2} \tag{6.12}$$

where a_j represents the quality-of-service *j*, d_{ij} is the distance from customer point *i* to bank *j*.

- *Selecting the best bank*

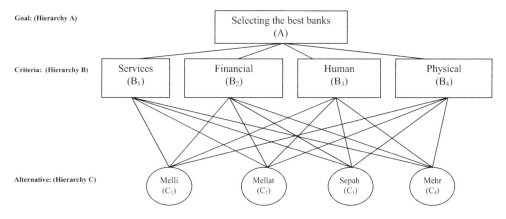

FIGURE 6.2

Hierarchy model of selecting the bank.

The major steps for selecting best bank are organized as follows:

Step 1 The first step is to define the goal and objectives.

Step 2 Selecting the bank parameters is the second step. Based on the collected information on important parameters that attract customers to the selected banks (Melli, Mellat, Sepah, and Mehr), it became clear that services, financial, human and physical parameters are most important parameters.

Step 3 The hierarchy from the top through the intermediate and the lowest levels should be structured. The goal is to attract customers to the bank. The criteria include four standard levels of important parameters for the four selected banks. Fig. 6.2 describes the hierarchy of a decision-making problem.

Step 4 The next step contains to design the format of questionnaire items as to process according to the hierarchy in Step 2. And then collect the input by a pairwise comparison of decision elements (according to Table 6.1).

Step 5 The eigenvalue method is used to estimate the consistence index. Then, it should be determined whether the input data satisfy a "consistence check" or not. If it does not, we should go back to Step 4 and redo the pairwise comparisons.

Criteria weights of all factors for the selected banks are calculated as presented in Table 6.2.

6.5 Model formulation

Given the diversity and complexity of issues in location–allocation problems, a number of classifications from various perspectives for this type of problems are available; however, one of the most common classifications is based on the number of service centers: (1) single source and (2) multisource problems. The latter is known as Multisource Weber Problem (MSWP), [29,30]. Yet, MSWP itself is classified into different types. One of them is called Capacitated MSWP where every service center has a certain service capacity [31]. Another form of this kind of problems is Multifacility Weber Problem (MFWP);

Table 6.1 The pairwise comparisons matrices.

A–B1–4. ($CR = 0.09 < 0.1$)

W	Physical	Human	Financial	Services	Selecting the best bank
0.16	2	3	1/7	1	Services
0.68	6	7	1		Financial
0.09	2	1			Human
0.07	1				Physical

B1–C1–4. ($CR = 0.03 < 0.1$)

W	Mehr	Sepah	Mellat	Melli	Services
0.39	4	3	1	1	Melli
0.38	3	4	1		Mellat
0.10	1/2	1			Sepah
0.13	1				Mehr

B2–C1–4. ($CR = 0.05 < 0.1$)

W	Mehr	Sepah	Mellat	Melli	Financial
0.47	2	9	2	1	Melli
0.26	2	3	1		Mellat
0.06	1/5	1			Sepah
0.21	1				Mehr

B3–C1–4. ($CR = 0.06 < 0.1$)

W	Mehr	Sepah	Mellat	Melli	Human
0.48	4	5	2	1	Melli
0.30	3	4	1		Mellat
0.08	1/3	1			Sepah
0.14	1				Mehr

B4–C1–4. ($CR = 0.05 < 0.1$)

W	Mehr	Sepah	Mellat	Melli	Physical
0.40	4	1	3	1	Melli
0.22	3	1	1		Mellat
0.30	4	1			Sepah
0.08	1				Mehr

CR, consistency ratio.

Table 6.2 Criteria weights of all factors.

Name of bank	Melli	Mellat	Sepah	Mehr
Attractiveness	0.45	0.28	0.09	0.18

in this case, each service center provides certain types of services. The MFWP is defined as locating simultaneously m facilities with particular service types satisfying the demand of n users to minimize the total transportation cost for each consumer from its closest facility [31].

Banks location–allocation problems under competitive condition assuming that each bank can have different type of services and there is no limit to the number of their customers can be classified as the Uncapacitated MFWP. In a competitive environment, selection of the best location for a new bank branch is an important decision that has significant effects on the efficiency of bank services. Because of the competitive environment, the goal is to attract more customers, so we must look for locations to establish new branches as far as possible from the branches of the same bank. Therefore, for solving this type of problem we need MO algorithms.

The assumptions for the defined problem can be expressed as the following statements:

1. We consider four different banks (Melli, Mellat, Sepah, and Mehr) in this study.
2. Each bank presents different levels of services.
3. Population densities in building blocks do not change during the study.
4. Population density (of people over 15 years of age) is available at the building block level.
5. Customers are interested in nearest bank branch with higher level of attraction to perform their financial tasks.
6. Banks have infinite capacity for accepting customers.
7. New bank branches should have maximum distance from branches of the same bank. This means that it should attract minimum number of customers from branches of the same bank.
8. Each customer refers to only one bank.
9. For customer convenience, sum of distance between customers and banks should be minimized.

According to the above-mentioned assumptions, mathematical model of the function for this problem is as follows:

First objective: The total distance needed for customers to travel to the newly established and existing branches should be minimized.

Second objective: The distance between newly established branch and other existing branches of the same bank should be maximized.

Regarding the aforementioned procedure, we have:

$$obj_1 = \min \sum_i \sum_j^{b+new} p_i d_{i,j} \qquad (6.13)$$

$$obj_2 = \max \sum_k \sum_{j_t}^{new} d_{k,j_t}, \quad \forall \ j_t \qquad (6.14)$$

where p_i is the number of customers in the building block i; new is the total number of bank branches to be established; b represents the total number of existing bank branches in the study area. $d_{i,j}$ is the distance between customer at block i and a bank branch j (existing or to be established) and d_{k,j_i} is the distance between a new branch k ($k = 1,...,new$) and an existing branch of the same bank j_t.

6.6 Implementation and results

Determining optimum location for bank branches under competitive conditions using spatial data often concerns multiple objectives. In this section, we implemented the MOCSS, MOPSO, and NSGA-II to solve the defined problem. A major part of the district III of Tabriz (612,850 m^2) was selected as the study area with a total population of 19,387 (over 15 years of age) with an average population density of 68 people per hectare. The number of existing bank branches in the study area are 13; 4, 3, 5, and

FIGURE 6.3

Banks spatial distribution and dispersion of the population in the study area.

1 branches for Melli, Mellat, Sepah, and Mehr banks, respectively as shown in Fig. 6.3. In this study we locate a new branch for Mellat bank in the study area.

Fig. 6.4 shows the graphical results produced by the MOCSS, the NSGA-II, and MOPSO for the defined problem. The Pareto Fronts of the problem are shown as circle points. It is also clear from figures that the distribution of solutions obtained by the MOCSS method is more better than the other methods.

Fig. 6.5 shows S metric obtained by the utilized algorithms. The MOCSS method shows better convergence than the other two methods. It is clear that the MOCSS algorithm indeed converges almost exponentially. Table 6.3 presents the comparison of results among the three algorithms considering the S metrics.

Fig. 6.6 shows the number of Pareto Front points obtained by the used algorithms at each iteration. As shown in Table 6.3, after 50 iterations, the MOCSS and NSGA-II methods obtained 50 solutions on the Pareto Front while the MOPSO method found only 33 solutions on the Pareto Front. Fig. 6.7 presents optimum locations of branch bank based on two objectives obtained by MOCSS, NSGA-II, and MOPSO methods. In this figure, red triangles show existing competitor branches, red triangles with blue stars show branches of the same bank, and yellow stars show proposed location for the new branches. Table 6.4 presents a comparison of set CS metrics and proves that the MOCSS can find better results and its results are with high probability dominate the results of the other two methods. The results of statistical and final tests indicate the accuracy and convergence speed of the MOCSS in finding optimal solutions for LBBCC.

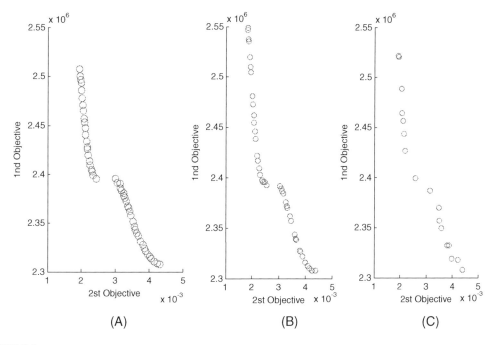

FIGURE 6.4

Pareto Front obtained by (A) MOCSS, (B) NSGA-II, and (C) MOPSO.

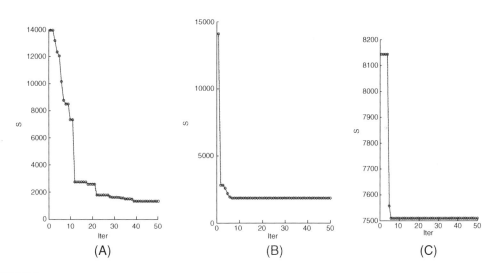

FIGURE 6.5

Convergence history of Spacing Index obtained by: (A) MOCSS, (B) NSGA-II, and (C) MOPSO.

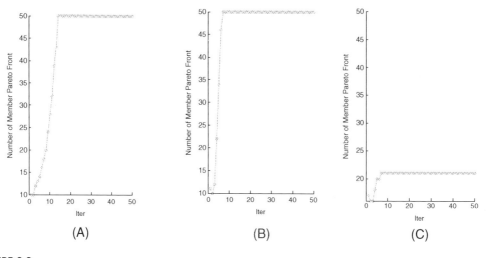

FIGURE 6.6

Number of Pareto Front solutions at each iteration obtained by (A) MOCSS, (B) NSGA-II, (C) MOPSO.

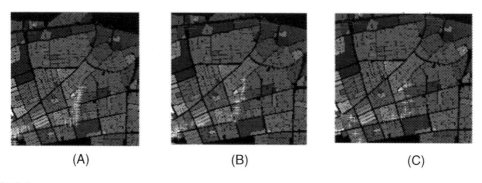

FIGURE 6.7

Optimum locations of branch bank obtained by (A) MOCSS, (B) NSGA-II, (C) MOPSO method.

Table 6.3 Comparison S index and number of Pareto Front ($n = 50$ and iteration $= 50$ for three approaches).			
Approach	**Spacing (Mean)**	**Spacing (SD)**	**Number of Pareto Front**
MOCSS	1344.8	193.5	50
NSGA-II	2074.0	304.4	50
MOPSO	4031.2	369.9	33

Table 6.4 Comparison of set coverage of two sets metrics ($n = 50$ and iteration $= 50$ for three approaches).

Approach	MOCSS		MOPSO		NSGA-II	
	NSGA-II	MOPSO	MOCSS	NSGA-II	MOCSS	MOPSO
Mean	0.14	0.26	0.04	0.08	0.10	0.26
SD	0.02	0.03	0.01	0.01	.01	0.02

6.7 Conclusions

This chapter aims to solve multiobjective problem of locating bank branches under competitive conditions by using the MOCSS method. On this issue, the following objectives were considered for the location of the branches under competitive conditions: the first objective is to minimize the total distance needed for customers to travel to the newly established and existing branches and the second objective contains maximizing the distance between newly established branch and other existing branches of the same bank. In this study, we compare the results of the MOCSS with the MOPSO and NSGA-II. From the obtained results for an LBBCC, it is clear that the MOCSS outperforms two other methods in terms of finding a diverse set of solutions, number of solutions on the Pareto Front, and the speed of convergence. The results of MOCSS method, with high probability, dominate the results the other two methods that prove the high-quality solutions obtained by the MOCSS method.

References

[1] N. Megiddo, K.J. Supowit, On the complexity of some common geometric location problems, SIAM J. Comput. 13 (1984) 182–196.
[2] E.G. Talbi, Metaheuristics: Form Design to Implementation, John Wiley and sons, New York, 2009.
[3] X. LI, A.G. YEH, Integration of genetic algorithms and GIS for optimal location search, Int. J. Geograph. Info. Sci. 19 (1) (2005) 581–601.
[4] H.Saadi Mesgari Aghamohammadi, M. Molaei, H. D. Aghamohammadi, Development a heuristic method to locate and allocate the medical centers to minimize the earthquake relief operation time, Iran. J. Publ. Health 42 (1) (2013) 63–71.
[5] S. Beheshtifar, A. Alimohammadi, Multi-Objective evolutionary algorithm for modeling of site suitability for health-care Facilities, Health Sci. J. 7 (2) (2013) 209–217.
[6] S. Talatahari, M. Azizi, Chaos Game Optimization: a novel metaheuristic algorithm, Artific. Intell. Rev. 54 (2020) 917–1004.
[7] A. Kaveh, S. Talatahari, N. Khodadadi, Stochastic paint optimizer: theory and application in civil engineering, Eng. Comput. (2020) 1–32.
[8] M. Hoard, J. Homer, W. Manley, P. Furbee, A. Haque, J. Helmkamp, System modeling in support evidence based disaster planning for rural areas, Int. J. Hyg. Environ. Health 208 (1-2) (2005) 117–125.
[9] D.L. Huff, Defining and estimating a trading area, J. Mark. 28 (1964) 34–38.
[10] M.N. Neema, A. Ohgai, Multi-objective location modeling of urban parks and open spaces: continuous optimization, Comput. Environ. Urban Syst. 34 (5) (2010) 359–376.

[11] S.H.A.Hajipoura Rahmati, S.T. V. Akhavan Niaki, A soft-computing Pareto-based meta-heuristic algorithm for a multi-objective multi-server facility location problem, Appl. Soft Comput. 13 (2013) 1728–1740.

[12] N. Xiao, D.A. Bennett, M.P. Armstrong, Using evolutionary algorithms to generate alternatives for multi-objective site-search problems, Environ. Plan. A 34 (4) (2002) 639–656.

[13] X. LI, J. HE, X. LIU, Intelligent GIS for solving high dimensional site selection problems using ant colony optimization techniques, Int. J. Geograph. Inform. Sci 23 (4) (2009) 399–416.

[14] Z. Masoomi, M.S. Mesgari, M. Hamrah, Allocation of urban land uses by multi-objective particle swarm optimization algorithm, Int. J. Geograph. Inform. Sci. 27 (3) (2013) 542–566.

[15] J.D. Duh, D.G. Brown, Knowledge-informed Pareto simulated annealing for multi-objective spatial allocation, Comput. Environ. Urban Syst. 31 (2007) 253–281.

[16] A. Gorener, H. Dinçer, Ü. Hacıoğlu, Application of multi-objective optimization on the basis of ratio analysis (MOORA) method for bank branch location selection, Int. J. Financ. Bank. Stud. 2 (2) (2013) 41–52.

[17] A. Ranjbar, S. Talatahari, F. Hakimpour, The application of multi-objective charged system search algorithm for optimization problems, Sci. Iran. 26 (3) (2019) 1249–1265.

[18] C.A. Coello, G.T. Pulido, M.S. Lechuga, Handling multiple objectives with particle swarm optimization, IEEE Trans. Evol. Comput. 8 (2004) 256–279.

[19] K.Pratap Deb, S. A. Agarwal, T. Meyarivan, A fast and elitist multiobjective genetic algorithm: NSGA–II, IEEE Trans. Evol. Comput. 6 (2) (2002) 182–197.

[20] C.A. Coello, M.S. Lechuga, MOPSO: A proposal for multiple objective particle swarm optimization, in: Proc. Congress on Evolutionary Computation, Honolulu, HI, 2002, pp. 1051–1056.

[21] K. Deb, Multi-objective Optimization Using Evolutionary Algorithms, John Wiley & Sons, New York, 2001.

[22] E. Zitzler, L. Thiele, M. Laumanns, C. Fonseca, et al., Performance assessment of multiobjective optimizer: an analysis and review, IEEE Trans. Evol. Comput. 7 (2003) 117–132.

[23] J.R. Schott, Fault tolerant design using single and multi criteria genetic algorithm optimization, in: M.S. thesis), Dept. Aeronautics and Astronautics, Massachusetts Inst. Technol., Cambridge, MA, 1995.

[24] E., Zitzler, L., Thiele. Multiobjective evolutionary algorithms: a comparative case study and the strength Pareto approach. IEEE Trans. Evol. Comput. 3 (4) (1999) 257–271

[25] E. Zitzler, K. Deb, L. Thiele, Comparison of multiobjective evolutionary algorithms: empirical results, Evol. Comput. 8 (2) (2000) 173–195.

[26] N. Srinivas, K. Deb, Multi objective optimization using non-dominated sorting in genetic algorithms, Evol. Comput. 2 (1994) 221–248.

[27] A. Kaveh, S. Talatahari, A novel heuristic optimization method: charged system search, Acta Mechanica 213 (3–4) (2010) 267–289.

[28] T.L. Saaty, The Analytic Hierarchy Process, McGraw-Hill International Book Company, New York, 1980.

[29] C. Singhtaun, P. Charnsethikul, Comparison of exact algorithms for rectilinear distance single-source capacitated multifacility Weber problems, J. Comput. Sci. 6 (2) (2010) 112–116.

[30] O. Berman, D. Krass, Facility location problems with stochastic demand and congestion, in: Z. Drezner, H.W. Hamacher (Eds.), Facility Location. Applications and Theory, Springer, Heidelberg, 2002, pp. 329–371.

[31] M. Bischoff, T. Fleischmann, K. Klamroth, The multi-facility location–allocation problem with polyhedral barriers, Comput. Oper. Res. 36 (2009) 1376–1392.

[32] D.A. Van Veldhuizen, Multi-objective evolutionary algorithms: classifications, analyses, and new innovations (PhD Thesis), Department of Electrical and Computer Engineering, Graduate School of Engineering, Air Force Institute of Technology, Wright-Patterson AFB, Ohio, 1991.

CHAPTER 7

Application of multiobjective Gray Wolf Optimization in gasification-based problems

Babak Talatahari[a], Siamak Talatahari[a] and Ali Habibollahzade[b]

[a]*Department of Civil Engineering, University of Tabriz, Tabriz, Iran,* [b]*School of Mechanical Engineering, College of Engineering, University of Tehran, Tehran, Iran*

7.1 Introduction

Gasification-based systems have been paid particular attention in recent years due to their great advantages such as benefiting from biomass renewable sources and achieving a high temperature required to reach desired products. Gasification systems work with an agent that can simply be air or even CO_2 or steam. The use of CO_2 and steam can potentially improve the performance of the system but the systems working on these agents should be optimized to enhance effectiveness of the gasifier and reduce char formation. Multiobjective optimization methods are quite handy in designing a gasification-based system to operate effectively. In this regard, conventional and advanced optimization techniques may be implemented to design the gasifier itself, or the bigger plant in which gasification is a smaller part of the control volume to which the optimization is applied. Usually, genetic algorithms are employed to conduct the optimization of the conventional gasification-based systems; however, researchers tend to optimize these systems following advanced optimization methods to achieve better solutions [23].

Some researchers investigated the gasification using air or other agents and focused on the gasifier itself or the gasification process to be optimized. Zhang et al. [26] investigated entrained flow gasification by employing an integrated array design to optimize the gasification process and Wang et al. [25] considered an entrained gasifier by using an artificial neural network technique after conducting computational fluid dynamic (CFD) calculations. They implemented machine learning algorithms to conduct the optimization methods based on the numerical analysis. Seçer et al. [22] focused on hydrogen production during the gasification process to be maximized using the conventional optimization methods. Hu et al. [13] performed an optimization for biomass gasification using a steam agent employing the response surface technique to maximize the higher heating value (HHV) of syngas. The optimization results showed that at the optimum solution point content of carbon monoxide, hydrogen, and methane—which have high HHVs is about 83%. Okolie et al. [19] focused on maximizing the valuable gas of hydrogen through a dedicated optimization scenario to optimize the effective factors for biomass gasification. Zhou et al. [27] used a dynamic reduced-order model to optimize a large-scale entrained-flow gasifier. Almost for all these works, conventional single-objective genetic-based algorithm was performed in these studies. Although very powerful optimization methods were developed until now (Stochastic Paint Optimizer [14], Chaos Game Optimization [24], Quantum Developed Swarm Optimizer [24] and many others), application of these methods for the defined problem is limited. As one of recent studies,

Habibollahzade et al. [8] investigated the gasification of the biomass using various gasification agents. They implemented the Gray Wolf Optimization (GWO) method to conduct triobjective optimization of the gasification process taking into account various biomass feedstocks and agent to determine the most suitable feedstock for each agent.

Some authors optimized the gasification applications for syngas/biofuel-driven systems. Sadeghi et al. [21] simulated a gasifier and subsequently examined and optimized a syngas fueled system from first and second laws of thermodynamics and economic viewpoints. They intended to maximize efficiency, while reducing cost of the system and CO_2 emissions using a multiobjective Genetic Algorithm optimization. They benefitted from the potentials of the optimization method to design an appropriate system driven by gasification process. Also, Mojaver et al. [17] optimized a gasification-based system using various feedstocks following a multiobjective optimization method. In addition, some other researchers examined gasification-based advanced systems in fuel cells and performed different multiobjective optimization methods to design appropriate systems [7,9]. Habibollahzade and Rosen [11] employed the Particle Swarm Optimization to conduct triobjective optimization of a gasification-based system operating on carbonaceous materials and air/O_2-enriched-air agents. In addition, more studies focused on the waste-to-energy plants (WtEPs) as the potential applications of syngas-driven systems.

Other applications of gasification-based systems would be WtEPs, which use the gasification to produce syngas to be burned and subsequently exploited for the power generation. Optimization of these systems has rarely been conducted in the literature. Holanda and Perrella Balestieri [12] optimized gas cleaning routes of a conventional WtEP and Alao et al. [1] designed a suitable waste-to-energy system through entropy-weighted, the technique for order of preference by similarity to ideal solution method and an optimization technique. Behzadi et al. [3] optimized a conventional WtEP via the Genetic Algorithm; the results showed that the efficiency of the system can be easily maximized through the optimization considering effective input parameters. In a recent study, Habibollahzade and Houshfar [10] examined a WtEP using a new type of agent and optimized the system using advanced triobjective GWO. They achieved a well-balanced solution in regard to thermodynamic, economic, and environmental criteria. In fact, the system can optimally operate at the well-balanced solution in which the system efficiency is satisfactory while carbon emissions and cost of the system are minimized.

7.2 Systems description

Optimization of the downdraft gasifier and WtEP consisting downdraft gasifier has been performed in the literature. Multiobjective optimization of the downdraft gasifier is performed to maximize its efficiencies while minimizing CO_2 emissions. However, at the bigger scales, where WtEPs are considered for the optimization, CO_2 emissions and efficiencies of the whole system are intended to be optimized which are not the same as CO_2 emissions of the gasifier itself. In the following subsections, a downdraft gasifier and WtEP working principles are described, briefly.

7.2.1 Downdraft gasifier

The gasifier used in WtEPs is usually a downdraft gasifier illustrated in Fig. 7.1. The figure shows that the gasifier has some operating zones and the equilibrium condition is applied to the reduction zone

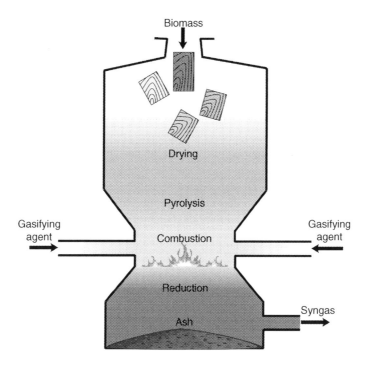

FIGURE 7.1

A downdraft gasifier with important gasification zones highlighted.

where higher temperatures are achieved. Initially, biomass (or any other carbonaceous material) is fed to the gasifier and after undergoing several chemical reactions and reaching thermodynamic equilibrium, syngas leaves the gasifier. Usually, the reduction zone temperature should be kept at higher values (higher than 700 °C) to avoid formation of tar and char. At such extreme values, the assumption of thermodynamic equilibrium is also reliable. The syngas composition is very important as it would be subsequently used in other applications such as WtEPs. Hence, optimization of the downdraft gasifiers is of particular importance to provide syngas with higher HHVs. It is also noted that, a gasifying agent is required to initiate and maintain gasification reactions throughout the process. The agents could be O_2-based agents or steam and even CO_2 whose advantages and disadvantages are listed in Table 7.1. The use of a mixture of these agents has also been investigated in the literature.

7.2.2 Waste-to-energy plant

Fig. 7.2 illustrates a conventional and advanced WtEP in which a downdraft gasifier with an air and a mixture of air and CO_2 agents are used. Both conventional and advanced systems have similar configurations where a gasification system and a Rankine cycle are the main parts. Initially, biomass is fed to the gasifier and the produced syngas is burned in the combustion chamber to provide sufficient heat for the Rankine cycle. The hot gases are still quite warm and they can be used in another cycle

Table 7.1 Main advantages and weaknesses of selected gasification agents [6,20].

Gasification agent	Advantages	Weaknesses
Air	1. Very inexpensive technology 2. Tar and char contents are relatively low 3. Temperature can be adjusted by equivalence ratio	1. Lower efficiency and HHV value of syngas 2. High nitrogen dilution
O_2	1. High efficiency and HHV value of syngas 2. Low nitrogen dilution 3. Very low tar and char 4. Temperature can be adjusted by equivalence ratio	1. Very expensive
Steam	1. High H_2 concentration in dry syngas compared to CO 2. High content of H_2 3. Less expensive than O_2 agent 4. High efficiency	1. Water and vaporization are needed 2. External heat source is required 3. High tar content
CO_2	1. Consumes CO_2 to produce CO 2. Efficiency can be high 3. High CO content in syngas	1. External heat source is required 2. Char content can be relatively high

called Organic Rankine Cycle system. It is noteworthy that the main difference between the advanced and conventional cycles lies on the use of gasification agents where some portion of CO_2 emitted is reused in the advanced cycle to minimize CO_2 emissions.

The main assumptions that are usually made to analyze WtEPs are as follows:

- Steady-state condition is assumed throughout the simulations [5].
- Content of CO_2, argon, and other minor gases are neglected (21% O_2 and 79% N_2) [5].
- A temperature and pressure of 101.3 kPa and 25 °C are applied with regard to a dead state [2].
- The pressure drop passing through heat exchangers would be 5% [4].
- All gases treated as ideal gases irrespective of temperature and pressure as higher pressures are not observed in WtEPs [18].
- All components operate adiabatically, that is, there is no heat transfer between the components and the environment [3].
- Gasifier and combustion chamber operate in thermodynamic equilibrium conditions [2].

7.3 Modeling

The gasifier and WtEP should be modeled to apply multiobjective optimization to optimize the objective functions. The modeling of the WtEP and the gasifier is initiated via modeling of the biomass and predicting its characteristics. Second, the gasification reactions are simulated to define the syngas content. Subsequently, the combustion chamber is modeled, before modeling the Rankine cycle and other minor parts. The modeling approach is provided for the gasifier and WtEP in two flowcharts shown in Fig. 7.3. Also, the corresponding equations needed to simulate these systems are provided in Table 7.2, while the related equations to model the Rankine cycle are not provided as they can be easily found elsewhere [10].

7.3 Modeling

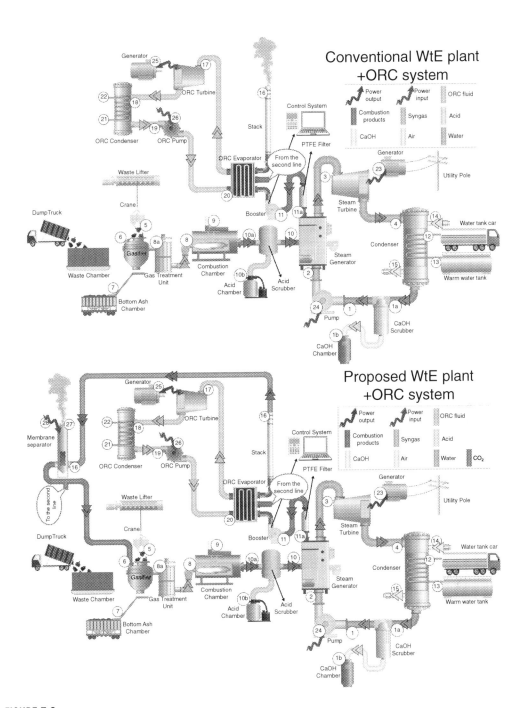

FIGURE 7.2

Schematics of a typical and advanced WtEPs [10].

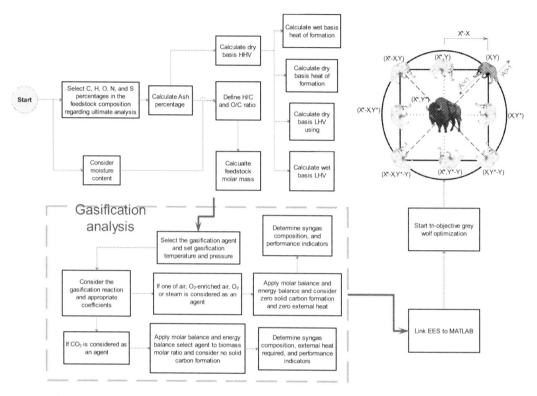

FIGURE 7.3

The modeling procedure of (A) the gasification system, (B) WtEP.

The economic assessment of the WtEP has been provided by Habibollahzade and Houshfar [10] in detail. However, the cost and exergy balance equations along with auxiliary equations are provided in Table 7.3. Also, the efficiencies, CO_2 emissions correlations for both gasification and WtEPs, and total product cost and cost rates of the systems of the WtEPs are explained in Habibollahzade et al. [8] and Habibollahzade and Houshfar [10].

7.4 Multicriteria Gray Wolf Optimization

The recently developed Multicriteria GWO (MCGWO) is a robust and advanced technique [16]. This optimization technique is stimulated by the societal leadership and hunting technique of gray wolves. The best solution is called alpha (α) referring to the alpha wolf. The next best solutions are beta (β) and delta (δ) which stand for beta and delta wolves inspired by the social hierarchy of wolves. The rest of the solution points are assumed as omega (ω) wolves that follow the other wolves. Through the MCGWO algorithm, the optimization is triggered and directed by the alpha, beta, and gamma wolves.

7.4 Multicriteria Gray Wolf Optimization

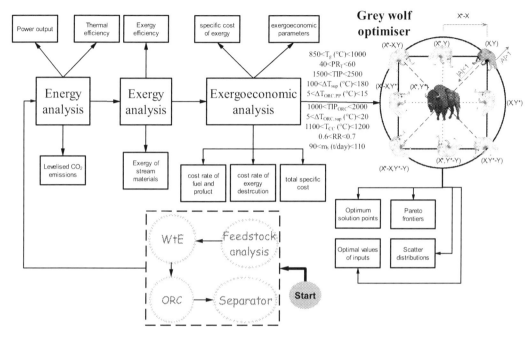

FIGURE 7.3 Continued

The encircling behavior of gray wolves during hunting is simulated using the following equations:

$$\vec{D} = \left| \vec{C}.\vec{X}_P(t) - \vec{X}(t) \right| \tag{7.1}$$

$$\vec{X}(t+1) = \vec{X}_P(t) - \vec{A}.\vec{D} \tag{7.2}$$

where t is the current iteration; $\vec{A}$ and $\vec{C}$ denote coefficient vectors and $\vec{X}_P$ denotes the position vector of the prey. Also, $\vec{X}$ is the position vector of a single gray wolf. Also, we have:

$$\vec{A} = 2\vec{a}.\vec{r}_1 - \vec{a} \tag{7.3}$$

$$\vec{C} = 2.\vec{r}_2 \tag{7.4}$$

where $\vec{a}$ reduces to within the range of 0~2 during the iteration and r_1, r_2 are random vectors in [0,1].

The MCGWO technique only stores the first three best solutions and neglects other ones. The algorithm also obliges other search agents to update their position according to the previous solution point. The following relations are used continually for each search agent during the optimization to simulate the position updating:

$$\vec{D}_i = \left| \vec{C}_m.\vec{X}_i - \vec{X} \right| \tag{7.5}$$

$$\vec{X}_m = \vec{X}_i - \vec{A}_m \cdot \left(\vec{D}_i \right) \tag{7.6}$$

Table 7.2 The whole modeling procedure of the gasifier and gasification system [8].

Equations/correlations	Description
Biomass analysis	
$CH_aO_bN_cS_d$	Chemical composition of the carbonaceous material
$MC_{CM} = \frac{\text{Mass of water}}{\text{Mass of wet biomass}}$	Moisture content of carbonaceous material
$Ash = 100 - (N + H + O + C + S)$	Ultimate analysis of biomass
$a = \frac{M_C \times H}{M_H \times C} \quad b = \frac{M_C \times O}{M_O \times C}$	Define the coefficients in chemical composition of the carbonaceous material
$HHV_{CM,dry} = 349.1C + 1178.3H + 100.5S - 103.40 - 15.1N - 21.1Ash$	Define the HHV of carbonaceous material on a dry basis
$h_{f,CM,dry} = [-3,935.00 - \frac{a}{2} \times HHV_{H_2} - d \times HHV_S + HHV_{CM,dry} \times M_{CM}]$	Define the heat of formation of the carbonaceous material on a dry basis
$hf,CM,wet = hf,CM,dry - v \times hf,water, 298\ K$	Define the heat of formation of the carbonaceous material on a wet basis
$M_{CM} = M_C + a \times M_H + b \times M_O + c \times M_N + d \times M_S$	Define the molar mass of carbonaceous material
$LHV_{CM,dry} = HHV_{CM,dry} - 2241.7 \times 9 \times \frac{H}{100}$	Define the lower heating value (LHV) of carbonaceous material on a dry basis
$LHV_{CM,wet} = HHV_{B,dry} \times (1 - MC_B) - 2241.7 \times \left(9 \times H \times \frac{(1-MC_B)}{100} + MC_B\right)$	Define the LHV of carbonaceous material on a wet basis
Gasification process	
$\dot{m}_{CM}CH_aO_bN_c + eH_2O + fO_2 + tN_2 + qCO_2$ $\rightarrow \dot{m}_1H_2 + \dot{m}_2CO + \dot{m}_3CO_2 + \dot{m}_4H_2O + \dot{m}_5CH_4 + \dot{m}_6N_2 + \dot{m}_7C$	Gasification reaction
$C: \dot{m}_B + q = \dot{m}_2 + \dot{m}_3 + \dot{m}_5 + \dot{m}_7$ $H: \dot{m}_B(a + 2v) + 2e = 2\dot{m}_1 + 2\dot{m}_4 + 4\dot{m}_5$ $O: \dot{m}_B(v + b) + e + 2f + 2q = \dot{m}_2 + 2\dot{m}_3 + \dot{m}_4$ $N: \dot{m}_B \times c + 2t = 2\dot{m}_6$	Molar balance
$K_1 = \frac{\dot{m}_2(\dot{m}_1)^3}{\dot{m}_4 \times \dot{m}_5} \left(\frac{P_g/P_0}{\dot{m}_{tot}}\right)^2 ; K_1 = \exp\left(-\frac{\Delta G_1^0}{RT_g}\right)$ $K_2 = \frac{\dot{m}_1 \times \dot{m}_3}{\dot{m}_2 \times \dot{m}_4} ; K_2 = \exp\left(-\frac{\Delta G_2^0}{RT_g}\right);$ $K_3 = \frac{\dot{m}_6}{(\dot{m}_2)^2} \left(\frac{P_g/P_0}{\dot{m}_{tot}}\right)^{-1} ; K_3 = \exp\left(-\frac{\Delta G_3^0}{RT_g}\right)$	Equilibrium constants
$\Delta G_1^0 = g_{CO}^0 + 3g_{H_2}^0 - g_{H_2O}^0 - g_{CH_4}^0$ $\Delta G_2^0 = g_{CO_2}^0 + g_{H_2}^0 - g_{H_2O}^0 - g_{CO}^0$ $\Delta G_3^0 = g_{CH_4}^0 - 2g_{H_2}^0$	Change of Gibbs free energies
$\dot{m}_{CM}h_{f,B,wet} + \underbrace{eh_{H_2O}}_{\text{at } T=100°C} + \underbrace{fh_{O_2} + th_{N_2} + qh_{CO_2}}_{\text{at } T=T_0, P=P} + \dot{Q}$ $= \dot{m}_1 h_{H_2} + \dot{m}_2 h_{CO} + \dot{m}_3 h_{CO_2} +$ $\underbrace{\dot{m}_4 h_{H_2O} + \dot{m}_5 h_{CH_4} + \dot{m}_6 h_{N_2}}_{\text{at } T=T} + \underbrace{\dot{m}_7 h_C}_{\text{solid carbon (char) at } T=T}$	Energy balance
Combustion process	
$\dot{m}_1H_2 + \dot{m}_2CO + \dot{m}_3CO_2 + \dot{m}_4H_2O + \dot{m}_5CH_4 + \dot{m}_6N_2$ $+ \dot{m}'(O_2 + 3.76N_2)$ $\rightarrow \dot{m}'_1CO_2 + \dot{m}'_2H_2O + \dot{m}'_3O_2 + \dot{m}'_4N_2$	Energy balance

Table 7.3 Exergy and cost conservation relations alongside cost functions and auxiliary equations [10].

Component	Exergy destruction rate	Cost balance, auxiliary equations, and cost function	Cost ref. year
Gasifier	$\dot{E}_{D,g} = \dot{E}_5 + \dot{E}_6 - \dot{E}_7 - \dot{E}_8$	$\dot{C}_5 + \dot{C}_6 + \dot{Z}_g = \dot{C}_7 + \dot{C}_8$ $c_5 = 2 \text{ US\$/GJ}$ $Z_g = 1600(\dot{m}_{F,daf}[\text{kg/h}])^{0.67}$	2013
Combustion chamber	$\dot{E}_{D,CC} = \dot{E}_8 + \dot{E}_9 - \dot{E}_{10}$	$\dot{C}_8 + \dot{C}_9 + \dot{Z}_{CC} = \dot{C}_{10}$ $Z_{cc} = 48.64 \dot{m}_9[\text{kg/s}] (1 + \exp(0.018 T_{10}[^\circ\text{C}] - 26.4)) \cdot \frac{1}{0.995 - \frac{P_{10}}{P_8}}$	2013
Pump	$\dot{E}_{D,pu} = \dot{W}_{pu} - (\dot{E}_2 - \dot{E}_1)$	$\dot{C}_1 + \dot{C}_{24} + \dot{Z}_{pu} = \dot{C}_2$ $c_{20} = c_{21}$ $Z_{pu} = 3540 (\dot{W}_{pu}[\text{kW}])^{0.71}$	2010
Steam turbine	$\dot{E}_{D,ST} = (\dot{E}_3 - \dot{E}_4) - \dot{W}_{ST}$	$\dot{C}_3 + \dot{Z}_{ST} = \dot{C}_4 + \dot{C}_{23}$ $c_3 = c_4$ $Z_{ST} = 6000 (\dot{W}_{ST}[\text{kW}])^{0.7}$	2010
Steam generator	$\dot{E}_{D,SG} = \dot{E}_2 + \dot{E}_{10} - \dot{E}_3 - \dot{E}_{11}$	$\dot{C}_2 + \dot{C}_{10} + \dot{Z}_{SG} = \dot{C}_{11} + \dot{C}_3$ $c_{11} = c_{10}$ $Z_{SG} = 6570 \left(\left(\frac{Q_{eco}[\text{kW}]}{\Delta T_{LMTD,eco}[K]} \right)^{0.8} + \left(\frac{Q_{eva}[\text{kW}]}{\Delta T_{LMTD,eva}[K]} \right)^{0.8} \right) + 21276 \dot{m}_3 [\text{kg/s}] + 1184.4 (\dot{m}_{10}[\text{kg/s}])^{1.2}$	1997
Condenser	$\dot{E}_{D,cond} = \dot{E}_{12} + \dot{E}_{14} + \dot{E}_4 - \dot{E}_1 - \dot{E}_{13} - \dot{E}_{15}$	$\dot{C}_4 + \dot{C}_{12} + \dot{C}_{14} + \dot{Z}_{cond} = \dot{C}_{15} + \dot{C}_{13} + \dot{C}_1$ $c_4 = c_1$ $Z_{cond} = 1773 \dot{m}_{12}[\text{kg/s}]$	2010
ORC Pump	$\dot{E}_{D,ORC,pu} = \dot{W}_{ORC,pu} - (\dot{E}_{20} - \dot{E}_{19})$	$\dot{C}_{19} + \dot{C}_{26} + \dot{Z}_{ORC,pu} = \dot{C}_{20}$ $c_{19} = c_{20}$ $Z_{ORC,pu} = 200 (\dot{W}_{ORC,pu}[\text{kW}])^{0.65}$	2010
ORC Turbine	$\dot{E}_{D,ORC,T} = (\dot{E}_{17} - \dot{E}_{18}) - \dot{W}_{ORC,T}$	$\dot{C}_{17} + \dot{Z}_{ORC,T} = \dot{C}_{25} + \dot{C}_{18}$ $c_{17} = c_{18}$ $Z_{ORC,T} = 4750 (\dot{W}_{ORC,T}[\text{kW}])^{0.75}$	2010
ORC Evaporator	$\dot{E}_{D,ORC,eva} = \dot{E}_{20} + \dot{E}_{11} - \dot{E}_{16} - \dot{E}_{17}$	$2\dot{C}_{11} + \dot{C}_{20} + \dot{Z}_{ORC,eva} = \dot{C}_{16} + \dot{C}_{17}$ $c_{11} = c_{16}$ $Z_{ORC,eva} = 309.14 (A_{ORC,eva}[\text{m}^2])^{0.85}$	2003
ORC Condenser	$\dot{E}_{D,ORC,cond} = \dot{E}_{18} + \dot{E}_{21} - \dot{E}_{22} - \dot{E}_{19}$	$\dot{C}_{21} + \dot{C}_{18} + \dot{Z}_{ORC,cond} = \dot{C}_{19} + \dot{C}_{22}$ $c_{18} = c_{19}$ $Z_{ORC,cond} = 516.62 (A_{ORC,cond}[\text{m}^2])^{0.6}$	2005
Separator	$\dot{E}_{D,sep} = \frac{1}{2}\dot{E}_{16} + \dot{W}_{sep} - (\dot{E}_{27} - \dot{E}_6)$	$\frac{1}{2}\dot{C}_{16} + \dot{C}_{28} + \dot{Z}_{sep} = \dot{C}_6 + \dot{C}_{27}$ $c_6 = c_{27}$ $Z_{sep} = 192000 \times \dot{m}_6[\text{kg/s}]$	2015

where i can be α, β, or δ, and m takes on a value of 1, 2, and 3, respectively, for α, β and δ. Thus, we have:

$$\vec{X}(t+1) = \frac{\vec{X}_1 + \vec{X}_2 + \vec{X}_3}{3} \tag{7.7}$$

The pseudo-code of the MCGWO algorithm is provided in Fig. 7.4.

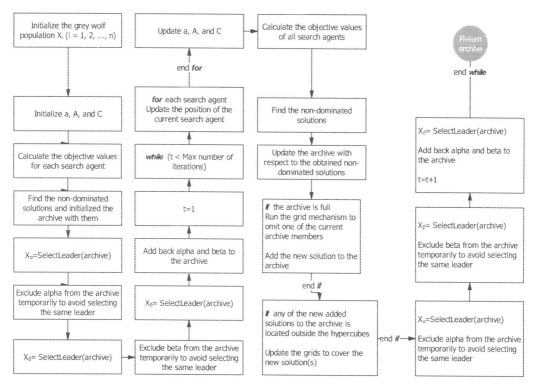

FIGURE 7.4

Pseudo-code of MCGWO [10].

The Linear Programming Technique for Multidimensional Analysis of Preference (LINMAP) and The Technique for Order of Preference by Similarity to Ideal Solution (TOPSIS) schemes are utilized to select the final optimal solution. Euclidean distances can be calculated using the normalized optimal solutions and normalized ideal solutions [15]:

$$ED^{+} = \sqrt{\sum_{i=1}^{n} \left(OF_i^N - OF_{i,\text{ideal}}^N\right)^2} \tag{7.8}$$

The normalized objective functions (OF) are determined as follows [15]:

$$OF_i^N = \frac{OF_i}{\sqrt{\sum_i (OF_i)^2}} \tag{7.9}$$

The lowest value of ED_i^+ is selected as the final optimal point from the perspective of the LINMAP method. Regarding the TOPSIS method, nonideal solutions are considered following a similar approach [15]:

$$ED^{-} = \sqrt{\sum_{i=1}^{n} \left(OF_i^N - OF_{i,\text{nonideal}}^N\right)^2} \tag{7.10}$$

Then, the relative closeness (RC) is defined as

$$RC_i = \frac{ED_i^-}{ED_i^- + ED_i^+} \tag{7.11}$$

The maximum value of RC is chosen as the final optimal point, for the TOPSIS scheme.

7.5 Results and discussion
7.5.1 Optimization at the gasifier level

The gasifier operating condition is optimized through MCGWO to maximize desired objective functions while minimizing unfavorable objectives. Accordingly, some Pareto frontiers are extracted to show a tradeoff between the objective functions as shown in Fig. 7.5 for the agents used. Thermodynamics efficiencies and emissions are considered to be the target functions.

Fig. 7.6 lists the outcomes of the optimization for all agents used for the optimization and the corresponding objective functions. It is noted that we can control the gasification temperature through adjusting biomass to agent ratio when O_2-based agents are used as the gasification process using such agents are exothermic. For other endothermic agents, the temperature should be controlled through external heat. Optimal range of conditions can be ascertained through defining the location of optimum points within the studied population. Further, when the local optimum points are dispersed in the studies range, it means there is no specific optimum operating condition.

The optimum ranges associated with various decision variables are shown in Fig. 7.7 for steam and CO_2 agent. When CO_2 agent is used, lower H/C ratios in the feedstock composition are suitable as the local optimum solutions are located at the bottom of the studied range. However, for the percentage of carbon in the feedstock, it is not possible to define a suitable local optimum range as the solutions are not located within a specific range. Therefore, the engineers should consider a suitable feedstock based on their own view. In addition, lower gasification pressures may be suitable in some conditions as the figure shows and higher gasification pressures can make the system useless. However, for the case of operating temperature, higher temperatures are preferred as lower temperatures result in lower efficiencies in some conditions. In addition, it is noted that moisture content (MC) is a contributing factor to efficiency and emissions and when steam is used as an agent, MC lower than 0.15 is suitable. In addition, external heat required for the gasification process should be maintained at lower values to make the system more effective. Higher external heat values can reduce the efficiency of the system as the system cannot produce syngas with high HHV to support the efficiency of the system. When CO_2 agent is used, very high efficiencies can be met if emissions are not the highest priority. For example, exergetic and cold gas efficiencies of 94.8% and 94.5% are expected if $CH_{1.3}O_{0.57}$ is used as the feedstock while the other decision variables are considered according to the values presented in Fig. 7.6. The problem is that cold gas efficiency and CO_2 emissions cannot be optimized as higher cold gas efficiency means higher CO_2 emissions as well. It can be seen that, when CO_2 agent is used to trigger the gasification process, the external heat used would be much higher as the reactions in the gasification are highly endothermic. The suitable feedstocks proposed to be used with different agents are listed in Table 7.4.

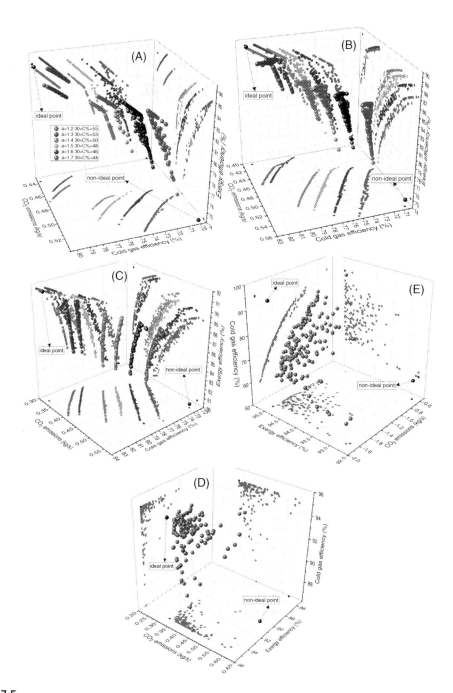

FIGURE 7.5

Pareto-optimal frontiers at several H/C ratios associated with various gasification agents: (A) air, (B) O_2-enriched air, (C) O_2, (D) steam, (E) CO_2 [8].

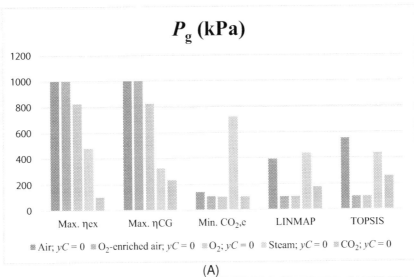

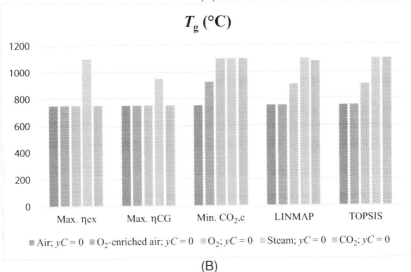

FIGURE 7.6

Values of objective functions and decision variables for biomass gasification using various agents [8].

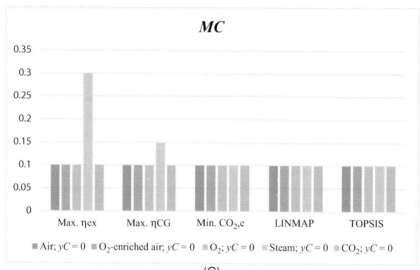

(C)

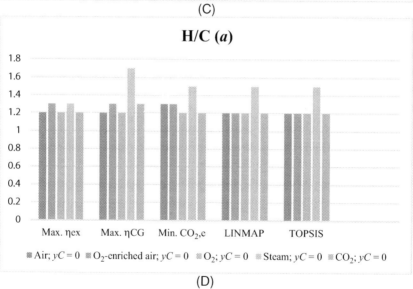

(D)

FIGURE 7.6

Continued

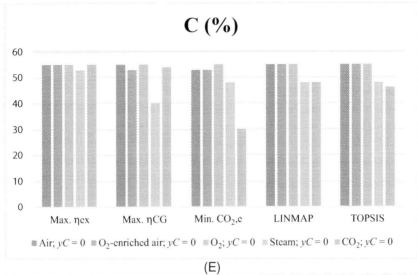

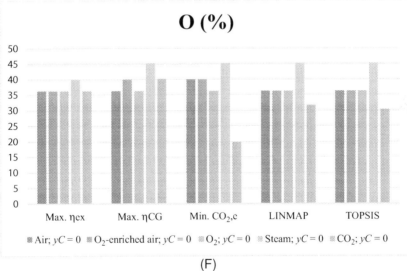

FIGURE 7.6

Continued

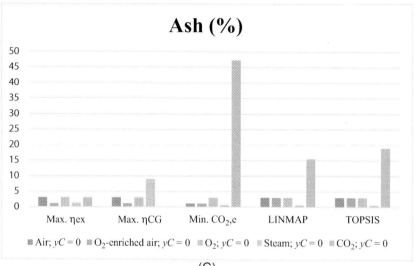

(G)

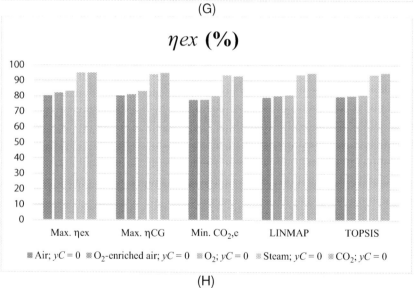

(H)

FIGURE 7.6

Continued

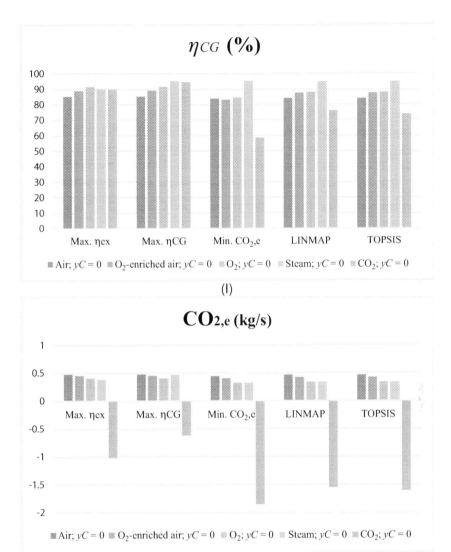

FIGURE 7.6

Continued

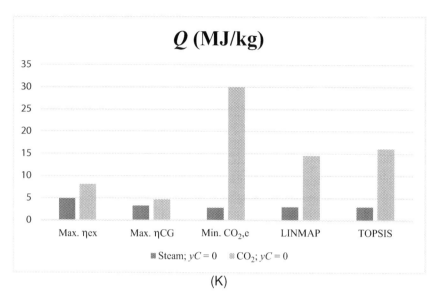

FIGURE 7.6

Continued

When steam agent is used, optimum operating ranges are shown in Fig. 7.7. According to this figure, it is impossible to define a specific optimum operating range within the studied range especially for H/C ratio and MC as external heat is required for steam agent. External heat is a contributing factor for efficiency and moist biomass may be more preferable in some conditions as less heat is required to trigger the gasification. It is also noted that, higher temperature and lower pressure values are preferred when steam is used in the gasification process as local optimum points are located around those values. Also, carbon-rich feedstocks preferred for the gasification using steam as optimum solutions are located at the bottom of the operating range. Regarding external heat values, lower external heat values are suitable as higher values can take the system away from its optimal conditions.

7.5.2 Optimization at the WtEP Level

Thermodynamics, environmental, and economic criteria are considered as the objective functions to optimize the system using MCGWO. Accordingly, some Pareto are extracted as presented in Fig. 7.8. As can be seen, with increasing one objective function, the other one increases as well, so it is necessary to determine an optimum solution as final solution point. Some solution points are highlighted in the figures indicating the maximum or minimum value of the objective functions. Also, the final solutions determined by TOPSIS and LINMAP methods are shown. In addition, the values of target functions alongside the decision variables are listed in Table 7.5. Additionally, the ideal and nonideal points are shown in the curves as well.

MCGWO is employed to optimize three objective functions at the same time as shown in Fig. 7.9. After conducting the optimization and optimizing three objective functions, the corresponding Pareto frontier and the results of the optimization are provided in Table 7.5 and Fig. 7.9. Fig. 7.9 shows

7.5 Results and discussion

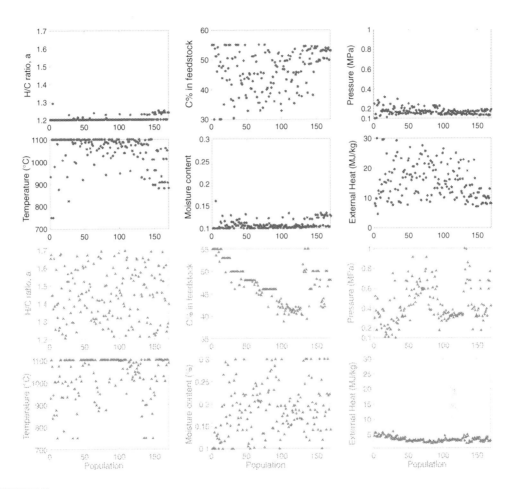

FIGURE 7.7

Scatter distribution of the local optimum points within the studied range of decision variables (Note: Gray circles refer to CO_2 agent and green triangles refer to the steam agent) [8].

Table 7.4 Feedstock for biomass gasification using various agents [8].

Solution point	Air; $y_C = 0$	O_2-enriched air; $y_C = 0$	O_2; $y_C = 0$	Steam; $y_C = 0$	CO_2; $y_C = 0$
Max. η_{ex}	$CH_{1.2}O_{0.495}$	$CH_{1.2}O_{0.495}$	$CH_{1.2}O_{0.495}$	$CH_{1.3}O_{0.57}$	$CH_{1.2}O_{0.495}$
Max. η_{CG}	$CH_{1.2}O_{0.495}$	$CH_{1.2}O_{0.495}$	$CH_{1.2}O_{0.495}$	$CH_{1.7}O_{0.85}$	$CH_{1.3}O_{0.57}$
Min. $CO_{2,e}$	$CH_{1.3}O_{0.566}$	$CH_{1.2}O_{0.495}$	$CH_{1.2}O_{0.495}$	$CH_{1.5}O_{0.71}$	$CH_{1.3}O_{0.50}$
LINMAP	$CH_{1.2}O_{0.495}$	$CH_{1.2}O_{0.495}$	$CH_{1.2}O_{0.495}$	$CH_{1.5}O_{0.71}$	$CH_{1.2}O_{0.50}$
TOPSIS	$CH_{1.2}O_{0.495}$	$CH_{1.2}O_{0.495}$	$CH_{1.2}O_{0.495}$	$CH_{1.5}O_{0.71}$	$CH_{1.2}O_{0.50}$

Table 7.5 Values of decision variables and performance indicators obtained by MCGWO [10]. PR, TIP, RR, LE and PP denote pressure ratio, turbine inlet pressure, recycling ratio, levelized emissions, and pinchpoit.

Operating condition	Decision variables										Performance indicators								
	T_g (°C)	PR_T	TIP (kPa)	ΔT_{sup} (°C)	$\Delta T_{ORC,PP}$ (°C)	TIP_{ORC} (kPa)	$\Delta T_{ORC,sup}$ (°C)	T_{CC} (°C)	RR	$\dot{m}_s$ (kg/s)*	η_{th} (%)	η_{ex} (%)	$\dot{m}_{CO_2}$ (kg/s)	LE (t/MWh)	$\dot{W}_{net}$ (kW)	$C_{P,tot}$ (US$/GJ)	$c_{P,tot}$ (US$/kWh)	PP (year)	$\dot{C}_{tot}$ (US$/h)
Objective functions: Power output vs total cost rate																			
Point A	1000.0	60.0	2500	180.0	7.3	1954	15.9	1200	0.6	1.2	31.1	16.9	2.2	1.7	4710	30.5	0.110	9.65	516.3
Point B	851.8	58.8	2181	173.6	9.0	1712	15.2	1200	0.6	1.0	27.9	15.1	1.8	1.8	3528	33.7	0.121	9.98	428.3
LINMAP	1000.0	60.0	2500	173.6	5.2	1448	15.2	1200	0.6	1.1	31.4	17.1	2.0	1.7	4343	30.5	0.110	9.86	476.9
TOPSIS	1000.0	60.0	2391	180.0	9.0	1516	12.4	1200	0.6	1.2	31.4	17.0	2.1	1.7	4535	30.4	0.109	9.68	496.4
Objective functions: Exergy efficiency vs CO_2 emissions																			
Point A	1000	60.0	2290	180.0	8.9	1247	15.6	1165	0.6	1.0	31.4	17.1	1.8	1.7	3973	30.9	0.111	10.09	441.9
Point B	850	60.0	2430	132.7	8.6	2000	7.2	1200	0.7	1.0	18.0	9.8	1.4	2.2	2286	52.8	0.190	10.40	434.7
LINMAP	1000	60.0	2500	180.0	7.6	1865	12.1	1200	0.6	1.0	28.4	15.4	1.7	1.7	3604	34.2	0.123	10.01	443.3
TOPSIS	1000	60.0	2111	173.7	6.0	1548	11.3	1200	0.6	1.0	31.2	17.0	1.8	1.7	3956	31.0	0.112	10.05	440.8
Objective functions: Exergy efficiency vs total cost rate																			
Point A	1000	60.0	2500	180.0	5.2	1866	12.3	1100	0.6	1.3	31.6	17.2	2.3	1.7	4895	30.2	0.109	9.67	532.8
Point B	870	46.2	2287	100.0	7.3	1979	6.7	1200	0.6	1.0	24.6	13.3	1.8	2.1	3114	37.5	0.135	9.97	420.8
LINMAP	1000	60.0	2500	168.4	7.6	1654	12.2	1200	0.6	1.0	31.1	16.9	1.8	1.7	3937	31.0	0.112	10.03	439.8
TOPSIS	1000	60.0	2500	180.0	9.6	1567	10.5	1200	0.6	1.0	31.4	17.0	1.8	1.7	3973	30.8	0.111	10.02	440.4
Objective functions: Exergy efficiency vs total cost rate vs CO_2 emissions																			
Point A	850	40.0	1993	100.0	5.0	1990	20.0	1200	0.6	1.0	23.2	12.6	1.8	2.2	2945	39.5	0.142	10.03	418.7
Point B	850	44.0	1974	100.0	14.7	2000	15.3	1200	0.7	1.0	15.3	8.3	1.4	2.6	1934	61.5	0.221	10.41	428.0
Point C	1000	60.0	2500	180.0	10.4	1000	14.4	1200	0.6	1.0	31.6	17.1	1.8	1.7	4001	30.7	0.111	10.04	441.4
LINMAP	1000	60.0	2405	180.0	7.0	1517	10.9	1200	0.6	1.0	28.9	15.7	1.7	1.7	3665	33.7	0.121	10.04	444.4
TOPSIS	1000	60.0	2500	180.0	5.0	1682	7.8	1100	0.6	1.0	31.3	17.0	1.8	1.7	3963	31.2	0.112	10.19	445.6

*Based on one operating line

7.5 Results and discussion

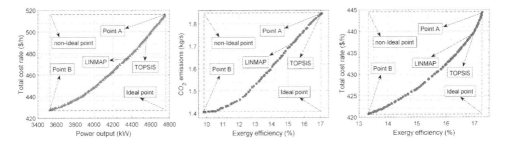

FIGURE 7.8

Pareto frontiers resulted from MCGWO considering various objective functions [10].

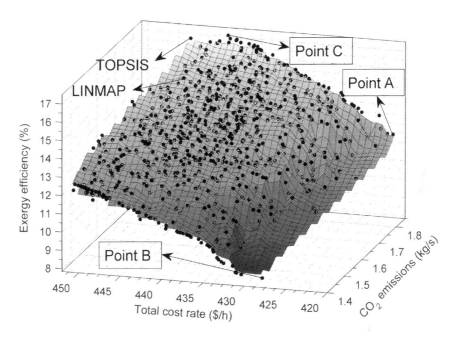

FIGURE 7.9

3D Pareto frontier resulted from MCGWO considering exergetic efficiency, total cost rate, and CO_2 emissions as the objective function [10].

the surface fitted on the optimal solutions alongside some optimum solutions indicating maximum efficiency, minimum cost and emissions in which the final optimum solutions are determined by the TOPSIS and LINMAP. The optimum values of the input parameters and objective functions associated with each operating point indicated in the figure are listed in Table 7.5. All of the economic parameters are considered according to the regulations of the Iranian government. Thermodynamics, economic, and environmental objective functions are valued according to various operating conditions.

References

[1] M.A. Alao, T.R. Ayodele, A.S.O. Ogunjuyigbe, O.M. Popoola, Multi-criteria decision based waste to energy technology selection using entropy-weighted TOPSIS technique: The case study of Lagos, Nigeria, Energy 201 (2020) 117675. https://doi.org/10.1016/j.energy.2020.117675.

[2] S. Azami, M. Taheri, O. Pourali, F. Torabi, Energy and exergy analyses of a mass-fired boiler for a proposed waste-to-energy power plant in Tehran, Appl. Therm. Eng. 140 (2018) 520–530. https://doi.org/10.1016/j.applthermaleng.2018.05.045.

[3] A. Behzadi, E. Gholamian, E. Houshfar, A. Habibollahzade, Multi-objective optimization and exergoeconomic analysis of waste heat recovery from Tehran's waste-to-energy plant integrated with an ORC unit, Energy 160 (2018a) 1055–1068. https://doi.org/10.1016/j.energy.2018.07.074.

[4] A. Behzadi, E. Houshfar, E. Gholamian, M. Ashjaee, A. Habibollahzade, Multi-criteria optimization and comparative performance analysis of a power plant fed by municipal solid waste using a gasifier or digester, Energy Convers. Manag. 171 (2018b) 863–878. https://doi.org/10.1016/j.enconman.2018.06.014.

[5] Y. Casas Ledón, L.E. Arteaga-Perez, J. Toledo, J. Dewulf, Exergoeconomic evaluation of an ethanol-fueled solid oxide fuel cell power plant. Energy 93 (2015) 1287–1295. https://doi.org/10.1016/J.ENERGY.2015.10.036

[6] N.D. Couto, V.B. Silva, E. Monteiro, A. Rouboa, P. Brito, An experimental and numerical study on the Miscanthus gasification by using a pilot scale gasifier, Renew. Energy 109 (2017) 248–261. https://doi.org/10.1016/j.renene.2017.03.028.

[7] E. Gholamian, P. Hanafizadeh, A. Habibollahzade, P. Ahmadi, Evolutionary based multi-criteria optimization of an integrated energy system with SOFC, gas turbine, and hydrogen production via electrolysis, Int. J. Hydrogen Energy 43 (2018). https://doi.org/10.1016/j.ijhydene.2018.06.130.

[8] A. Habibollahzade, P. Ahmadi, M.A. Rosen, Biomass gasification using various gasification agents: optimum feedstock selection, detailed numerical analyses and tri-objective grey wolf optimization, J. Clean. Prod. 284 (2021) 124718. https://doi.org/10.1016/j.jclepro.2020.124718.

[9] Habibollahzade, A., Gholamian, E., Behzadi, A., 2019. Multi-objective optimization and comparative performance analysis of hybrid biomass-based solid oxide fuel cell/solid oxide electrolyzer cell/gas turbine using different gasification agents. Appl. Energy 233–234, 985–1002. https://doi.org/10.1016/j.apenergy.2018.10.075

[10] A. Habibollahzade, E. Houshfar, Improved performance and environmental indicators of a municipal solid waste fired plant through CO_2 recycling: exergoeconomic assessment and multi-criteria grey wolf optimisation, Energy Convers. Manag. 225 (2020) 113451. https://doi.org/10.1016/j.enconman.2020.113451.

[11] A. Habibollahzade, M.A. Rosen, Syngas-fueled solid oxide fuel cell functionality improvement through appropriate feedstock selection and multi-criteria optimization using Air/O_2-enriched-air gasification agents, Appl. Energy 286 (2021) 116497. https://doi.org/10.1016/j.apenergy.2021.116497.

[12] M.R. Holanda, J.A. Perrella Balestieri, Optimisation of environmental gas cleaning routes for solid wastes cogeneration systems. Part II - Analysis of waste incineration combined gas/steam cycle, Energy Convers. Manag. 49 (2008) 804–811. https://doi.org/10.1016/j.enconman.2007.07.020.

[13] Q. Hu, Y. Dai, C.H. Wang, Steam co-gasification of horticultural waste and sewage sludge: product distribution, synergistic analysis and optimization, Bioresour. Technol. 301 (2020) 122780. https://doi.org/10.1016/j.biortech.2020.122780.

[14] A. Kaveh, S. Talatahari, N. Khodadadi, Stochastic paint optimizer: theory and application in civil engineering, Eng. Comput. (2020) 1–32. https://doi.org/10.1007/s00366-020-01179-5.

[15] Z.K. Mehrabadi, F.A. Boyaghchi, Thermodynamic, economic and environmental impact studies on various distillation units integrated with gasification-based multi-generation system: comparative study and optimization, J. Clean. Prod. 241 (2019) 118333. https://doi.org/10.1016/j.jclepro.2019.118333.

[16] S. Mirjalili, S. Saremi, S.M. Mirjalili, L.D.S. Coelho, Multi-objective grey wolf optimizer: a novel algorithm for multi-criterion optimization, Expert Syst. Appl. 47 (2016) 106–119. https://doi.org/10.1016/j.eswa.2015.10.039.

[17] P. Mojaver, S. Khalilarya, A. Chitsaz, Multi-objective optimization and decision analysis of a system based on biomass fueled SOFC using couple method of entropy/VIKOR, Energy Convers. Manag. 203 (2020) 112260. https://doi.org/10.1016/j.enconman.2019.112260.

[18] O.J. Ogorure, C.O.C. Oko, E.O. Diemuodeke, K. Owebor, Energy, exergy, environmental and economic analysis of an agricultural waste-to-energy integrated multigeneration thermal power plant, Energy Convers. Manag. 171 (2018) 222–240. https://doi.org/10.1016/j.enconman.2018.05.093.

[19] J.A. Okolie, S. Nanda, A.K. Dalai, J.A. Kozinski, Optimization and modeling of process parameters during hydrothermal gasification of biomass model compounds to generate hydrogen-rich gas products, Int. J. Hydrogen Energy 36 (17) (2019) 18275–18288. https://doi.org/10.1016/j.ijhydene.2019.05.132.

[20] P. Parthasarathy, K.S. Narayanan, Hydrogen production from steam gasification of biomass: influence of process parameters on hydrogen yield – a review, Renew. Energy 66 (2014) 570–579. https://doi.org/10.1016/j.renene.2013.12.025.

[21] M. Sadeghi, A.S. Mehr, M. Zar, M. Santarelli, Multi-objective optimization of a novel syngas fed SOFC power plant using a downdraft gasifier, Energy 148 (2018) 16–31. https://doi.org/10.1016/j.energy.2018.01.114.

[22] A. Seçer, E. Fakı, Ş. Türker Üzden, A. Hasanoğlu, Hydrothermal co-gasification of sorghum biomass and çan lignite in mild conditions: an optimization study for high yield hydrogen production, Int. J. Hydrogen Energy 45 (2020) 2668–2680. https://doi.org/10.1016/j.ijhydene.2019.11.196.

[23] S. Talatahari, M. Azizi, Chaos Game Optimization: a novel metaheuristic algorithm, Artific. Intell. Rev. 54 (2020) 917–1004.

[24] S. Talatahari, M. Azizi, Optimal design of real-size building structures using quantum-behaved developed swarm optimizer, Struct. Des. Tall Spec. Build. 29 (11) (2020) e1747.

[25] H. Wang, D. Chaffart, L.A. Ricardez-Sandoval, Modelling and optimization of a pilot-scale entrained-flow gasifier using artificial neural networks, Energy 188, (2019) 116076. https://doi.org/10.1016/j.energy.2019.116076.

[26] J. Zhang, J. Hou, Z. Feng, Q. Zeng, Q. Song, S. Guan, Z. Zhang, Z. Li, Robust modeling, analysis and optimization of entrained flow co-gasification of petcoke with coal using combined array design, Int. J. Hydrogen Energy 45 (2020) 294–308. https://doi.org/10.1016/j.ijhydene.2019.10.153.

[27] H. Zhou, T. Xie, F. You, On-line simulation and optimization of a commercial-scale shell entrained-flow gasifier using a novel dynamic reduced order model, Energy 149 (2018) 516–534. https://doi.org/10.1016/j.energy.2018.02.031.

CHAPTER 8

A VDS-NSGA-II algorithm for multiyear multiobjective dynamic generation and transmission expansion planning

Ali Esmaeel Nezhad[f], Mohammad Sadegh Javadi[b], Alberto Borghetti[a], Morteza Taherkhani[c], Alireza Heidari[d] and João P.S. Catalão[b,e]

[a]*Department of Electrical, Electronic, and Information Engineering, University of Bologna, Bologna, Italy,* [b]*Institute for Systems and Computer Engineering, Technology and Science (INESC TEC), Porto, Portugal,* [c]*Department of Electrical Engineering, West Tehran Branch, Islamic Azad University, Tehran, Iran,* [d]*School of Electrical Engineering and Telecommunications (EE&T), The University of New South Wales (UNSW), Sydney NSW, Australia,* [e]*Faculty of Engineering of the University of Porto, (FEUP), Porto, Portugal,* [f]*Department of Electrical Engineering, School of Energy Systems, LUT University, Lappeenranta, Finland*

8.1 Introduction

Steady-state problems of power systems are generally studied in four time horizons as real-time studies, short-term studies, mid-term studies, as well as long-term studies. In this respect, short-term and real-time horizons usually relate to the operation of electric power systems, while the mid-term horizon is usually devoted to maintenance scheduling and fuel allocation. Long-term horizon, which is usually defined from one year to 10–20 years, mainly includes the expansion planning problems, among which the three problems known as generation expansion planning or "GEP," transmission expansion planning or "TEP," besides the substation expansion planning or "SEP" are well known. The power system expansion planning models can be generally tackled by using centralized models, decentralized models, or even semi-decentralized models. The modeling strategy mainly relies on the economic priorities and aspects as well as the market conditions. In this respect, Park et al. [1] included a broad review of the planning problems in the regulated environment.

The problem of power system planning is proposed and solved to decide on the capacity, type, time, and site of new assets in power systems. An optimal planning model would guarantee the desired performance of the system [2]. The methods presented so far to tackle the mentioned problem may be generally divided into two main groups. The first group is based on the mathematical optimization and includes methods such as linear programming, Lagrangian relaxation, and branch-and-bond [3–6]. The second group comprises heuristic methods such as genetic algorithm, known as genetic algorithm (GA) and particle swarm optimization, known as "PSO" [7,8]. Billinton and Allan [9] investigated the coordination of GEP and TEP problems. Billinton and Zhang [10] used a game theory–based method to characterize the relationship between the generation sector investment and transmission sector investments. It is noteworthy that the capacity expansion of the system is done economically in line with the future requirements of the system, while guaranteeing the system's desired reliability level [11–13]. In this regard, there are too many documents available thus far, investigating the system reliability at HL-II [14–16]. As the research paper in [2] emphasized, any long-term planning problem of power systems would comprise three main sections as follows: the input data, the modeling and computations, and results analysis. The amount of load demand, the technoeconomic data of power plants, the location of assets as well as climate conditions forecasts are categorized into input data of the problem. The second section includes other subsections such as the cost due to power system operation, the cost due to the new assets' investment, which form the planning model. Usually, tackling the mentioned problem as a coordinated expansion planning problem would face various difficulties due to the lack of enough information and solution intractability [16]. The problem is also aimed at minimizing the total cost (TC), including the operating and investment costs, neglecting construction sites of plants, that is, all load centers and units are placed at the same bus. However, by introducing the power system deregulation, the fundamentals of integrated planning were significantly changed. A distributed expansion planning model has been proposed in Botterrud et al. [17], where a hybrid centralized and decentralized decision maker has been used. It is noteworthy that the uncertainties due to the load demand and price have also been taken into account. A coordinated expansion planning framework has been presented in [18], taking into consideration market conditions. In this regard, different players submit their expansion plans to the planning entity, while seeking to maximize their profits. The best plan would then be selected by taking into consideration system expansion planning priorities, such as reliability constraints and stability indexes. Nevertheless, it should be noted that excessively increasing the number of players would balance the price across the power system which in turn reduces profit of the players [19]. Seddighi and Ahmadi-Javid [20] proposed an environmental-friendly technique for the composite dynamic expansion planning problem to expand the transmission and generation systems' capacity. Moreover, a clustering model based on a bilevel technique and objective-based scenario choice was proposed in [21]. The clustering variables include the variables, relating to the investment stage and power flow problem. Moreover, a k-means clustering technique was deployed in [22] within a two-stage planning model, where the clusters are the decisions of the wind power investment stage. An optimal model has been suggested in [23] to address generation companies' (GenCos') profit through bilateral as well as multilateral contracts to trade electricity and for haggling as a function of price [12]. El-Khattam et al. [24] presented an optimal long-term planning model for the markets that are not fully competitive. In this regard, the equilibrium would be obtained by iteratively maximizing profit. A game theory–based dynamic multiobjective optimization framework has been developed in [25] for the distributed GEP problem, ensuring the profits of prevalent power plants within the restructured environment. A coordinated microgrids expansion model has been designed in [26] while addressing the impacts of wind power and storage devices.

The integrated GEP and TEP problem is solved to determine the most desired generation units and transmission lines to be installed and added to the power system over the planning horizon, generally intended to minimize the TC and ensuring the system's reliability. The existing restructured power systems have given rise to several challenges caused by the impact of transmission systems on the reliability issues [27,28]. In this regard, concurrently solving both GEP and TEP would be a very difficult task. As a result, first the GEP problem is tackled and the units to be installed are specified. Then, the problem of TEP is solved to determine the lines to be added to the transmission system [29]. The deterministic N-1 and N-2 contingency analyses are carried out and deterministic load balance constraint is considered to alleviate the computational effort of the problem [27].

As the TEP is highly nonconvex, weighted sum approaches could not ensure Pareto-optimal solutions. Accordingly, this chapter shows the use of a posterior approach, named "nondominated sorting genetic algorithm II (NSGA II)" [9] to derive the Pareto set, which effectively exploits previously provided knowledge. Afterward, the most appropriate solution is selected by using the fuzzy decision maker [30]. The presented planning framework comprises two stages: a master problem (MP), and a slave problem composed by two subproblems. The MP is devoted to producing the binary decision variables by NSGA-II which show the status of candidate assets. The reliability index and the TC of the future system are determined by applying an iterative mixed-integer linear programming using the enumeration approach. Furthermore, a virtual database is applied along with the NSGA-II to reduce the calculation effort, significantly boosting the rate of convergence of the optimization method.

The main characteristic of this procedure is a novel point of view to the reliability-oriented integrated GEP and TEP problem. The developed framework makes the independent system operator (ISO) able to measure the composite reliability issues of the existing and future configurations of the power system. The novelties of this chapter are listed below.

1. Developing a dynamic multiobjective optimization framework for the integrated GEP and TEP problem with the capability to be utilized by the ISO for the optimal system expansion.
2. Presenting a virtual database as an asset to boost the convergence rate of the NSGA-II by mitigating the computational effort.
3. Developing a dynamic multiobjective optimization framework for the composite GEP and TEP problem with the capability to be utilized by the ISO for the optimal system expansion.
4. Presenting a virtual database as an asset to boost the convergence speed of the NSGA-II by mitigating the computational load of the problem.

The remainder of this chapter has been prepared as follows. Section 8.2 gives a comprehensive review of the mathematical formulation of the studied problem. The fundamentals of the multiobjective optimization and the descriptions of the NSGA-II algorithm, together with the descriptions of VIKOR decision maker are given in Sections 8.3 and 8.4. The results, obtained from simulating the proposed problem on a six-bus test system are included in Section 8.4, and Section 8.5 draws some relevant conclusions.

8.2 Problem formulation

The main objective of the multiyear multiobjective dynamic generation and TEP (MMDGTEP) problem is minimizing the TC and total expected energy not supplied (EENS) while meeting all constraints of the system. The first objective covers the cost due to the investment in new capacity additions plus the system operating cost. The second objective is introduced to minimize the composite generation and transmission risk index, that is the EENS at the hierarchy level II, $EENS_{HL-II}$. As mentioned in the Introduction, the MMDGTEP problem is formulated as a hybrid two-level optimization model. The MP deals with the problem of adding new capacities to the system. In other words, the assessment of first-level decision variables is done in the MP from the economic perspective. In the slave problem, the operating cost and reliability assessment of the proposed plans are evaluated. The solution obtained from the described problem specifies the capacity, the site, and the time new assets should be added to the system, that is, new generation and transmission assets, in an economic and secured manner. In this respect, the forecasted load demand according to the predicted growth rate should be supplied and the reliability of the system should be at a desired level.

8.2.1 Master problem

The MP's objective is set as minimizing the investment cost associated with new capacity additions in the generation and transmission sectors. The MP is modeled as (Eq. 8.1), subject to the planning constraints:

$$\text{Min } MP = \sum_{y=1}^{NY} \sum_{i=1}^{NG} \sum_{k=1}^{NCU} \frac{GI_{kiy}(Gn_{kiy} - Gn_{ki(y-1)})}{(1+d)^{y-1}} + \sum_{y=1}^{NY} \sum_{j=1}^{NCL} \frac{TI_{jy}(Ln_{jy} - Ln_{j(y-1)})}{(1+d)^{y-1}} \quad (8.1)$$

where NY is planning horizon, NG is number of generation buses, NCU is number of candidate units, GI is generation unit's investment cost, Gn is generation unit's decision variable, NCL is number of candidate lines, Ln is transmission line's decision variable, y is index for planning year, j is index for transmission line, k is index for candidate unit, i is index for generation bus, TI is transmission line's investment cost, d is annual discount rate.

TC comprises the investment cost plus the system operating cost. In the MP, the optimal plan would be determined, provided that the subproblems have been optimally solved. In other words, the MP determines the investment cost while the subproblems of the slave problem deal with the operating cost. The capital investment costs in the generation and transmission sectors in a year are constrained as Eqs. (8.2) and (8.3). Besides, the predicted capacity in a year in the generation and transmission sectors is constrained as Eqs. (8.4) and (8.5), and the construction time of the candidate investment in the generation and transmission sectors are applied by Eqs. (8.6) and (8.7), respectively. Constraints (8.6) and (8.7) also specify that once an asset is added to the power system, its investment status remains "1" until the end of the horizon.

$$\sum_{i=1}^{NG} \sum_{k=1}^{NU} GI_{kiy}(Gn_{kiy} - Gn_{ki(y-1)}) \leq TGI_y \quad (8.2)$$

$$\sum_{j=1}^{NL} TI_{jy}\left(Ln_{jy} - Ln_{j(y-1)}\right) \leq TTI_y \quad (8.3)$$

$$\sum_{i=1}^{NG} \sum_{k=1}^{NU} PG_{ki}^{\max,C}(Gn_{kiy} - Gn_{ki(y-1)}) \leq TGC_y \quad (8.4)$$

$$\sum_{j=1}^{NL} PL_{jy}^{\max,C}\left(Ln_{jy} - Ln_{j(y-1)}\right) \leq TTC_y \quad (8.5)$$

$$Gn_{ki(y-1)} \leq Gn_{kiy}, \ Gn_{kiy} = 0 \ if \ y < MTGI_{ki} \quad (8.6)$$

$$Ln_{j(y-1)} \leq Ln_{jy}, \ Ln_{jy} = 0 \ if \ y < MTTI_j \quad (8.7)$$

where NU is the total number of generation units, TGI is total number of new generation units can be installed for each year, TTI is total number of new transmission lines can be installed for each year, TGC is the maximum annual budget for the generation units' investment, TTC is the maximum annual budget for the transmission lines' investment, $MTGI$ is minimum time required for the generation unit installation, $MTTI$ is minimum time required for the transmission line installation.

8.2.2 Slave problem

The slave problem includes two subproblems: one minimizes the operating cost and the other minimizes the EENS. The objective function of the first problem is associated with the operating cost of the given plan and installed assets on the basis of the optimal power flow (OPF). An annual load duration curve (LDC) is exploited to calculate the yearly operating cost. The objective of the second subproblem represents the reliability index of the prospective expansion plans $EENS_{HL-II}$.

The investment plan determined by the MP is given to the associated subproblems. The investment cost of the prospective plan is integrated in the first objective while the investment plans and the status of capacity additions are assigned to the second subproblem. The next section provides the mathematical modeling of the two subproblems.

8.2.3 TC assessment objective of the MMDGTEP problem

The first subproblem considers the operational constraints, such as the power balance at each bus and the maximum power that can be transmitted in the lines and transformers, while addressing the annual operating limitations, and finds the minimum total operation cost over the planning years by using a DC-OPF technique.

The model of the first subproblem is as follows:

$$\text{Min } f_1 = TOC$$
$$TOC = \sum_{y=1}^{NY} \sum_{b=1}^{NB} \sum_{i=1}^{NG} \sum_{k=1}^{NU} \frac{DT_{by} \cdot OC_{kiby} \cdot PG_{kiby}}{(1+d)^{y-1}} \tag{8.8}$$

Subject to:

$$\sum_{i=1}^{NG}\sum_{k=1}^{NEU} PG^E_{kiby} + \sum_{i=1}^{NG}\sum_{k=1}^{NCU} PG^C_{kiby} - PD_{lby} - \sum_{j=1}^{NEL} PL^E_{jby} - \sum_{j=1}^{NCL} PL^C_{jby} = 0 \tag{8.9}$$

$$0 \le PG^E_{kiby} \le PG^{max,E}_{ki} IG^E_{kiby} \tag{8.10}$$

$$0 \le PG^C_{kiby} \le PG^{max,C}_{ki} Gn_{kiy} IG^C_{kiby} \tag{8.11}$$

$$PL^E_{jby} = B_j(\delta^E_{mby} - \delta^E_{nby}) \tag{8.12}$$

$$-PL^{max,E}_j \le PL^E_{jby} \le PL^{max,E}_j \tag{8.13}$$

$$PL^C_{jby} - B_j(\delta^C_{mby} - \delta^C_{nby}) - M^C_j(1 - Ln_{jy}) \le 0 \tag{8.14}$$

$$PL^C_{jby} - B_j(\delta^C_{mby} - \delta^C_{nby}) + M^C_j(1 - Ln_{jy}) \ge 0 \tag{8.15}$$

$$-PL^{max,C}_j Ln_{jy} \le PL^C_{jby} \le PL^{max,C}_j Ln_{jy} \tag{8.16}$$

$$\delta_{ref} = 0 \tag{8.17}$$

where *TOC* is total operational cost of generation units, *NB* is number of load buses, *DT* is duration of load blocks, *OC* is operational cost, *PG* is power generated by generation units, *b* is Index for load bus, *B* susceptance matrix of power grid *PD* is demand power, *PL* is power transmitted throughout transmission lines, *C* is index for candidate assets, *E* is index for existing assets, δ is bus voltage angle. Besides, NU is the total number of generation units, NEU is the total number of existing generation units, NEL is the total number of existing transmission lines, IG shows a binary variable representing the operation status of the asset, M is a sufficiently big value, m is the sending bus, and n is the receiving bus.

The power balance at bus *l* is modeled by (8.9), in which superscript *E* show the existing assets and superscript *C* denote the candidate assets. It should be noted that the status of each candidate asset is given by the solution of the MP and the selection procedure is not done here. In this stage, the generation level of the generation units, and transmission lines flow are the decision variables. Set *j* comprises the lines that are connected to bus *l* and labeled by "to bus" or "from bus." The constraints represent the capacity of the existing and candidate generation units (8.10)–(8.11), the line flows of existing transmission lines (8.12)–(8.13), the line flows of candidate lines (8.14)–(8.16), and the phase angle of the slack bus (8.17).

8.2.4 EENS$_{HL-II}$ evaluation procedure of the MMDGTEP problem

The annualized value of EENS at HL-I is derived utilizing the reduced scenarios as

$$EENS_{HL-I} = \sum_{k=1}^{N} E_k P_K \qquad (8.18)$$

where the EENS at the hierarchical level I, that includes the generation sector only, is denoted by $EENS_{HL-I}$. In this regard, it is supposed that all generation units are at one bus. The probability of contingency *k* and the corresponding energy curtail are indicated by P_k and E_k, respectively. Moreover, the annualized value of $EENS_{HL-II}$ which is defined as the energy curtailed at each bus is obtained as [31]

$$EENS_k = \sum_{j \in (x,y)} L_{kj} D_{kj} F_j \qquad (8.19)$$

$$EENS_{HL-II} = \sum_{k=1}^{NL} EENS_k \qquad (8.20)$$

The load curtailed at bus *k* for the sake of mitigating the overload as a result of contingency *j* is denoted by L_{k_j}. Besides, the duration of the curtailment and the occurrence frequency of this contingency are represented by D_{k_j} and F_j, respectively. The amount of load curtailment is evaluated by incorporating a virtual generator at each load bus by using an incidence matrix based DC optimal power flow (IM-DCOPF) method [32,33], considering the noninterconnectivity feature of distant generation units in the case of line outages.

The second subproblem aims at minimizing the total annual EENS of the associated prospective network:

$$Min \; f_2 = \sum_{y=1}^{NY} EENS_{HL-II}^y \qquad (8.21)$$

A probabilistic index using the enumeration technique along with a minimum cost evaluation model has been employed as described in the following five steps:

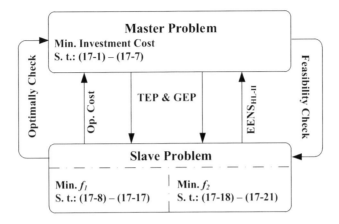

FIGURE 8.1

The proposed expansion planning framework.

1. First, the multistage annual load model should be produced, omitting the chronology and aggregating the load states by means of the data of the hourly load. This curve would be obtained from the annual LDC.
2. The second step is to choose the system states at the load level utilizing the mentioned enumeration technique. Generation units are usually modeled, utilizing multistate random variables, comprising the up, down, and derated states. On the other hand, transmission assets are modeled by means of two-state variables, comprising only the up and down states and the derated state has been neglected to more simplify the model. It is noted that transmission lines are assumed 100% reliable, which is case-dependent. The transmission system constraints must be considered as the generation capacity may be limited by the transmission system topology.
3. The solution of a cost minimization problem described in the subsequent section allocates the generation, determines the associate cost, and the amount of load curtailed together with the associated cost at each bus. With respect to the fact that the objective is to specify the amount of the load curtailed, and afterward, specify the "EENS," the identical value for the significance weighting factor of the load bus, and the identical unit generation cost are used to avoid the merit order of commitment of generation units.
4. Iteration of the second and the third steps until convergence is reached for every load level.
5. The corresponding probabilities are used to weight the obtained results for each load level to determine the annual indices of the expected cost of generation and risk.

Resuming, the conceptual model of the mentioned procedure illustrated in Fig. 8.1 shows the master and slave problems and their associated decision variables and outputs.

8.3 Multiobjective optimization principle

In case the optimization problem is aimed at optimizing more than one objective functions with conflicting nature, multiobjective optimization tools should be used. Unlike single-objective optimization problems, solving a problem with multiple objectives would give more than one optimal solution, known as Pareto set. It is noted that all the solutions obtained are nondominant and a better value of each objective in each Pareto-optimal solution cannot be derived except at the cost of deteriorating the values of other objectives [34–36]. Expression (8.22) shows a typical multiobjective optimization problem, subject to different constraints.

$$\text{Minimize } f_i(x) \quad i = 1, 2, \ldots, N_{obj}$$
$$\text{Subject to} \begin{cases} g_k(x) = 0 \quad k = 1, 2, \ldots, K \\ h_l(x) \leq 0 \quad l = 1, 2, \ldots, L \end{cases} \quad (8.22)$$

The objective function i and a decision vector are denoted by f_i and x, respectively. It should be noted that all solutions are optimal and the most adequate solution should be specified by the planning entity taking into account the preferences of the problem. Various optimization methods have been applied to cope with the multiobjective optimization problem [36–39], among which weighted sum method, epsilon-constraint method, and goal programming transform the primary multiobjective problem into a single-objective one and then, solve the problem. The main drawback of these methods relates to generating nonoptimal solutions in the Pareto set [39], while needing relatively complete information of the problem and a relatively high number of runs [32]. However, some well-established approaches are already available to solve multiobjective problems, utilizing the concept of nondominancy with respect to all objectives [32]. NSGA-II, presented in the next section, is known as an effective algorithm, with the capability to tackle nonconvex and mixed-integer problems [40].

8.4 Nondominated sorting genetic algorithm-II
8.4.1 Computational flow of NSGA-II

This section investigates the computational procedure of the NSGA-II together with a flowchart, showing the detail. In general, the following stages are taken by the NSGA-II [39]:

(Step 1) Initialization: first, a parent population with size N_P is randomly generated.
(Step 2) Non-dominated sorting of parent population: this stage sorts the generated population with respect to the non-domination level. In this regard, every population would be given a rank, showing its non-domination level or front number in a way that "1" shows the most desired level and "2" is the next desired one, etc. The crowding distance of populations at every non-domination level is determined and the population is put in order in a declining manner on the basis of the crowding distance.
(Step 3) Choosing tournament: two members are picked randomly and their front number and crowding distance are compared. The more desired one would be picked and copied to the mating pool.
(Step 4) Crossover and mutation: the simulated binary crossover and polynomial mutation are adopted in the procedure described in this chapter.
(Step 5) Merging the parent and child populations with the size $2N_P$.
(Step 6) Non-dominated sorting of the merged population: the non-domination and crowding distance indices are used to sort the new merged population. In this regard, elitism would be guaranteed

as all members of the two populations are used. The population of the most superior non-dominated set, shown by F_1 within the new merged population, would be highlighted among all other members. In case N_P is larger than F_1, the entire population F_1 would be selected to be used in the new population, while the other members will be selected from the next non-dominated fronts according to their ranks. Hence, F_2 would be subsequently selected and followed by the solution from F_3, etc. It is noteworthy that this process would be carried out up to the time it is not possible to accommodate any set. Assume F_1 shows the last non-dominated set, after which it would not be possible to accommodate any other sets. Generally, the count of solutions within the whole sets will be greater than N_P.

(Step 7) Termination criterion: the procedure will be terminated following a given number of generations, or at the time no considerable enhancement in the solution is observed. The NSGA-II would be run for a given number of generations in this chapter. Step 8 would be taken, provided that the termination criterion is met; otherwise, a new population is copied to the parent population and the third step is processed.

(Step 8) Choose the first member of the population of the first front.

(Step 9) Termination.

The conceptual flowchart of the presented NSGA-II is depicted in Fig. 8.2.

8.4.2 VDS-NSGA-II

The VDS-NSGA-II is developed and utilized in this chapter to avoid the monotonous cases in determining the reliability, which uses the enumeration technique. In this respect, the total number of cases to be used for calculating $EENS_{HLII}$ will be significantly mitigated by only accepting the cases associated with a probability greater than a predetermined threshold. It should be noted that the number of solutions that must be exploited in the enumeration technique is of very high order, particularly once a heuristic method such as NSGA-II is used. As the total number of assets, both existing and new ones, are determined by the ISO, and EENS has been calculated for some of the cases, the database can be used to restore those states and recalculating the system risk again would be redundant. For a particular case, the EENS would not vary. Accordingly, utilizing a virtual database for the presented expansion planning problem, including binary decision variables, can substantially help enhance the search capability and search speed of the NSGA-II.

8.4.3 Methodology

This section describes the implementation of VDS-NSGA-II in the MMDGTEP problem. As the problem of MMDGTEP is a dynamic mixed-integer programming problem of a very large size, a virtual mapping procedure (VMP) is taken into account together with the proposed VDS-NSGA-II to improve the effectiveness of aforementioned soft computing algorithm. In the proposed MMDGTEP problem, transforming the combination of candidate units' and lines' statuses into a dummy variable for every stage of MMDGTEP allows to use the virtual database.

The three stages of the VMP are described as below.

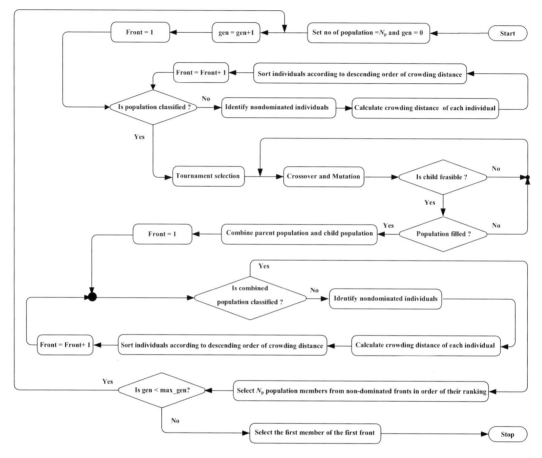

FIGURE 8.2

Computational flowchart of NSGA-II.

1. Form the prospective plan using the corresponding integer decision vector. As mentioned above, the investment status of an asset changes to "1" immediately after it is installed and remains unchanged to the end of the horizon.
2. Extract the annually available assets for the planning horizon.
3. Represent the status of each combination as a decimal number to submit the plan number to the associated objective functions.

Hence, a multivariable decision vector would be mapped to a single variable one for each year. The proposed VDS-NSGA-II incorporates the aforementioned VMP to manage the address cell of the virtual database in both operating cost and $EENS_{HL-II}$ objective functions. Fig. 8.3 illustrates the proposed procedure.

	Generation Expansion Planning				Transmission Expansion Planning			
	G.4	G.5	G.6	G.7	T.8	T.9	T.10	T.11
	Decision Vector							
Year	0	7	6	0	7	0	9	4
1	0	0	0	0	0	0	0	0
2	0	0	0	0	0	0	0	0
3	0	0	0	0	0	0	0	0
4	0	0	0	0	0	0	0	1
5	0	0	0	0	0	0	0	1
6	0	0	1	0	0	0	0	1
7	0	1	1	0	1	0	0	1
8	0	1	1	0	1	0	0	1
9	0	1	1	0	1	0	1	1
10	0	1	1	0	1	0	1	1

FIGURE 8.3

The virtual mapping procedure technique adopted in VDS-NSGA-II.

	Generation Expansion Planning				Transmission Expansion Planning			
	G.4	G.5	G.6	G.7	T.8	T.9	T.10	T.11
	Decision Vector							
Year	0	7	6	0	8	0	9	4
1	0	0	0	0	0	0	0	0
2	0	0	0	0	0	0	0	0
3	0	0	0	0	0	0	0	0
4	0	0	0	0	0	0	0	1
5	0	0	0	0	0	0	0	1
6	0	0	1	0	0	0	0	1
7	0	1	1	0	0	0	0	1
8	0	1	1	0	1	0	0	1
9	0	1	1	0	1	0	1	1
10	0	1	1	0	1	0	1	1

FIGURE 8.4

Illustrative framework of eliminating the repetitive calculations.

As mentioned above, the single variable decision vector, the so-called "decision address," has been considered for each year. By considering this fact that the proposed expansion planning framework is a dynamic one, decomposition of the problem into the associated planning year and considering the dynamic feature of the expansion planning accelerates the simulation process.

For example, if the commitment of transmission line "T.8" is postponed from the seventh year to the eighth one, recalculation of entire cases in the objective functions is avoided. As the yearly calculations have been carried out in the previous stages, it is only needed to evaluate the recent seventh year plan. It means that the first six years and also last three years calculations are available from the database and only the evaluation of the recent expansion plan should be carried out. Fig. 8.4 illustrates this procedure.

The proposed virtual database in line with the virtual mapping procedure is depicted in Fig. 8.5.

168 Chapter 8 A VDS-NSGA-II algorithm for multiyear multiobjective dynamic

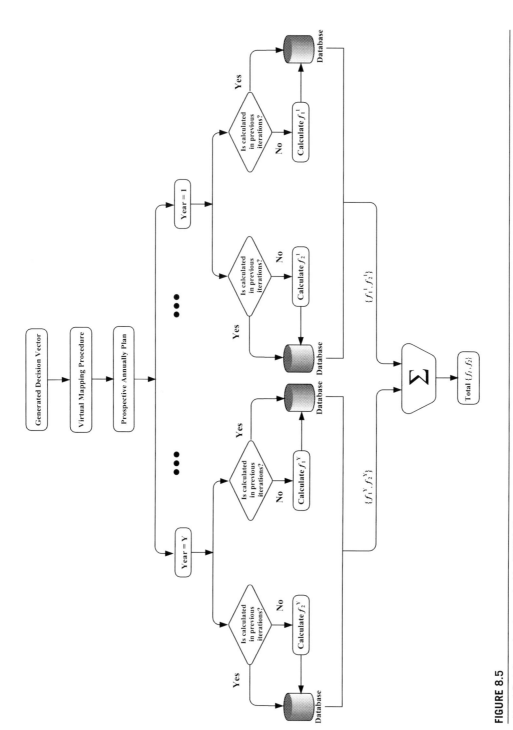

FIGURE 8.5

virtual Database Implemented in NSGA-II

As it is shown in the flowchart, an annual virtual database is considered for each objective function. It means that the numbers of virtual databases are $k \times Y$, in which k and Y denote the number of objective functions and planning horizon, respectively.

8.4.4 VIKOR decision making

The VIKOR decision maker was first developed by Oprikovic in 1998 [30] and it performs on the basis of allocating positive ideal values and negative ideal values to effectively specify the relative interval between every solution and Pareto-optimal solution. Afterwards, the significance of all Pareto solutions would be specified through a ranking, indicated by x_j, where j stands for the members of Pareto set and it is up to P [42]:

1. Using f_{ij}, showing the rating functions, computing the value, pertaining to criterion i for the solution x_j. After that, the best value of the objective function, indicated by f_i^+ and the worst value of that objective function, indicated by f_i^-, would be determined by using relationships (8.23) and (8.24).

$$f_i^+ = \max\left[(f_{ij})|j = 1, 2, \ldots, m\right] \quad (8.23)$$

$$f_i^- = \min\left[(f_{ij})|j = 1, 2, \ldots, m\right] \quad (8.24)$$

2. Specifying the value of group utility measure, shown by S_j and the value of individual regret measure, shown by R_j by employing relationships (8.25) and (8.26).

$$S_j = \sum_{i=1}^{n} w_i \frac{(f_i^+ - f_{ij})}{(f_i^+ - f_i^-)} \quad (8.25)$$

$$R_j = \max_i \left[w_i \frac{(f_i^+ - f_{ij})}{(f_i^+ - f_i^-)}\right] \quad (8.26)$$

where w_i shows the weight of each objective so that the sum would be equal to 1 [43]. Q_j is also computed based on the relationship (8.27):

$$Q_j = w_j \left[\frac{S_j - S^+}{S^- - S^+}\right] + (1 - w_j)\left[\frac{R_j - R^+}{R^- - R^+}\right] \quad (8.27)$$

where

$$S^+ = \text{Min}\left[(S_j)|j = 1, 2, \ldots, m\right] \quad (8.28)$$

$$S^- = \text{Max}\left[(S_j)|j = 1, 2, \ldots, m\right] \quad (8.29)$$

$$R^+ = \text{Min}\left[(R_j)|j = 1, 2, \ldots, m\right] \quad (8.30)$$

$$R^- = \text{Max}\left[(R_j)|j = 1, 2, \ldots, m\right] \quad (8.31)$$

3. Ranking the Pareto set with respect to the values of the group utility measure, the individual regret measure, and also Q_j in an ascending order where the solution with the least value of Q_j would be picked as the most preferred optimal solution [44].

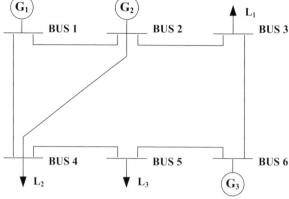

FIGURE 8.6

The studied six-bus system.

Table 8.1 Data of generation units.

Unit	G.1	G.2	G.3	G.4	G.5	G.6	G.7
Node	1	2	6	1	2	2	3
Size (MW)	100	100	50	100	80	60	20
Operating Cost ($/MWh)	Installed	Installed	Installed	200	270	250	250
Investment Cost ($/kW)	60	72	92	60	84	96	96

8.5 Simulation results

This chapter uses a six-bus test system to evaluate the performance of the developed optimization framework to solve the MMDGTEP problem. Fig. 8.6 demonstrates the studied system, [41], which is comprised of six nodes and seven transmission lines.

Tables 8.1 and 8.2 represent the data of the studied system, including the generating units and transmission system [41]. In this regard, four candidate transmission lines and four candidate generation units will be taken into consideration. It is assumed that the time, required to construct a generation unit is three years, while a transmission line can be installed within a year. The problem is solved for a planning horizon of 10 years, while Fig. 8.7 indicates the annual peak load demand. The shares of buses 3, 4, and 5 in the load demand are 40%, 30%, and 30%, respectively. The load duration of each year has been divided into four blocks as shown in Fig. 8.8 for the sake of more simplification. It should be noted that no constraint has been considered for the annual investments or the number of assets to be installed. Besides, the discount rate has been assigned to the model zero [41].

Table 8.2 Installed and candidate transmission lines' data [41].

Line	T.1	T.2	T.3	T.4	T.5	T.6	T.7	T.8	T.9	T.10	T.11
F_bus	1	2	1	2	4	5	3	1	2	1	5
To_bus	2	3	4	4	5	6	6	2	3	4	6
Reactance (per unit)	0.170	0.037	0.258	0.197	0.037	0.140	0.018	0.170	0.037	0.258	0.140
Capacity (MW)	80	70	140	100	50	140	130	80	70	140	140
Investment cost ($/kW)	Installed	Installed	Installed	Installed	Installed	Installed	Installed	80	96	120	56

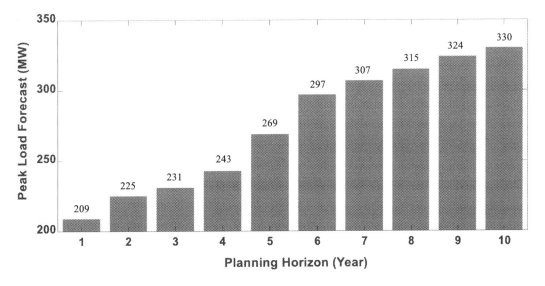

FIGURE 8.7

Yearly peak load forecast of six-bus test system.

The results of the MMDGTEP problem are shown in Table 8.3. All 25 reported plans are nondominated plans of the MMDGTEP problem in the six-bus test system.

The load demand during the first year of the horizon is supplied using the existing generation units and the power flow of lines is feasible. In this respect, the generation units, associated with lower costs, that is G.1 and G.2 are scheduled to operate at their maximum power and G.3 is supposed to meet the remaining load demand. Nonetheless, it is noted that T.2 will be congested and it prevents G.2 from operating at its rated capacity. Thus, G.3 as a more costly unit would operate at a higher level that in turn leads to an increased operating cost. Therefore, the candidate transmission line T.9 is added at the first year in all nondominated plans, resulting in raising the transmission capacity between nodes 2 and 3. This increased transmission capacity enables G.2 to operate at a higher generation level over the following years. Consequently, the overall load will be smaller than the overall installed generation capacity over the first 4 years of the planning horizon. Taking into account the zero value of the discount

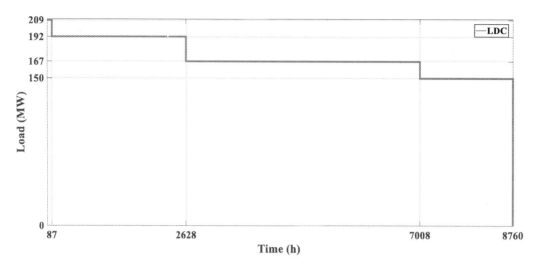

FIGURE 8.8

Load blocks in the first year.

rate and the minimum time to install a generation unit, a new generation unit will be added to the system in year 3. Thus, G.4 would be installed at bus 2 as higher power can be transferred by installing line T.9 in all prospective plans. These expansion plans are seen in all nondominated plans. However, the system reliability enforces the installation of more capacity.

The sensitivity analysis based on different values of w for this test system are represented in Table 8.4. Based on the reported best and worst plans and the planner's desired values for the objective functions, the final expansion plan would be extracted. For example, if the planner decides to avoid the risk, P14 is the best option. The best plans, {P1, P12, P14}, have the same transmission and generation investments, {G.4, T.8, T.9, T.10}. Due to zero discount rate and elimination of capacity and budget constraints, the aforementioned capacity additions would be considered the most efficient prospective projections.

Implementing the virtual database in line with the NSGA-II accelerates the convergence time and mitigates the computational load, particularly in the next simulation iterations. As the number of elements in the search space is countable, the size of virtual database would be determined. By decomposing the dynamic framework to a static one in the proposed approach, the size of database reduces. Furthermore, by considering the construction time of generation units and transmission lines in the proposed decoding framework of decision vector in an annually manner, the size of virtual database is reduced. For example, as the construction time for generating units and transmission lines are considered 3 years and less than 1 year, respectively, the maximum acceptable expansion plans for the first 3 years would be 16 . Thus, the size of database for the first three years is limited to 16 cells. The same analysis is carried out for the committed plans. It is noteworthy that the size of population and the upper bound of generations are assigned to the method as 5 and 100, respectively. Moreover, the probabilities associated with the crossover and mutation are considered 0.8 and 0.2, respectively.

Table 8.3 The obtained Pareto set and VIKOR decision-maker's results.

Plan	f_1 Total cost (M$*)	f_2 EENS$_{HL-II}$ (MWh*)	R	S	Q
P01	675.147	40.322	0.31615	0.3668	0.40268
P02	669.940	58.060	0.5	0.5	0.98622
P03	691.339	26.083	0.20817	0.37672	0.25361
P04	685.147	27.574	0.18401	0.33194	0.11702
P05	689.565	26.885	0.19091	0.36778	0.20667
P06	689.479	27.016	0.19007	0.3683	0.20649
P07	689.388	27.159	0.18919	0.3689	0.20638
P08	689.293	27.311	0.18826	0.36955	0.20634
P09	689.194	27.449	0.1873	0.37002	0.20586
P10	678.522	28.653	0.19519	0.27868	0.01769
P11	689.568	26.499	0.19093	0.36379	0.19794
P12	691.289	26.478	0.20768	0.38033	0.26078
P13	705.147	14.135	0.34249	0.38720	0.48916
P14	706.883	14.128	0.35937	0.40401	0.55280
P15	721.339	9.821	0.50000	0.50000	0.98622
P16	715.147	11.414	0.43976	0.45627	0.79483
P17	719.566	10.882	0.48274	0.49374	0.94516
P18	719.479	11.009	0.48191	0.49422	0.94490
P19	719.388	11.146	0.48102	0.49475	0.94465
P20	719.293	11.291	0.48009	0.49533	0.94445
P21	708.522	12.574	0.37532	0.40385	0.57770
P22	719.568	10.498	0.48276	0.48978	0.93649
P23	712.602	12.573	0.41500	0.44353	0.72766
P24	714.930	12.556	0.43766	0.46601	0.81289
P25	721.289	10.473	0.49951	0.50627	0.99923

* Calculated based on a 10-year horizon.

8.6 Conclusion

This chapter has described the implementation of a virtual database together with the use of heuristic optimization algorithms within the countable searching space of a mixed-integer optimization problem. The method is useful to reduce the computational effort. In particular, the VDS-NSGA-II soft computing algorithm is implemented to tackle the MMDGTEP problem, which is a dynamic and mixed-integer optimization problem associated that includes the time-consuming procedure for the minimization of EENS$_{HL-II}$.

This chapter also highlighted the requirements of considering the two conflicting objective functions, that is the investment cost and the risk.

Table 8.4 Sensitivity analysis results for the VIKOR decision maker.

w_{cost}	w_{EENS}	Best plan	R	S	Q
0.50	0.50	P10	0.19519	0.27868	0.01769
0.20	0.80	P21	0.15013	0.19579	0.00990
0.80	0.20	P01	0.12646	0.20750	0.00623

The results obtained for an illustrative case show the effectiveness and efficiency of the presented optimization framework. In this respect, the procedure provides diverse optimal expansion plans to the ISO as the decision maker. The VIKOR decision maker can be used to pick the most adequate plan.

Acknowledgment

João P.S. Catalão acknowledges the support by FEDER funds through COMPETE 2020 and by Portuguese funds through FCT under POCI-01-0145-FEDER-029803 (02/SAICT/2017).

References

[1] Y.M. Park, J.R. Won, J.B. Park, D.G. Kim, Generation expansion planning based on an advanced evolutionary programming, IEEE Trans. Power. Syst. 14 (1999) 299–305. https://doi.org/10.1109/59.744547.

[2] S.A. Mansouri, M.S. Javadi, A robust optimisation framework in composite generation and transmission expansion planning considering inherent uncertainties, J. Exp. Theor. Artif. Intell. 29 (2017). https://doi.org/10.1080/0952813X.2016.1259262.

[3] T. Akbari, M.T. Bina, Approximated MILP model for AC transmission expansion planning: global solutions versus local solutions, IET Gener. Transm. Distrib. 10 (2016) 1563–1569. https://doi.org/10.1049/iet-gtd.2015.0723.

[4] S. Binato, M.V.F. Pereira, S. Granville, A new Benders decomposition approach to solve power transmission network design problems, IEEE Trans. Power. Syst. 16 (2001) 235–240. https://doi.org/10.1109/59.918292.

[5] L.L. Garver, Transmission network estimation using linear programming, IEEE Trans. Power. Appar. Syst. (1970) 1688–1697 PAS-89. https://doi.org/10.1109/TPAS.1970.292825.

[6] G.B. Shrestha, P.A.J. Fonseka, Congestion-driven transmission expansion in competitive power markets, IEEE Trans. Power. Syst. 19 (2004) 1658–1665. https://doi.org/10.1109/TPWRS.2004.831701.

[7] J.H. Roh, M. Shahidehpour, Y. Fu, Market-based coordination of transmission and generation capacity planning, IEEE Trans. Power. Syst. 22 (2007) 1406–1419. https://doi.org/10.1109/TPWRS.2007.907894.

[8] S.K.K. Ng, J. Zhong, C.W. Lee, A game-theoretic study of the strategic interaction between generation and transmission expansion planning, In: Proc. IEEE/PES Power Syst. Conf. Expo. PSCE (2009) 2009. https://doi.org/10.1109/PSCE.2009.4840024.

[9] R. Billinton, R.N. Allan, Reliability Evaluation of Power Systems, Springer US, 1996. https://doi.org/10.1007/978-1-4899-1860-4.

[10] R. Billinton, W. Zhang, Adequacy equivalent development of composite generation and transmission systems using a D.C. load flow, Reliab. Eng. Syst. Saf. 65 (1999) 295–305. https://doi.org/10.1016/S0951-8320(99)00012-5.

[11] A. Akhavein, M.F. Firuzabad, R. Billinton, D. Farokhzad, Review of reduction techniques in the determination of composite system adequacy equivalents, Electr. Power. Syst. Res. 80 (2010) 1385–1393. https://doi.org/10.1016/j.epsr.2010.06.002.

[12] M.S. Javadi, M. Saniei, H. Rajabi Mashhadi, An augmented NSGA-II technique with virtual database to solve the composite generation and transmission expansion planning problem, J. Exp. Theor. Artif. Intell. 26 (2014). https://doi.org/10.1080/0952813X.2013.815280.

[13] M. Sadegh Javadi, M. Saniei, H. Rajabi Mashhadi, G. Gutiérrez-Alcaraz, Multi-objective expansion planning approach: Distant wind farms and limited energy resources integration, IET Renew. Power. Gener. 7 (2013) 652–668. https://doi.org/10.1049/iet-rpg.2012.0218.

[14] A. Mazer, Electric Power Planning for Regulated and Deregulated Markets, John Wiley & Sons, Inc., Hoboken, NJ, USA, 2007. https://doi.org/10.1002/9780470130575.

[15] A. Motamedi, H. Zareipour, M.O. Buygi, W.D. Rosehart, A transmission planning framework considering future generation expansions in electricity markets, IEEE Trans. Power Syst. 25 (2010) 1987–1995. https://doi.org/10.1109/TPWRS.2010.2046684.

[16] Wang X (Xifan), McDonald JR, James R. Modern Power System Planning. London, New York McGraw-Hill 1994.

[17] A. Botterrud, M.D. Ilic, I. Wangensteen, Optimal investments in power generation under centralized and decentralized decision making, IEEE Trans. Power Syst. 20 (2005) 254–263. https://doi.org/10.1109/TPWRS.2004.841217.

[18] J.H. Roh, M. Shahidehpour, L. Wu, Market-based generation and transmission planning with uncertainties, IEEE Trans. Power Syst. 24 (2009) 1587–1598. https://doi.org/10.1109/TPWRS.2009.2022982.

[19] M.S. Javadi, Esmaeel Nezhad A. Intelligent particle swarm optimization augmented with chaotic searching technique to integrate distant energy resources, Int. Trans. Electr. Energy Syst. 27 (2017). https://doi.org/10.1002/etep.2447.

[20] A.H. Seddighi, A. Ahmadi-Javid, Integrated multiperiod power generation and transmission expansion planning with sustainability aspects in a stochastic environment, Energy 86 (2015) 9–18. https://doi.org/10.1016/j.energy.2015.02.047.

[21] M. Sun, F. Teng, I. Konstantelos, G. Strbac, An objective-based scenario selection method for transmission network expansion planning with multivariate stochasticity in load and renewable energy sources, Energy 145 (2018) 871–885. https://doi.org/10.1016/j.energy.2017.12.154.

[22] S. Zolfaghari, T. Akbari, Bilevel transmission expansion planning using second-order cone programming considering wind investment, Energy 154 (2018) 455–465. https://doi.org/10.1016/j.energy.2018.04.136.

[23] S.L. Chen, T.S. Zhan, M.T. Tsay, Generation expansion planning of the utility with refined immune algorithm, Electr. Power. Syst. Res. 76 (2006) 251–258. https://doi.org/10.1016/j.epsr.2005.06.005.

[24] W. El-Khattam, K. Bhattacharya, Y. Hegazy, M.M.A. Salama, Optimal investment planning for distributed generation in a competitive electricity market, IEEE Trans. Power Syst. 19 (2004) 1674–1684. https://doi.org/10.1109/TPWRS.2004.831699.

[25] Budi R.F.S. Sarjiya, S.P. Hadi, Game theory for multi-objective and multi-period framework generation expansion planning in deregulated markets, Energy 174 (2019) 323–330. https://doi.org/10.1016/j.energy.2019.02.105.

[26] R. Hemmati, H. Saboori, P. Siano, Coordinated short-term scheduling and long-term expansion planning in microgrids incorporating renewable energy resources and energy storage systems, Energy 134 (2017) 699–708. https://doi.org/10.1016/j.energy.2017.06.081.

[27] J. Choi, T. Tran, A.A. El-Keib, R. Thomas, H.S. Oh, R. Billinton, A method for transmission system expansion planning considering probabilistic reliability criteria, IEEE. Trans. Power. Syst. 20 (2005) 1606–1615. https://doi.org/10.1109/TPWRS.2005.852142.

[28] D. Kalyanmoy, Multi-Objective Optimization using Evolutionary Algorithms, Wiley, NJ, 2021, https://www.wiley.com/en-us/Multi+Objective+Optimization+using+Evolutionary+Algorithms-p-9780471873396.

[29] A.E. Nezhad, A. Ahmadi, M.S. Javadi, M. Janghorbani, Multi-objective decision-making framework for an electricity retailer in energy markets using lexicographic optimization and augmented epsilon-constraint, Int. Trans. Electr. Energy. Syst. 25 (2015). https://doi.org/10.1002/etep.2059.

[30] S. Opricovic, G.-H. Tzeng, Compromise solution by MCDM methods: a comparative analysis of VIKOR and TOPSIS, Eur. J. Oper. Res. 156 (2004) 445–455. https://doi.org/10.1016/S0377-2217(03)00020-1.

[31] M. Moeini-Aghtaie, A. Abbaspour, M. Fotuhi-Firuzabad, Incorporating large-scale distant wind farms in probabilistic transmission expansion planning-part I: theory and algorithm, IEEE Trans. Power Syst. 27 (2012) 1585–1593. https://doi.org/10.1109/TPWRS.2011.2182363.

[32] M. Basu, Dynamic economic emission dispatch using nondominated sorting genetic algorithm-II, Int. J. Electr. Power Energy Syst. 30 (2008) 140–149. https://doi.org/10.1016/j.ijepes.2007.06.009.

[33] M. Basu, Economic environmental dispatch of hydrothermal power system, Int. J. Electr. Power Energy Syst. 32 (2010) 711–720. https://doi.org/10.1016/j.ijepes.2010.01.005.

[34] I.Y. Kim, O.L. De Weck, Adaptive weighted-sum method for bi-objective optimization: Pareto front generation, Struct. Multidiscip. Optim. 29 (2005) 149–158. https://doi.org/10.1007/s00158-004-0465-1.

[35] I.Y. Kim, O.L. De Weck, Adaptive weighted sum method for multiobjective optimization: a new method for Pareto front generation, Struct. Multidiscip. Optim. 31 (2006) 105–116. https://doi.org/10.1007/s00158-005-0557-6.

[36] D. Jones, M. Tamiz, Advanced topics in goal programming formulation, Int. Ser. Oper. Res. Manag. Sci. 141 (2010) 53–75. https://doi.org/10.1007/978-1-4419-5771-9_4.

[37] C.A. Coello Coello, A.D. Christiansen, Moses: a multiobjective optimization tool for engineering design, Eng. Optim. 31 (1999) 337–368. https://doi.org/10.1080/03052159908941377.

[38] M. Varadarajan, K.S. Swarup, Solving multi-objective optimal power flow using differential evolution, IET Gener. Transm. Distrib. 2 (2008) 720–730. https://doi.org/10.1049/iet-gtd:20070457.

[39] N. Srinivas, K. Deb, Muiltiobjective optimization using nondominated sorting in genetic algorithms, Evol. Comput. 2 (1994) 221–248. https://doi.org/10.1162/evco.1994.2.3.221.

[40] M.S. Javadi, A. Esmaeel Nezhad, Multi-objective, multi-year dynamic generation and transmission expansion planning- renewable energy sources integration for Iran's National Power Grid, Int. Trans. Electr. Energy Syst. 29: (2019) e2810. https://doi.org/10.1002/etep.2810.

[41] A. Khodaei, M. Shahidehpour, S. Kamalinia, Transmission switching in expansion planning, IEEE Trans. Power Syst. 25 (2010) 1722–1733. https://doi.org/10.1109/TPWRS.2009.2039946.

[42] M. Tavana, R. Kiani Mavi, F.J. Santos-Arteaga, E. Rasti Doust, An extended VIKOR method using stochastic data and subjective judgments, Comput. Ind. Eng. 97 (2016) 240–247. https://doi.org/10.1016/j.cie.2016.05.013.

[43] R.N. Kackar, Off-line quality control, parameter design, and the Taguchi Method, J. Qual. Technol. 17 (1985) 176–188. https://doi.org/10.1080/00224065.1985.11978964.

[44] M.K. Sayadi, M. Heydari, K. Shahanaghi, Extension of VIKOR method for decision making problem with interval numbers, Appl. Math. Model. 33 (2009) 2257–2262. https://doi.org/10.1016/j.apm.2008.06.002.

CHAPTER 9

A multiobjective Cuckoo Search Algorithm for community detection in social networks

Shafieh Ghafori and Farhad Soleimanian Gharehchopogh

Department of Computer Engineering, Urmia Branch, Islamic Azad University, Urmia, Iran

9.1 Introduction

Over the past 20 years, a new multidisciplinary research field called social network analysis has attracted researchers' attention [1,2]. As social networks grow, their structure and coherence become more complex, so new ways of analyzing and discovering relationships must be defined. Community detection (CD) is essential for understanding the structure and function of nodes in social networks and has attracted a great deal of attention due to its wide range of applications [3]. For example, nodes belonging to a community have useful nodes, and the network graph moves in the direction of magnification, leading to other analyses such as link prediction, information dissemination, and more [4].

Although there are different definitions for a CD, recognizing a group of densely connected nodes or based on a sufficiently similar pattern [5], most community structures on social networks overlap with other communities on the network, making the CD analysis more complex [6]. The purpose of the CD is to identify groups of nodes that have many edges between them. Using this definition, two essential factors must be considered in a CD. One to minimize links between communities and the other to maximize internal links in communities [7]. The goal is to optimize these two functions simultaneously.

As most engineering problems have multiple goals for optimization, and the goals are generally contradictory, it is almost impossible to find an optimal answer that simultaneously minimizes all functions [8,9]. However, it is possible to find answers that create the best interactions between goals [10]. The answer to an multi objective optimization (MOO) problem is a set of decision-making vectors that neither member overcomes nor each of these decision-making vectors can be selected as the answer to the problem [11,12]. Finding the best optimal answer should be done according to the CD's limitations and needs [13].

In this study, Multiobjective Cuckoo Search Algorithm (MOCSA) [14] is used for CD. This algorithm is a metaheuristic algorithm and has various capabilities in producing optimal solutions and finding optimal points. Yang and Deb [15] introduced the Cuckoo Search Algorithm (CSA). CSA uses levy flight, which detects search space better than the standard Gaussian trend. There is also a balance between local search and global exploration. CSA has fewer dependent parameters compared to some optimization algorithms. Cross-curricular algorithms can use the search process to achieve a set of Pareto answers, the best of which is the optimal answer, or one can hope that it is not too far from the optimal answer set [16]. In conventional methods, the decision-making function may not have enough information to determine priorities and constraints before or during the search process, which is less common in metaheuristic algorithms. The objectives for this study are NMI maximization and modularity in CD.

The CD issue is a multipart optimization issue derived from subgraph separation. In this chapter, the optimal global solution is obtained through the multiple objective functions based on the MOCSA model. The contribution of this book chapter is as follows:

- Two objective functions are designed for CD. The advantages of using the multiobjective method compared to the single objective function are that it is possible to optimize multiple criteria simultaneously, and instead of a single solution, a set of solutions is produced. Each solution has a different number of communities that create an optimal balance.
- In the MOCSA model, several solutions are generated randomly to scan the search space. As local spaces get better than others, the chances of producing a new population increase, and optimal solutions are obtained. Local tourism helps search for nearby neighbors.
- Experimental results on several datasets show that the MOCSA model has performed significantly better in most cases.

This chapter's general structure is organized as follows: Section 9.2 will overview related research. In Section 9.3, we explain the CD and the proposed model. In Section 9.4, we will evaluate and compare the proposed model results with other models. In Section 9.5, we present the conclusions and future work.

9.2 Related works

Over the years, various models for CD have been proposed. Most of these models use mass intelligence and machine learning algorithms. These algorithms have been more accurate than conventional and linear models. In this section, it is reviewed by some models of CD algorithms.

In [17], a model based on the Multiobjective Genetic Algorithm (MOGA) is proposed for CD. For this purpose, MOGA uses a coding type of information based on edge information. The MOGA model consists of two objective functions (the first function is to maximize the internal connection of communities; the second function is to minimize external connections). Evaluations of various datasets showed that the overall efficiency of the algorithm was 80%.

In [18], the Multiobjective Evolutionary Algorithm (MOEA) is used for CD. In this method, solution vectors are routed based on the local method that causes rapid convergence and diversity. The results showed that the NMI value on the Football was 0.9254. The multiobjective adaptive fast evolutionary algorithm model has been proposed to maximize NMI [19]. The probability of mutation and crossover's probability vary according to population fit and correlational characteristics between individuals. Finally, to produce a set of independent subnets, an optimal Pareto solution with the highest modularity is selected and decoded. The evaluation has shown that the NMI value on the four datasets (Karate, Dolphin, Football, and Polbooks) averages to 0.9200.

The Multiobjective Discrete Backtracking Search Algorithm (MODBSA/D) model [20] is proposed for CD based on minimizing the Negative Ratio Association (NRA) and Ratio Cut (RC) functions. In general, the NRA function indicates the internal link density of communities, and RC represents the link density between communities. Therefore, by minimizing NRA and RC, it can be ensured that the connections within a community are dense, and the connections between communities are dispersed. In this way, the CD in a complex network with these two objective functions can be formulated to maximize the number of internal links and minimize external links. The relationship between these two objective functions must strike a balance in the direction of minimization. By balancing these two objective functions, the right partition can be obtained for a complex network. NMI and modularity criteria have also been used for evaluation. An evaluation has shown that the NMI value on the four datasets (Karate, Dolphin, Football, and Polbooks) averages 0.8876.

The Multiobjective Discrete Biogeography Based Optimization (MODBBO) model [21] has been proposed based on the balance between the density of the connections and the nodes' similarity to find the communities of a network. In the MODBBO model, two objective functions are assigned to the CD. Modularity is designed as the first objective function to find communities according to a social network's topological structure. The second function is to find communities according to the standard features of a social network's nodes. The results on the various datasets show that the MODBBO model had low time complexity and high accuracy.

In [22], Multiobjective Enhanced Firefly Algorithm (MOEFA) is recommended for CD. In the MOEFA model, a chaotic parameter is used to adjust the mutation strategy to improve its overall performance. The results obtained on the Dolphins dataset showed that the MOEFA model for NMI and modularity obtained the best values of 0.9888 and 0.5242, respectively.

In the WOCDA [23] model, a new strategy and three actions (shrinking encircling, spiral updating, and random search) are designed to mimic CD's whaling behavior. In the first stage, the new strategy aims to identify the optimal solutions based on the objective function (finding the degree of nodes) and continuing the search. The operation to reduce the siege based on the optimal solution was discovered to find the neighboring nodes. A one-way crossover operator performs spiral update operations to maintain good communities. Finally, a random search operation was performed to randomly select a neighboring node's label and update the current node tag to increase the nationwide search capability. The results on four datasets (Karate, Dolphin, Football, and Polbooks) showed that the NMI value averaged 0.8951.

In [24], a multiobjective community recognition algorithm called MOCD-ACO is presented with the hybrid of Ant Colony Optimization (ACO) exploration operator and MOEA (MOEA/D). MOCD-ACO simultaneously optimizes two objective functions, NRA and RC. Each ant is responsible for discovering the optimal solution for an underlying problem. Ants are divided into several groups, and each group shares a pheromone matrix. Ants use random sampling models to build solutions. Each ant can update its current solution if it has a better goal. For the algorithm not to be easily optimized locally, the local annealing search operator has been used to expand the integrated search domain. The results on four different datasets show that the NMI value is acceptable. However, the disadvantage of MOCD-ACO has been the high execution time.

The Modularity-based Discrete State Transition Algorithm (MDSTA) model [25] has been proposed to achieve optimal solutions. The vertex substitute transformation operator and community substitute transformation operator have been proposed for global search based on the network's exploratory information. Each primary person evolves through these two alternative operators. In the next step, the elite population, which includes high fitness people, is selected from these evolved individuals. Finally, a two-way crossover operation is being conducted among elite individuals for local search. The MDSTA framework is straightforward and easy to implement. Evaluations on four datasets (Karate, Dolphin, Football, and Polbooks) demonstrated that Karate's modularity value was 0.4198 and for Football was 0.6046.

In [26], the Genetic Algorithm for Community Diagnosis (GACD) with a new local search strategy has been proposed to accelerate convergence and improve accuracy. In the GACD model, a random probability is created to determine whether crossover or mutation is used, and if the value of this probability is less than the probability of crossover, the crossover is used. Otherwise, a mutation is used. After producing as many children as the original population, all of these solutions are deciphered, and the best solution is chosen based on modularity. Crossover and mutation operators require computational costs O (N) and O (1), respectively. The modularity calculation requires O (m) 's time complexity where m is the number of edges. Evaluations on the four primary datasets showed that the mean NMI was 0.8132.

9.3 Proposed model
9.3.1 Community diagnosis

A social network can be represented by a nondirectional diagram $G = (V, E)$, as V and E indicate the set of nodes (network components) and edges (network relationships), respectively. Society is defined as a group of nodes that have more intralinks than interlinks. The purpose of this chapter is to do CD in unguided and nondirectional networks [27].

An adjacent matrix is a method for displaying graphs. If A's adjacent matrix of a graph is considered, the first-degree node in the graph with k_i is shown to be equal to $k_i = \sum_j A_{ij}$, considering an S subgraph as $S \subseteq G$.

$$SGk_i(S) = k_i^{in}(S) + k_i^{out}(S) \tag{9.1}$$

In Eq. (9.1), the degree of each node for internal and external communities is defined as $k_i^{in}(S) = \sum_{j \in S} A_{ij}$ and $k_i^{out}(S) = \sum_{j \notin S} A_{ij}$. This means that the first-degree node i, which is a member of the community j, is defined as the set of internal communities by $k_i^{in}(S) = \sum_{j \in S} A_{ij}$ and the first-degree node is called i, which is not a member of the community j can be defined as the sum of the external communities $k_i^{out}(S) = \sum_{j \notin S} A_{ij}$. Communities whose total internal nodes are much larger than the sum of its external nodes are considered a subset of the entire input graph, and the identification of this subroutine is called CD, and the discovery of the most optimal and accurate communities are defined as [27]

$$\sum_{jS} k_i^{in}(S)'' \sum_{jeS} k_i^{out}(S) \tag{9.2}$$

According to Eq. (9.2), $\sum_{j \in S} k_i^{in}(S)$ is the sum of the degrees of all nodes in the community or subscript S, and $\sum_{j \notin S} k_i^{out}(S)$. The sum of the degrees of all nodes do not exist in the community or subroutine S. For CD, the neighborhood matrix is usually calculated using graphs of the relationships between people on social networks. Then different results are obtained by applying CD algorithms to this matrix on different data.

9.3.2 Multiobjective optimization

Each optimization problem has three main parts: optimization variables, cost functions (objective functions), and constraints. An optimization problem can generally be expressed as [28]

$$\begin{aligned} &x = [x_1, x_2, x_3, \ldots, x_n] \\ &\text{to minimize } f(x) \\ &\text{subjected to } \begin{cases} g_j(x) \leq 0, j = 1, 2, \ldots, J \\ h_k(x) = 0, k = 0, 1, 2, \ldots, K \end{cases} \end{aligned} \tag{9.3}$$

In Eq. (9.3), x is a vector with n variables called the variable vector. The $F(x)$ parameter is the objective function of the optimization problem. $g_j(x)$ and $h_k(x)$ are unequal and equal constraints, respectively. The criterion by which the optimization process is performed is called the cost function, the value function, or the objective function.

In MOO problems, it is often impossible to find a single answer that can simultaneously optimize all objective functions. However, it is possible to achieve a set of answers that best interact between goals. The goal of an MOO problem is to discover a set of solutions in the search space. This set of answers is called the Pareto answer. The Pareto is, in fact, the optimal answer that any other set of answers cannot defeat. In MOO problems, there is a set of answers that makes a proper compromise between all functions. This set of answers is called the Pareto answer. The mathematical definition of the defeat of an answer is as follows:

In an optimization problem, multi-$mf_i, i = 1, 2, \ldots, m$ when the answer x_1 defeats the answer x_2, only if the following two conditions are met:

1. For all f_i, $f_i(x_1)$ is not worse than $f_i(x_2)$.
2. For at least one of the f_i, $f_i(x_1)$ is better than $f_i(x_2)$.

If one of the above two conditions is not met, then the answer x_1 does not defeat the answer x_2. Of course, if the answer x_1 does not overcome the answer x_2, it does not necessarily mean that x_2 overcomes x_1. According to the above definitions, the optimal $x^* \in X$ answer is in Pareto mode if, for each of the objective functions, none of the $x \in X$ answers can defeat x^*. That is, $f_i(x^*)$ is not worse than $f_i(x)$ for all x, and $f_i(x^*)$ is better than $f_i(x)$ for at least one of the f_i.

9.3.3 CD based on MOCSA

In the first stage of MOCSA [14], the initial habitat (nest) is produced for the cuckoos. In the CD, the habitats will be an $N \times 1$ array of graph nodes that show the current position of the cuckoo's life. This array is shown in Eq. (9.4).

$$Habitat = [x_1, x_2, \ldots, x_n] \tag{9.4}$$

Every egg that is laid in the nest is a representative of a solution. New solutions are also being developed with the help of levy flight. Levy flight is an accidental step in which levy distribution follows the steps. Studies have shown that flying Levy can increase the efficiency of the search process in uncertainty. In MOCSA, new answers are generated using levy flight, according to the following equation:

$$x_i(t+1) = x_i(t) + \alpha \times Levy() \times (x_i(t) - x_{best}(t)) \tag{9.5}$$

$$Levy = r^-: (1 << 3) \tag{9.6}$$

In Eq. (9.5), $x_i(t)$ indicates the position of ith nest in the repetition stage t, $x_{best}(t)$ is the position of the best nest (best answer) in the t repetition stage, and $Levy$ (λ) is the distribution of the levy flight. The step's length should also be determined in proportion to the problem's scale and the answer's space. According to Eq. (9.5), if cuckoo eggs are more similar to host bird eggs, then the chances of nondiagnosis and survival are high. Some of the new solutions are selected by the Louis distribution around the best solution obtained at each step ($x_{best}(t)$) to increase the local search speed. Several other solutions are also generated randomly to scan the search space. As local spaces get best than others, the chances of producing a new population increase, and optimal solutions are obtained. Local tourism helps search for nearby neighbors. The distribution of levy flight is also calculated according to Eq. (9.6). An r parameter is a random number in the range of zero to one. Fig. 9.1 shows the MOCSA flowchart for CD.

MOO models usually use several functions for optimization. This chapter uses two objective functions for CD optimization to optimize and explore communities among nodes. The goal of the $f1$ function is minimization, and the goal of the $f2$ function is maximization.

$$\begin{cases} \min f1 = \sum_{j=1}^{k} p(S_j) \text{ where } p(S_j) = \sum_{i \in s} \frac{k_i^{in}(S)}{k_i^{in}(S) + k_i^{out}(S)} \\ \max f2 = p(v_i, v_j) = \frac{1}{1 + \sqrt{\sum_n (v_i - v_j)^2}} \end{cases} \qquad (9.7)$$

In Eq. (9.7), the k parameter is the number of communities, the $k_i^{in}(S)$ parameter is the number of nodes at the node i connected to the other S subroutine nodes. The $k_i^{out}(S)$ parameter is the number of nodes i connected to other nodes outside the S subgraph. In Eq. (9.7), vi and vj are two neighboring nodes. Fig. 9.2 shows an example of a CD based on the $f2$ function with 11 nodes. Fig. 9.2A shows the graph structure. For each element, the value of the function $f2$ is calculated based on the nodes' nodes. As shown in Fig. 9.2B, the nodes are clustered based on the value obtained from $f2$. Three communities are created from the graph structure. After the distances' value defines the vector elements, the number of communities is formed according to the three major vector elements.

In the MOCSA model, the matrix is first defined as the array matrix. An archive matrix is a matrix in which Pareto's optimal answers are placed. Firstly, that is responses that do not defeat any of the individuals, and second, improvements in an objective function cause the CD. At the beginning of the algorithm, this matrix is empty, and there is no answer. After performing the first iteration, all the answers are compared, and the answers that any of the population members have not defeated are transferred to this matrix. In subsequent repetitions and with a new population's production, all the new individuals are compared, and the unwanted individuals are added to the archive matrix.

Given that new people transferred to the archive matrix may be defeated by one or more people in the previous matrix or vice versa, several former archive matrix members may be defeated by these new individuals. Each must be repeated after new people are added to the list. The archive matrix has been refined, and the losers have been removed. This process will be repeated until the end of the algorithm is met.

Another point to keep in mind is that in single-objective optimization algorithms, the best answer, which is the optimal answer, is usually unique. Therefore, the $x_{best}(t)$ parameter in x_{gbest} will be the best algorithm answer in step t and the best answer up to that point, respectively. However, in matters of MOO problems, none of Pareto's answers outweighs each other, as mentioned earlier. In this chapter, the congestion distance concept is used to maintain the answers' diversity and quality. The distance between each person in the archive matrix and the two adjacent people (also in the archive matrix) is determined using the congestion distance. The larger the congestion distance of a person, the better the optimal answer of Pareto. Therefore, that range is considered part of the new range, and more members must converge toward this range so that the density of Pareto members in the response space has a uniform distribution.

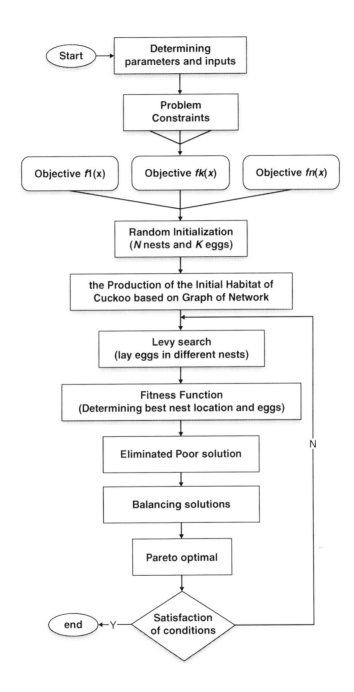

FIGURE 9.1

MOCSA model flowchart for CD.

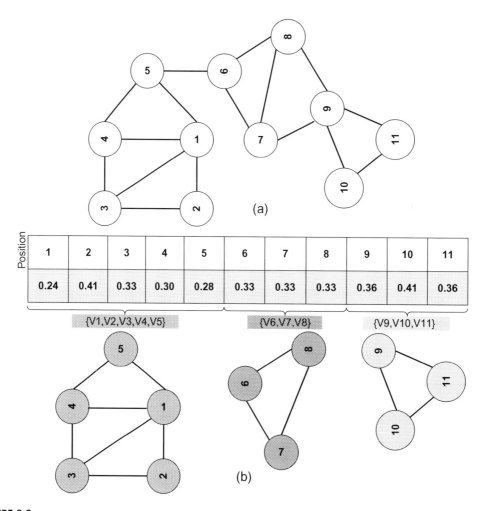

FIGURE 9.2

Graph chart structure and community discovery.

9.3.4 Fitness function

Social networks' structure is highly complex due to the relatively close links in communities and the relatively scattered links between different communities. Two essential features should be considered in the structure of social networks: the first is to minimize links between communities, and the other is to maximize links with leaders in society. In Eq. (9.8), the I_{ij} parameter indicates the connected edges between nodes i and j, and d_i and d_j represent the degree of nodes i and j.

$$Fit = \frac{I_{ij}}{m} - \frac{d_i d_j}{2m^2} \qquad (9.8)$$

Table 9.1 Details of Social Networks (datasets).

Dataset	Node	Edges	Format	Edge weights	Maximum degree	Category	Clusters
Zachary's Karate	34	78	Undirected	Unweighted	17 edges	Human social	2
Dolphin	62	159	Undirected	Unweighted	12 edges	Animal	2
Football	115	613	Undirected	Unweighted	12 edges	Interaction network	12
Political books	105	441	Undirected	Unweighted	25 edges	Miscellaneous network	3
Email network	1133	5451	Undirected	Unweighted	71 edges	Communication	Unknown
Geom network	7343	11898	Undirected	Unweighted	102 edges	Geometry collaboration	Unknown
NetScience	1589	2742	Undirected	Unweighted	19 edges	Infrastructure	Unknown
Power grid	4941	6594	Undirected	Unweighted	–	coauthorship network	Unknown

9.4 Evaluation and results

The proposed model is implemented on a system with 6 GB of memory, an Intel Corei5-2540M 2.6 GHz processor, the operating system of Windows 8, and MATLAB 2017 software. Evaluations were performed on eight datasets Karate [29], Dolphin [29], Football [29] and Polbooks [29] Email [30], Geom [30], NetScience [30], and Power Grid [30]. All results are reported based on an average of 10 executions. The initial population and the number of iterations in the MOCSA model are 100 and 200, respectively.

NMI [31] and modularity criteria [31] have been used for evaluation. The NMI criterion is defined according to Eq. (9.9). In Eq. (9.9), the parameters A and B are two segments separated from a network, C is the entanglement matrix, C_{ij} is equal to the number of nodes in common i in section A and community j in section B. The NMI index shows the overall relationship between nodes. This index represents the correlation of the network. In a high-density interconnected network, there are many connections between nodes. The NMI index is between 0 and 1. A value of 1 indicates, all nodes in the network are connected to each other. A density of 0 indicates that there is no connection between the nodes in the network.

$$NMI = \frac{-2\sum_{i=1}^{C_A}\sum_{j=1}^{C_B} C_{ij}\log\left(C_{ij}\cdot\frac{n}{C_i C_j}\right)}{\sum_{i=1}^{C_A} C_i \log(C_i/n) + \sum_{j=1}^{C_B} C_j \log(C_j/N)} \quad (9.9)$$

A modularity criterion has been proposed to evaluate CD algorithms' performance in social networks [31]. How to calculate this criterion is shown in the following equation:

$$Q = \sum_{s=1}^{k}\left[\frac{l_s}{m} - \left(\frac{d_s}{2m}\right)^2\right] \quad (9.10)$$

In Eq. (9.10), l_s is the total number of edges connecting the vertices within the s cluster, and the d_s parameter is the sum of the nodes in s, and m is the total number of network edges. The possible values of the modularity criterion are in the range [1 to −0.5]. (Table 9.1)

Table 9.2 shows the MOCSA model results based on NMI and modularity criteria with 200 iterations cycles on different datasets. The NMI for the Karate, Dolphin, Football, and Polbooks datasets are

Table 9.2 MOCSA model results on different datasets.

Datasets	NMI			Modularity		
	Best	Average	Worst	Best	Average	Worst
Karate	1.0000	1.0000	1.0000	0.4192	0.4163	**0.4146**
Dolphin network	0.9984	0.9972	0.9935	0.5262	0.5238	**0.5245**
College football	0.9486	0.9425	0.9414	0.6025	0.5996	**0.5881**
Books politics (Polbooks)	0.7455	0.7384	0.7352	0.5264	0.5237	**0.5215**
Email network	–	–	–	0.5362	0.5337	**0.5318**
Geom network	–	–	–	0.7025	0.7017	**0.7008**
NetScience	–	–	–	0.9477	0.9456	**0.9442**
Power grid	–	–	–	0.8382	0.8364	**0.8336**

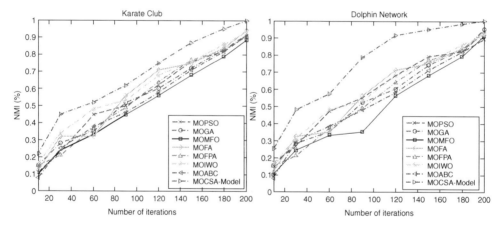

FIGURE 9.3

NMI comparison chart based on different models on Karate and Dolphin.

1.0000, 0.9984, 0.9486, and 0.7455, respectively. Karate, Dolphin, Football, and Polbooks datasets' modularity value was 0.4192, 0.5262, 0.6025, and 0.5264, respectively. The modularity for the Email, Geom, NetScience, and Power datasets are 0.5362, 0.7025, 0.9497, and 0.8382, respectively.

Fig. 9.3 shows a comparison chart of a round of iterations based on Karate and Dolphin's different models. The MOCSA model has more NMI than other models. To demonstrate the performance of the MOCSA model, other multipurpose algorithms such as Multiobjective Particle Swarm Optimization (MOPSO) [32], Multiobjective Genetic Algorithm (MOGA) [33], Multiobjective Moth-Flame Optimization (MOMFO) [34], Multiobjective Firefly Algorithm (MOFA) [35], Multiobjective Discrete Flower Pollination Algorithm (MOFPA) [36], Multiobjective Discrete Invasive Weed Optimization (MOIWO) [37], Multiobjective Artificial Bee Colony (MOABC) [38] used for comparison.

The NMI values in the MOPSO, MOGA, MOMFO, MOFA, MOFPA, MOIWO, MOABC models on the Karate dataset for 200 iteration cycles were 0.9115, 0.9221, 0.8858, 0.9414, 0.9096, 0.9314,

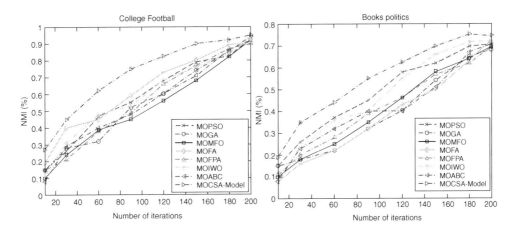

FIGURE 9.4

NMI comparison chart based on different models on Football and Polbooks.

and 0.9105, respectively. At best, the MOCSA has been able to improve the NMI in Karate by about 12.89%.

NMI values in the MOPSO, MOGA, MOMFO, MOFA, MOFPA, MOIWO, MOABC models on the Dolphin for 200 iterations cycles were 0.8925, 0.9525, 0.9120, 0.9341, 0.9278, 0.9196 and 0.9321, respectively. At best, the MOCSA has improved the NMI on Dolphin by about 11.87%.

Fig. 9.4 shows a comparison chart of a round of iterations based on different models for Football and Polbooks. The MOCSA model has more NMI than other models. NMI values in MOPSO, MOGA, MOMFO, MOFA, MOFPA, MOIWO, MOABC models on the Football dataset for 200 iteration cycles were 0.9205, 0.9425, 0.9225, 0.9041, 0.8978, 0.9247 and 0.9212, respectively.

At best, the MOCSA has improved NMI in Football by about 5.66%. The NMI values in the MOPSO, MOGA, MOMFO, MOFA, MOFPA, MOIWO, MOABC models on the Polbooks dataset for 200 iteration cycles were 0.7084, 0.6824, 0.6945, 0.7198, 0.6914, 0.7198, and 0.7048, respectively. At best, the MOCSA has improved the NMI on Polbooks by as much as 9.25%.

Fig. 9.5 shows a comparison chart of a round of iterations based on modularity for Email and Geom. Fig. 9.6 shows a comparison chart of a round of iterations based on modularity for NetScience and Power Grid. It can be observed from Figs. 9.5 and 9.6 that MOCSA-Model has the best performance in terms of the values of Modularity (Q) on the Email, Geom, NetScience, and Power Grid.

Table 9.3 shows the comparison of MOCSA model with MOGA-Net [39], MOCD [40], MOEA/D-Net [41], MODPSO [42], DMOPSO [43], MODTLBO/D [44] models on four real datasets. The NMI values in the MOGA-Net and MOCD models on Karate are 0.8388 and 0.8371, respectively. Table 9.3 shows the highest NMI on Karate belonging to MOEA/D-Net, MODPSO, MODTLBO/D, and MOCSA. The values of NMI in the MOEA/D-Net and MODPSO models on Dolphin are 1.0000 and 1.0000, respectively. The highest NMI on Football belongs to MOCSA. The highest NMI value on Polbooks belongs to MOCSA, and its value is 0.7455. Discrete inverse modelling-based multi-objective evolutionary algorithm with decomposition (DIM-MOEA/D) [45] and MODPSO models have also been used to compare modularity on the Email, Geom, NetScience, and Power Grid datasets.

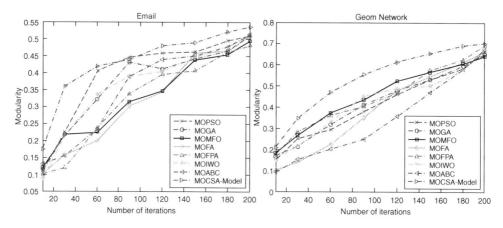

FIGURE 9.5

Comparison chart based on modularity on Email and Geom.

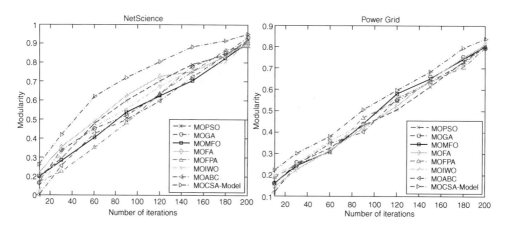

FIGURE 9.6

Comparison chart based on modularity on NetScience and Power Grid.

In Table 9.3, the highest modularity values in Email and Geom are 0.5362 and 0.7025, respectively, which belong to MOCSA. Also, the modularity value in NetScience and Power is 0.9503 and 0.8382, respectively. The results showed that MOCSA performed well on CD, and the results were mostly successful. The NMI value results in the MOCSA model were higher than in the three datasets compared to the other models, and only in Dolphin, the amount was meager. In general, the efficiency of the MOCSA model for CD in different datasets is 92.23%, which is a high degree of accuracy for node recognition and node communication.

Table 9.3 Comparison of MOCSA model with different MOO Models.

Datasets	Models	NMI	Modularity
Karate club	MOGA-Net [39]	0.8388	0.4147
	MOCD [40]	0.8371	0.4187
	MOEA/D-Net [41]	1.0000	0.4197
	MODPSO [42]	1.0000	0.4197
	DMOPSO [43]	0.8610	0.3948
	MODTLBO/D [44]	1.0000	0.4197
	MOCSA model	1.0000	0.4192
Dolphin network	MOGA-Net [39]	0.9918	0.5110
	MOCD [40]	0.9972	0.5213
	MOEA/D-Net [41]	1.0000	0.5190
	MODPSO [42]	1.0000	0.5190
	DMOPSO [43]	0.8934	0.5071
	MODTLBO/D [44]	1.0000	0.5220
	MOCSA model	0.9984	0.5262
College football	MOGA-Net [39]	0.7863	0.5163
	MOCD [40]	0.8613	0.5847
	MOEA/D-Net [41]	0.9265	0.6034
	MODPSO [42]	0.9289	0.6035
	DMOPSO [43]	0.8381	0.5861
	MODTLBO/D [44]	0.9366	0.6045
	MOCSA model	0.9486	0.6025
Poll books	MOGA-Net [39]	0.5834	0.5137
	MOCD [40]	0.5861	0.5205
	MOEA/D-Net [41]	0.5885	0.5259
	MODPSO [42]	0.5910	0.5263
	DMOPSO [43]	0.5809	0.5186
	MODTLBO/D [44]	0.6172	0.5269
	MOCSA model	0.7455	0.5264
Email network	MODTLBO/D [44]	–	0.5105
	DIM-MOEA/D [45]	–	0.5355
	MOCSA model	–	0.5362
Geom network	MODTLBO/D [44]	–	0.7018
	DIM-MOEA/D [45]	–	0.7007
	MOCSA model	–	0.7025
NetScience	DIM-MOEA/D [45]	–	0.9268
	MODPSO [42]	–	0.9503
	MOCSA model	–	0.9497
Power grid	DIM-MOEA/D [45]	–	0.8247
	MODPSO	–	0.8299
	MOCSA model	–	0.8382

9.5 Conclusion and future works

This chapter proposed a multiobjective algorithm based on a local updating called MOCSA for CD on social networks. In MOCSA, a local space–based cuckoo update method was used that improved evolution and increased population convergence toward optimal responses. The MOCSA model can detect high-quality communities with frequent use of local space–based updates. In MOO problems, there are two or three objective functions that the goals should be designed according to the type of problem so that it is possible to find a final optimal solution and satisfy the problem answer. On the other hand, solving these problems usually finds solutions (known as a set of Pareto solutions) that create the right balance between different goals.

The MOCSA model results on eight datasets showed that the amount of NMI on Karate has improved up to about 12.89%, Dolphin up to 11.87%, Football up to 5.66% Polbooks up to 9.25%, in comparison with different models. The Email, Geom, NetScience, and Power Grid datasets' modularity was 0.5362, 0.7025, 0.9497, and 0.8382. The results also showed that the MOCSA model performed better than most MOGA-Net, MOCD, MOEA/D-Net, MODPSO, DMOPSO, and MODTLBO/D models. The MOCSA model results on Email, Geom, NetScience, and Power showed that the modularity value was satisfactory, and the MOCSA model performed well compared to other models. Based on the results, it can be concluded that the MOCSA model was suitable and efficient for CD and community discovery among different nodes.

With this procedure, we can use metaheuristic algorithms multiobjective for future work for CD based on the hybrid of different operators such as crossover and mutation and for improving the power of an exploration and exploitation algorithm.

References

[1] J. Chu, et al., Social network community analysis based large-scale group decision making approach with incomplete fuzzy preference relations, Inf. Fusion 60 (2020) 98–120.
[2] M. Abedi, F.S. Gharehchopogh, An improved opposition based learning firefly algorithm with dragonfly algorithm for solving continuous optimization problems, Intell. Data Anal. 24 (2) (2020) 309–338.
[3] X. Li, et al., Communities detection in social network based on local edge centrality, Phys. A Stat. Mech. Appl. 531 (2019) 121552.
[4] A. Singh, et al., Probabilistic data structure-based community detection and storage scheme in online social networks, Fut. Gen. Comput. Syst. 94 (2019) 173–184.
[5] F. Hu, et al., Community detection in complex networks using Node2vec with spectral clustering, Phys. A Stat. Mech. Appl. 545 (2020) 123633.
[6] H. Jiang, et al., Community detection in complex networks with an ambiguous structure using central node based link prediction, Knowledge-Based Syst. 195 (2020) 105626.
[7] D.-Y. Nan, et al., A framework of community detection based on individual labels in attribute networks, Phys. A Stat. Mech. Appl. 512 (2018) 523–536.
[8] F.S. Gharehchopogh, H. Shayanfar, H. Gholizadeh, A comprehensive survey on symbiotic organisms search algorithms, Artific. Intell. Rev. (2019) 1–48.

[9] F.S. Gharehchopogh, H. Gholizadeh, A comprehensive survey: whale optimization algorithm and its applications, Swarm Evol. Comput. 48 (2019) 1–24.
[10] S. Talatahari, M. Azizi, Optimization of constrained mathematical and engineering design problems using chaos game optimization, Comput. Ind. Eng. 145 (2020) 106560.
[11] M.A. Javed, et al., Community detection in networks: a multidisciplinary review, J. Netw. Comput. Appl. 108 (2018) 87–111.
[12] I. Messaoudi, N. Kamel, A multi-objective bat algorithm for community detection on dynamic social networks, Appl. Intell. 49 (6) (2019) 2119–2136.
[13] R. Shang, H. Liu, L. Jiao, Multi-objective clustering technique based on k-nodes update policy and similarity matrix for mining communities in social networks, Phys. A Stat. Mech. Appl. 486 (2017) 1–24.
[14] X.-S. Yang, S. Deb, Multi-objective cuckoo search for design optimization, Comput. Oper. Res. 40 (6) (2013) 1616–1624.
[15] X.-S. Yang, S. Deb, Cuckoo Search via Levy flights, Proc. World Congress on Nature & Biologically Inspired Computing (NaBIC) 2009 (2009) 1–15.
[16] H. Shayanfar, F.S. Gharehchopogh, Farmland fertility: a new metaheuristic algorithm for solving continuous optimization problems, Appl. Soft Comput. 71 (2018) 728–746.
[17] G. Bello-Orgaz, S. Salcedo-Sanz, D. Camacho, A Multi-objective genetic algorithm for overlapping community detection based on edge encoding, Inform. Sci. 462 (2018) 290–314.
[18] F. Cheng, et al., A local information based multi-objective evolutionary algorithm for community detection in complex networks, Appl. Soft Comput. 69 (2018) 357–367.
[19] Q. Li, et al., A multi-objective adaptive evolutionary algorithm to extract communities in networks, Swarm Evol. Comput. 52 (2020) 100629.
[20] F. Zou, et al., Community detection in complex networks: multi-objective discrete backtracking search optimization algorithm with decomposition, Appl. Soft Comput. 53 (2017) 285–295.
[21] A. Reihanian, M.-R. Feizi-Derakhshi, H.S. Aghdasi, Community detection in social networks with node attributes based on multi-objective biogeography based optimization, Eng. Appl. Artific. Intell. 62 (2017) 51–67.
[22] B. Amiri, et al., Community detection in complex networks: multi–objective enhanced firefly algorithm, Knowledge-Based Syst. 46 (2013) 1–11.
[23] Y. Zhang, et al., WOCDA: A whale optimization based community detection algorithm, Phys. A Stat. Mech. Appl. 539 (2020) 122937.
[24] P. Ji, S. Zhang, Z. Zhou, A decomposition-based ant colony optimization algorithm for the multi-objective community detection, J. Amb. Intell. Hum. Comp. 11 (1) (2020) 173–188.
[25] X. Zhou, et al., A novel modularity-based discrete state transition algorithm for community detection in networks, Neurocomputing 334 (2019) 89–99.
[26] M. Moradi, S. Parsa, An evolutionary method for community detection using a novel local search strategy, Phys. A Stat. Mechan. Appl. 523 (2019) 457–475.
[27] W. Zhang, et al., Groups make nodes powerful: Identifying influential nodes in social networks based on social conformity theory and community features, Exp. Syst. Appl. 125 (2019) 249–258.
[28] H. Zare, M. Hajarian, Determination of regularization parameter via solving a multi-objective optimization problem, Appl. Numer. Math. 156 (2020) 542–554.
[29] dataset1, http://www-personal.umich.edu/~mejn/netdata/. (Accessed on 2020.8.30).
[30] dataset2, http://konect.uni-koblenz.de/networks/. (Accessed on 2020.8.30).
[31] K. Steinhaeuser, N.V. Chawla, Identifying and evaluating community structure in complex networks, Patt. Recogn. Lett. 31 (5) (2010) 413–421.
[32] M. Amoozegar, B. Minaei-Bidgoli, Optimizing multi-objective PSO based feature selection method using a feature elitism mechanism, Exp. Syst. Appl. 113 (2018) 499–514.

[33] D. Dutta, J. Sil, P. Dutta, Automatic clustering by multi-objective genetic algorithm with numeric and categorical features, Exp. Syst. Appl. 137 (2019) 357–379.
[34] Z. Zhang, et al., Improved Multi-objective Moth-flame Optimization Algorithm based on R-domination for cascade reservoirs operation, J. Hydrol. 581 (2020) 124431.
[35] L. Lv, et al., Multi-objective firefly algorithm based on compensation factor and elite learning, Fut. Gen. Comput. Syst. 91 (2019) 37–47.
[36] K. Wang, X. Li, L. Gao, A multi-objective discrete flower pollination algorithm for stochastic two-sided partial disassembly line balancing problem, Comput. Ind. Eng. 130 (2019) 634–649.
[37] Z. Shao, D. Pi, W. Shao, A multi-objective discrete invasive weed optimization for multi-objective blocking flow-shop scheduling problem, Exp. Syst. Appl. 113 (2018) 77–99.
[38] A. Sheikhahmadi, A. Zareie, Identifying influential spreaders using multi-objective artificial bee colony optimization, Appl. Soft Comput. 94 (2020) 106436.
[39] C. Pizzuti, A multi-objective genetic algorithm to find communities in complex networks, IEEE Trans. Evol. Comput. 16 (3) (2012) 418–430.
[40] C. Shi, et al., Multi-objective community detection in complex networks, Appl. Soft Comput. 12 (2) (2012) 850–859.
[41] M. Gong, et al., Community detection in networks by using multi-objective evolutionary algorithm with decomposition, Phys. A Stat. Mech. Appl. 391 (15) (2012) 4050–4060.
[42] M. Gong, et al., Complex network clustering by multiobjective discrete particle swarm optimization based on decomposition, IEEE Trans. Evol. Comput. 18 (1) (2014) 82–97.
[43] L. Li, et al., Quantum-behaved discrete multi-objective particle swarm optimization for complex network clustering, Patt. Recogn. 63 (2017) 1–14.
[44] D. Chen, et al., Multi-objective optimization of community detection using discrete teaching–learning-based optimization with decomposition, Inform. Sci. 369 (2016) 402–418.
[45] F. Zou, et al., Inverse modelling-based multi-objective evolutionary algorithm with decomposition for community detection in complex networks, Phys. A Stat. Mech. Appl. 513 (2019) 662–674.

CHAPTER 10

Finding efficient solutions of the multicriteria assignment problem

Emmanuel Kwasi Mensah[a], Esmaeil Keshavarz[b] and Mehdi Toloo[c]

[a]Department of Economics, University of Insubria, Italy, [b]Department of Mathematics, Sirjan Branch, Islamic Azad University, Sirjan, [c]Department of Business Transformation, Surrey Business School, University of Surrey, Guildford, United Kingdom; Department of Systems Engineering, Faculty of Economics, Technical University of Ostrava, Ostrava, Czech Republic; Department of Operations Management & Business Statistics, College of Economics and Political Science, Sultan Qaboos University, Muscat, Oman

10.1 Introduction

One of the most important and applicable problems in the field of Multicriteria Combinatorial Optimization (MCCO) is the Multicriteria Assignment Problem (MCAP). It is an extension of the well-known assignment problem (AP) that deals with the problem of assigning n elements (e.g., machines, students) to n jobs (e.g., work, courses) such that all jobs are completed in minimum total time (cost) [9,13]. The problem arises in practice owing to a varying degree of efficiency, cost, or time of machines or persons for performing different jobs, thus the need to optimize the assignment by either minimizing cost or maximize profit or both. The AP is useful to a wide range of applications, such as the vehicle routing and signal processing [2] including as subproblem to complicated combinatorial problems such as the transportation problem, traveling salesman problem, and the distribution problem.

Generally, the single-objective AP is known to be polynomially solvable and there exist several methods that can be used to obtain the solution. Heuristic approaches such as the well-known Hungarian method [15], the simplex based methods (see [4]), and primal-dual combinatorial methods are among few popular ones. The Hungarian method that explores the total unimodularity of the assignment constraint matrix is one of the most efficient methods for finding the optimal solution of the AP.

The AP extension to include multiple criteria is however a complex problem to deal with. Finding an efficient solution while fulfilling all related criteria typifies an assignment with a difficult combinatorial structure. It can be verified that enlisting all feasible solutions for the MCAP in the objective space for instance is a herculean task. In fact, it is not economical even for the small number of assignments. For example, an assignment with $n = 8$ only amounts to 40,320 distinct feasible solutions. Nonetheless, just like other problems in MCCO, several diverse methods in the literature can be used to obtain the nondominated solutions without enumerating all feasible solutions. For example, the scalarization method, the fuzzy method, the goal programming method, search-based, and hybrid algorithm. The scalarization method, a popular method among them, transforms the multiobjective problem into a single objective with additional constraints. The most common techniques among the scalarization methods are the weighted sum, Chebyshev, and the $\varepsilon -$ constraint methods [9]. Belhoul et al. [5] including a host of researchers have considered scalarizing methods to determine the best compromise solution for the MCAP.

Although the scalarization methods are straightforward, they can only find supported efficient solutions [13]. Supported solutions are optimal solutions of a model with a weighted sum scalarization objective function. Nonsupported solutions on the other hand are not. Finding all supported and nonsupported efficient solutions in practice is a major challenge in MCCO. Mattias Ehrgott [10] and most researchers thus resort to heuristics methods such as the branch and bound, primal-dual algorithms, two-phase approaches (see [5,7,18,20]), and some metaheuristic algorithms such as the genetic algorithm and simulated annealing (see [1,22]).

In this chapter, a two-phase exact solution method based on the data envelopment analysis (DEA) approach is considered to find both supported and nonsupported efficient solutions of the MCAP. By informal definition, the DEA is a linear programming tool that measures the relative efficiency of homogeneous decision-making units (DMUs) with multiple inputs and multiple outputs. Its concept of relative efficiency is based on Pareto optimality to establish the efficiency status for Multicriteria Decision Making (MCDM) problems particularly when one considers DMUs as a set of alternatives and data as the value of criteria [17,21]. The DEA efficiency partitions all the DMUs into two sets: efficient and inefficient, where an efficient DMU achieves a score of 1 and inefficient DMU achieves a score of less than 1 [11,24]. One of the general approaches to finding the efficient solutions of MCCO is a two-phase method that was initially proposed by Ulungu and Teghem [25,26]. Based on this approach, supported efficient solutions are found in the first phase by solving some weighted sum problems and nonsupported efficient solutions are detected in the second phase. We propose a different two-phase method in which a minimal complete set (MCS) of efficient assignments is obtained in phase I, and all the obtained efficient assignments are classified into the supported and nonsupported solutions in phase II. Each of the employed models in the proposed approach is formulated based on the traditional DEA models.

The rest of the chapter is structured as follows: The next section provides an introduction to the assignment in the one objective case. Section 10.3 reviews MCAP and DEA models, and then introduces the relationships between the DEA and MCAP models. We employ these relations and discuss efficient solution approaches and algorithms to solve the MCAP in Section 10.4. In Section 10.5, we present the details of an illustrative example to demonstrate the applicability of the proposed algorithm. Finally, we conclude the chapter in Section 10.6.

10.2 The basic AP

The linear AP is a well-known combinatorial optimization problem and one of the most studied problems in mathematical programming. Informally speaking, we are given n items (e.g., jobs) to n machines (or processing units) such that each machine is assigned exactly one job with nonnegative cost (c_{ij}) associated with assigning machine i ($i = 1, \ldots n$) for performing job j ($j = 1, \ldots n$). The primary objective is to minimize the cost for all n assignments. This problem is a special case of the transportation problem that involves a network structure of m nodes and n sinks. Thus, the problem can be solved with the transportation method (see [4]).

An alternative formulation and solution to the AP can be through graph theory. In this case, for a given combinatorial structure of the assignment network, the AP is seen as the problem of finding a minimum weighted matching on a bipartite graph $G = (U \cup V, E)$ with $|U| = |V| = n$ such that any edge $(i, j) \in E$ links a vertex i in U to a vertex j in V with cost c_{ij}. The graph-based combinatorial AP is then to find a perfect matching of minimum cost. [7]. Mathematically, the AP is a 0-1 integer linear programming given below:

$$\min \sum_{i=1}^{n} \sum_{j=1}^{n} c_{ij} t_{ij}$$
s.t.
$$\sum_{j=1}^{n} t_{ij} = 1 \qquad i = 1, \ldots, n \qquad (10.1)$$
$$\sum_{i=1}^{n} t_{ij} = 1 \qquad j = 1, \ldots, n$$
$$t_{ij} \in \{0, 1\} \qquad i, j = 1, \ldots, n$$

where t_{ij} is a 0-1 decision variable that is 1 if machine i performs job j and 0 otherwise. We will name any feasible solution of model (10.1) a feasible assignment. On this note, there are $n!$ distinct feasible assignment in an $n \times n$ AP where the constraints of the combinatorial structure admit exactly n nonzero variables at each feasible solution. Efficient solutions of the model can be obtained by using primal-dual combinatorial algorithms, simplex-like methods, or genetic algorithms. The Hungarian method is one of the earliest methods that utilize the primal-dual approach based on the shortest path algorithm. Its solution technique involves first, solving step-by-step iteratively the continuous AP problem where $t_{ij} \geq 0$ ($i, j = 1, \ldots, n$) in model (10.1) and its corresponding dual formulation:

$$\max \sum_{i=1}^{n} u_i + \sum_{j=1}^{n} v_j$$
s.t. $\qquad (10.2)$
$$u_i + v_j \leq c_{ij} \qquad i = 1, \ldots, n$$
$$u_i, v_j \in \mathbb{R} \qquad j = 1, \ldots, n$$

where u_i and v_j are dual variables associated with the first and second constraints. At each step, an assignment is added to a current partial primal solution and the dual feasible solution is determined such that the complementary slackness condition $\{\bar{c}_{ij} : c_{ij} - u_i - v_j, x_{ij}\bar{c}_{ij} = 0, i, j = 1, \ldots, n\}$ is satisfied. The steps are updated and iterated until the primal-dual (solution) pair is optimal. There are a plethora of algorithms that use the primal-dual approach to solve the AP. The class of algorithms generally runs in $O(n^4)$ time although the complexity can be reduced to $O(n^3)$. We refer the reader to Dell'Amico and Toth [7] for a survey on some algorithms and codes for the AP. A complete golden survey on the subject can also be found in Pentico [19].

10.3 Restated MCAP and DEA: models and relationship

In this section, we provide the background to some MCDM problems. We focus our attention on the MCAP, the DEA, and their integrated model.

10.3.1 The multicriteria assignment problem (MCAP)

Consider a real-life assignment network structure that contains more than one objective or criteria that needs to be satisfied. In this case, the multicriteria optimization becomes the domain that is applied when optimal decisions have to be taken. Examples of such MCAP include scheduling on parallel machines, allocation, and machine set-up problems. Let us consider an MCAP with cost vectors $C_{ij} = (c_{ij}^{(1)}, c_{ij}^{(2)}, \ldots, c_{ij}^{(k)})$ and the profit vectors $P_{ij} = (p_{ij}^{(1)}, p_{ij}^{(2)}, \ldots, p_{ij}^{(s)})$ where $c_{ij}^{(m)}$ and $p_{ij}^{(r)}$ are nonnegative integer numbers for $i, j = 1, \ldots, n$. The MCAP is stated as the following model:

$$\min C_m(x) = \sum_{i=1}^{n} \sum_{j=1}^{n} c_{ij}^m t_{ij} \quad m = 1, \ldots, k$$

$$\min P_r(x) = \sum_{i=1}^{n} \sum_{j=1}^{n} p_{ij}^r t_{ij} \quad r = 1, \ldots, s$$

s.t.

$$\sum_{j=1}^{n} t_{ij} = 1 \quad i = 1, \ldots, n \quad (10.3)$$

$$\sum_{i=1}^{n} t_{ij} = 1 \quad j = 1, \ldots, n$$

$$t_{ij} \in \{0, 1\} \quad i, j = 1, \ldots, n$$

Denote $\mathcal{X}$, the set of all feasible assignments. The set $\mathcal{X} \subset \{0, 1\}^{n^2} \subset \mathbb{R}^{n^2}$ is known as the feasible set in decision space. Similarly, $\mathcal{Z} = \{(C(t), P(t)) : t \in \mathcal{X}\} \subset \mathbb{N}^{k+s} \subset \mathbb{R}^{k+s}$ is called the feasible set in the objective space where $C(t) = (C_1(t), C_2(t), \ldots, C_k(t))$ and $P(x) = (P_1(t), P_2(t), \ldots, P_s(t))$ are, respectively, the minimum (cost) and maximum (profit) type objective function value vectors. Notice that the set $\mathcal{X}$ is nonconvex that translates also to the set $\mathcal{Z}$. Usually, due to the conflicting nature of the different criteria, optimal solution in the MCAP is meaningless, indeed it is common knowledge that a feasible assignment that optimizes simultaneously all the objectives in model (10.3) does not exist. The general concept that provides optimality conditions for these multiobjective combinatorial optimization (MOCO) problems belongs to the concept of Pareto optimality or Pareto efficiency. An assignment is efficient in MCAP if and only if it cannot improve some criteria without sacrificing others. To understand this concept, we need to introduce the following definitions.

Definition 1. Let $t, t' \in \mathcal{X}$. A feasible assignment t dominates t' if $C_m(t) \leq C_m(t'), m = 1, \ldots, k, P_r(x) \geq P_r(t'), r = 1, \ldots, s$ and $(C_m(t), P_r(t)) \neq (C_m(t'), P_r(t'))$ and equivalently, $(C(t), P(t))$ dominates $(C(t'), P(t'))$. If $C_m(t) < C_m(t') \ \forall m = 1, \ldots, k$, and $P_r(t) > P_r(t') \ \forall r = 1, \ldots, s$, then t strictly dominates t'.

Definition 2. A feasible assignment $t^* \in \mathcal{X}$ is said to be efficient or Pareto-optimal solution of model (3) if there is no other feasible assignment $t \in \mathcal{X}$ such that t dominates t^*, otherwise the assignment t^* is not efficient.

Here, if $t, t' \in \mathcal{X}$ such that $(C(t), P(t))$ dominates $(C(t'), P(t'))$, then the point t dominates t' or the latter is inferior and we write $t' \preccurlyeq t$. The set of all efficient solution denoted by $\mathcal{X}_E$ is generally called a complete set of efficient solutions and the image of $\mathcal{X}_E$ in $\mathcal{Z}$ forms the nondominated frontier $\mathcal{Z}_E$. Fig. 10.1 shows the Pareto front that indicates the image of the Pareto-optimal set in the objective

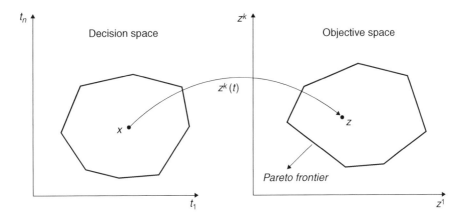

FIGURE 10.1

Mapping of decision space into objective space.

space from the decision space for a general multicriteria decision problem. For the nonconvexity of the feasible set of the MCAP, the set of efficient solutions is partitioned into two subsets: supported and nonsupported efficient solutions [10,13]. To define the supported efficient solutions, we consider the following weighted sum single objective model:

$$\max \left\{ \sum_{r=1}^{s} u_r P_r(t) - \sum_{m=1}^{k} v_m C_m(t) : t \in \mathcal{X} \right\} \quad (10.4)$$

where u_r, $r = 1, \ldots, s$ and v_m, $m = 1, \ldots, k$ are positive weights parameters. The definitions below follows.

Definition 3. An efficient solution $t^* \in \mathcal{X}$ is a *supported efficient solution* if it is the optimal solution of model (10.4) for at least one positive vector (u, v). If $\mathcal{X}_{SE}$ denotes the set of all supported efficient solutions, the corresponding point $(C(t^*), P(t^*))$ is called a *supported nondominated point* for all $t^* \in \mathcal{X}_{SE}$.

Definition 4. An efficient solution $t^* \in \mathcal{X}$ is a *nonsupported efficient solution* if there exists no positive vector (u, v) such that t^* is an optimal solution of model (10.4). If $\mathcal{X}_{NE}$ denotes the set of all nonsupported efficient solutions, the corresponding point $(C(t^*), P(t^*))$ is called the *nonsupported, nondominated point* for all $t^* \in \mathcal{X}_{NE}$.

The nonsupported efficient solutions are efficient solutions that are not optimal solutions of model (10.4) for any (u, v) and accordingly their points. We have used the notations Z_{SE} and Z_{NE} to denote the set of all supported and nonsupported, nondominated points, respectively. Fig. 10.2 shows a hypothetical feasible assignment represented by a minimized cost vector and a maximized profit vector in objective space. The points A, B, C, D, and F are the supported nondominated points whose solution is the extreme solutions in the decision space. Obviously, the points define extreme points of the convex hull of Z. The points E, F, H, and I are nondominated points, however, nonsupported. Unlike the supported nondominated points, the latter is in the interior of the convex hull of Z while G, J, L, and M are dominated points spanned by inefficient solutions in the decision space. From Fig. 10.2, it is obvious

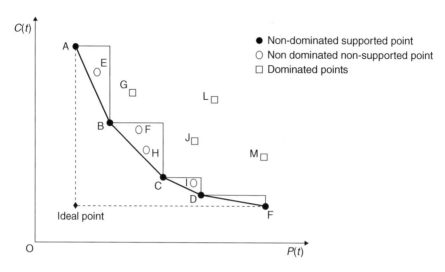

FIGURE 10.2

Supported and nonsupported points in biobjective space.

that the ideal point is obtained by solving model (10.3). An ideal point unlike an anti-ideal point is a point with the best solution in the objective space. Note however that, an ideal point here is not a feasible solution.

10.3.2 Data envelopment analysis

The DEA is a linear programming and an MCDM tool built on the concept of free disposability and the convexity of the inputs and outputs. It is mainly used to assess the relative efficiency of a set of DMUs with multiple max- and min-type factors. Structurally, the DEA formulation is quite similar to the multiobjective linear programming (MOLP) based on the reference direction approach to generate efficient solutions [12]. In this direction, efficient points in the objective space of the MOLP are considered as efficient DMUs in the DEA framework.

Let the performance evaluation problem be such that there are n DMUs, DMU$_j$ ($j=1, \ldots, n$) where each DMU consumes m amounts of inputs x_{ij} ($i=1, \ldots, m$) to produce s amounts of output y_{rj} ($r=1, \ldots, s$). It is supposed that these inputs and outputs are semipositive, that is, $x_j \geq \mathbf{0}_m$, $x_j \neq \mathbf{0}_m$ and $y_j \geq \mathbf{0}_s$, $y_j \neq \mathbf{0}_s$ for $j=1, \ldots, n$. The DEA efficiency is a dominance-based concept that follows similar Pareto optimality conditions as in the MOLP.

Definition 5. DMU$_j$ is dominated by DMU$_k$ if $x_{ij} \geq x_{ik}$ for all i and $y_{rj} \leq y_{rk}$ for all r. An immediate conclusion follows that DMU$_k$ is more efficient than DMU$_j$.

Stewart [23] provides an alternative definition using the weighted sum of the inputs-outputs $\sum_{r=1}^{s} u_r y_{rj} - \sum_{i=1}^{m} v_i x_i$. That is, DMU$_j$ is dominated by DMU$_k$ if

$$\sum_{r=1}^{s} u_r y_{rj} - \sum_{i=1}^{m} v_i x_{ij} < \sum_{r=1}^{s} u_r y_{rk} - \sum_{i=1}^{m} v_i x_{ik} \tag{10.5}$$

for any semipositive weight vector $(\boldsymbol{u}, \boldsymbol{v})$. It follows also then that, DMU_o is Pareto optimal or efficient if there is at least one positive weights vector $(\boldsymbol{u}, \boldsymbol{v})$ such that

$$\sum_{r=1}^{s} u_r y_{rj} - \sum_{i=1}^{m} v_i x_{ij} \leq \sum_{r=1}^{s} u_r y_{ro} - \sum_{i=1}^{m} v_i x_{io} \tag{10.6}$$

for all other DMUs j.

The idea of enforcing DEA efficiency in the MOLP context appears with the Pareto efficiency concept that both approaches share. In the DEA case, the means of achieving this Pareto optimality or efficient DMUs is by solving a linear programming model, the first of which was proposed by Charnes et al. [6]. This model is called the Charnes, Cooper and Rhodes (CCR) model. It is one of two traditional models that measures the relative efficiency of DMUs based on the ratio of the weighted sum of outputs to the weighted sum of inputs under constant returns to scale technology. For our later purpose, we will consider also an extended CCR model known as the Banker, Charnes and Cooper (BCC) model, proposed under variable returns-to-scale (VRS) technology by Banker et al. [3]. Furthermore, the Free Disposable Hull (FDH), which unlike the CCR and BCC assumes a nonconvex production frontier, will be considered. There are two forms of the BCC model which are mutually dual: envelopment and multiplier. The models are formulated as follows:

BCC envelopment form

$$\min \theta - \varepsilon \left(\sum_{i=1}^{m} s_i^- + \sum_{r=1}^{s} s_r^+ \right)$$

s.t.
$$\sum_{j=1}^{n} \lambda_j x_{ij} + s_i^- = \theta x_{io} \qquad i = 1, \ldots, m$$
$$\sum_{j=1}^{n} \lambda_j y_{rj} - s_r^+ = y_{ro} \qquad r = 1, \ldots, s \tag{10.7}$$
$$\sum_{j=1}^{n} \lambda_j = 1$$
$$\lambda_j \geq 0 \qquad j = 1, \ldots, n$$
$$s_i^- \geq 0 \qquad i = 1, \ldots, m$$
$$s_r^+ \geq 0 \qquad r = 1, \ldots, s$$

BCC multiplier form

$$\max \theta = \sum_{r=1}^{s} u_r y_{ro} + u_o$$

s.t.
$$\sum_{i=1}^{m} v_i x_{io} = 1$$
$$\sum_{r=1}^{s} u_r y_{rj} - \sum_{i=1}^{m} v_i x_{ij} + u_o \leq 0 \qquad j = 1, \ldots, n \tag{10.8}$$
$$v_i \geq \varepsilon \qquad i = 1, \ldots, m$$
$$u_r \geq \varepsilon \qquad r = 1, \ldots, s$$
$$u_o \text{ is free}$$

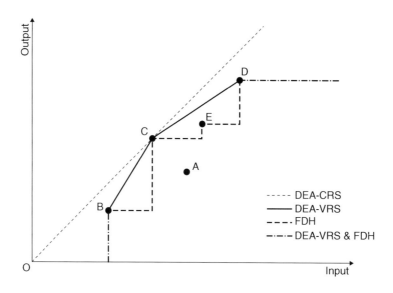

FIGURE 10.3

DEA-VRS and FDH frontiers.

where u_o is the returns to scale variable and ε is the non-Archimedean infinitesimal added essentially to keep the multipliers (weights) away from zero. The FDH model proposed in Deprins et al. [8] is formulated as the following:

FDH model

$$\min \theta - \varepsilon \left(\sum_{i=1}^{m} s_i^- + \sum_{r=1}^{s} s_r^+ \right)$$

s.t.

$$\sum_{j=1}^{n} \lambda_j x_{ij} + s_i^- = \theta x_{io} \quad i = 1, \ldots, m$$

$$\sum_{j=1}^{n} \lambda_j y_{rj} - s_r^+ = y_{ro} \quad r = 1, \ldots, s \quad (10.9)$$

$$\sum_{j=1}^{n} \lambda_j = 1$$

$$\lambda_j \in \{0, 1\} \quad j = 1, \ldots, n$$
$$s_i^- \geq 0 \quad i = 1, \ldots, m$$
$$s_r^+ \geq 0 \quad r = 1, \ldots, s$$

The main difference between the BCC and FDH models lies in the structure of their efficient frontier that envelops the DMUs in different ways depending on the underlying assumptions. As depicted in Fig. 10.3, it can be observed that DMUs B, C, E, and D are efficient from the FDH point of view, whereas DMU E is not efficient from the DEA-VRS view-point. The main reason for the difference here results from the convexity assumption by the DEA-VRS frontier that considers only linear combinations of feasible and efficient input and output transformation as efficient DMUs. Thus, the set of nondominated units on the DEA-VRS frontier is always less or equal to the nondominated units on the FDH frontier and as a result, if DMU_k is BCC efficient, then it is FDH efficient as well. Note that, DMU A is dominated by units on both frontiers. In models (10.7)–(10.9), the set of efficient and inefficient DMUs are obtained by solving the respective model for each DMU. The efficiency of DMUs for the BCC and FDH models are defined according to the following:

Definition 6. DMU_o is BCC efficient in the envelopment model (10.7) if and only if there is an optimal solution $(\theta^*, \lambda^*, s^{-*}, s^{+*})$ that satisfies $\theta^* = 1$, $\lambda^* = e_o$, $s^{-*} = 0_m$, and $s^{-*} = 0_s$ where e_o is a vector having zero components, except for a 1 in the oth position; otherwise, DMU_o is called BCC inefficient.

Definition 7. DMU_o is BCC efficient in multiplier model (10.8) if $\theta^* = 1$ and there exist at least one optimal solution (v^*, u^*, u_o^*) with $v^* > 0_m$ and $u^* > 0_s$ otherwise DMU_o is called BCC inefficient.

Definition 8. DMU_o is FDH efficient in model (10.9) if there exist at least an optimal solution $(\theta^*, \lambda^*, s^{-*}, s^{+*})$ that satisfies $\theta^* = 1$, $\lambda^* = e_o$, $s^{-*} = 0_m$, and $s^{-*} = 0_s$, otherwise, DMU_o is called FDH inefficient.

As noted from Fig. 10.3, the production possibility set of the FDH is a subset of the BCC; however, observing from models (10.7) and (10.9), the treatment of slacks in the input and outputs for the radial efficiency is not much different, rather pervasive in the case of the FDH. A similar treatment exists if we consider the models in the additive (ADD) form. ADD models are formulated based on the slacks of input excesses and output shortfalls. They are particularly useful because of their unit-invariant property that is an important characteristic for subjective unit of measurement in efficiency measurement. Various ADD models, herein, ADD-BCC and ADD-FDH models, can be formulated by eliminating θ from models (10.7) and (10.9). We favor the ADD model because of the stated advantage. Hence, we consider the following ADD-FDH model:

$$\min \sum_{i=1}^{m} s_i^- + \sum_{r=1}^{s} s_r^+$$

s.t

$$\sum_{j=1}^{n} \lambda_j x_{ij} + s_i^- = x_{i0} \quad i = 1, \ldots, m$$

$$\sum_{j=1}^{n} \lambda_j y_{rj} - s_r^+ = y_{r0} \quad r = 1, \ldots, s \quad (10.10)$$

$$\sum_{j=1}^{n} \lambda_j = 1$$

$$\lambda_j \in \{0, 1\} \quad j = 1, \ldots, n$$
$$s_i^- \geq 0 \quad i = 1, \ldots, m$$
$$s_r^+ \geq 0 \quad r = 1, \ldots, s$$

ADD-BCC model can be achieved by replacing constraint $\lambda_J \in \{0, 1\}$ with $\lambda_J \geq 0$ in the above model.

An efficient DMU in the ADD models is defined as follows [27]:

Definition 9. DMU_o is ADD-FDH efficient (ADD-BCC efficient) if there exists at least one optimal solution $(\lambda^*, s^{-*}, s^{+*})$ for ADD-FDH (ADD-BCC) model with $\lambda^* = e_p$, $s^{-*} = 0_k$, and $s^{+*} = 0_s$; otherwise, DMU_o is named ADD-FDH inefficient (ADD-BCC inefficient).

Esmail Keshavarz and Toloo [14] proved that a DMU is FDH efficient if and only if is ADD-FDH efficient. In the next section, we will apply this note to design our proposed approach.

10.3.3 An integrated DEA and MCAP

Although the DEA and MCDM differ in terms of their objectives, generally they coincide if we view DMUs as alternatives, inputs as criteria to be minimized, and outputs as criteria to be maximized. The methodological connection between DEA and MCDM has been established in Stewart [23] including studies on combined multicriteria DEA models [11,16]). In this chapter, the unique relationship of the DEA and the MCAP is of utmost interest to establish the status of feasible solutions. Note *a priori* that the DEA input–output space is equivalent to the objectives space of the MCAP. That is $\mathcal{Z} = \{(C(t)), P(t) : t \in \mathcal{X}\}$ is the set of all DMUs under evaluation. Hence, any efficient assignment $t* \in \mathcal{X}$ in model (10.3) can be considered as a DMU with input–output vector $(C(t^*), P(t^*))$. Considerably, the question of whether an efficient solution is supported or not and whether a feasible solution is efficient or not can be explored using this unique relationship. Esmail Keshavarz and Toloo [14] establish theorems that exploit this relationship to answer these questions. For the reason of brevity, these theorems are only restated here.

Theorem 1. A feasible solution $t^* \in \mathcal{X}$ of MCAP (3) is a supported efficient solution, if and only if the vector $(C(t^*), P(t^*))$ is BCC efficient

Theorem 2. A feasible solution $t^* \in \mathcal{X}$ of the MCAP (3) is an efficient solution if and only if the vectors $(C(t^*), P(t^*))$ is FDH efficient in model (10.9) and equivalently ADD-FDH efficient in model (10).

We refer the reader to the systematic proof of these theorems in reference therein. The intuition to the theorems can be inferred geometrically from Fig. 10.3. Considering the input and output axes as $C(t)$ and $P(t)$, it is not difficult to see the BCC- and FDH-efficient points from the objective space, equally from Fig. 10.2. It is sufficient to see the points B, C, D, and E are nondominated and FDH efficient. These points form the efficient solutions $t^* \in \mathcal{X}$ in MCAP as specified in Theorem 2. It is worth noting that only the points B, C, and D are the supported nondominated. Indeed, to specify the facet of the efficient frontier of the BCC model. Quite intuitive from Theorem 1, if $t^* \in \mathcal{X}$ is a supported solution in the input-output space, then $(C(t^*), P(t^*))$ is a supported nondominated point in the objective space for the MCAP.

10.4 Finding efficient solutions using DEA

The existence of a nonsupported efficient solution in MOCO implies the problem is difficult to solve. As mentioned above, it is not economical to find all feasible solutions (DMUs) that are also feasible solutions of the MCAP. Neither is it satisfactory to list all $n!$ feasible assignment to determine efficient assignment. The purpose of this section is therefore to employ the DEA technique in a two-phase algorithm to find efficient solutions to the MCAP without listing all the assignments.

10.4.1 The two-phase algorithm

We present a two-phase algorithm to find and classify an MCS of efficient solutions for the MCAP. This approach here is based on the DEA models presented in the previous section. The two-phase algorithm here has the advantage of exploiting the unimodularity property including the ability to find efficient assignment(s) without necessarily enlisting all feasible assignment as compared to other methods in the literature (see for example, [10]).

Phase I of the algorithm utilizes the FDH model to find an MCS of efficient solutions. As noted in Theorem 2, a feasible assignment is an efficient assignment if and only if it is FDH efficient. In this phase, an ADD-FDH based model is formulated for finding efficient assignments. First, we compute the anti-ideal point denoted $(c_{AI}^1, c_{AI}^2, \ldots, c_{AI}^k, p_{AI}^1, p_{AI}^2, \ldots, p_{AI}^s)$. As stated before, this point is dominated by all feasible points in the objective space. The anti-ideal points are obtained by solving the following single-objective problem:

$$
\begin{aligned}
c_{AI}^m = \quad & \max \sum_{i=1}^n \sum_{j=1}^n c_{ij}^m t_{ij} && m = 1, \ldots, k \\
\text{s.t.} \quad & \sum_{j=1}^n t_{ij} = 1 && i = 1, \ldots, n \\
& \sum_{i=1}^n t_{ij} = 1 && j = 1, \ldots, n \\
& t_{ij} \in \{0, 1\} && i, j = 1, \ldots, n \\
p_{AI}^r = \quad & \min \sum_{i=1}^n \sum_{j=1}^n p_{ij}^r t_{ij} && r = 1, \ldots, s \\
\text{s.t.} \quad & \sum_{j=1}^n t_{ij} = 1 && i = 1, \ldots, n \\
& \sum_{i=1}^n t_{ij} = 1 && j = 1, \ldots, n \\
& t_{ij} \in \{0, 1\} && i, j = 1, \ldots, n
\end{aligned}
\tag{10.11}
$$

Practically selecting the inefficient anti-ideal points for this model ensures that the associated assignment of the optimal solution, as a project point in the objective space, is efficient. We can also formulate a mixed-integer linear programming (MILP) that is a manipulative version of the ADD-FDH model to obtain efficient assignments as follows:

$$
\begin{aligned}
& \max \sum_{i=1}^m s_i^- + \sum_{r=1}^s s_r^+ \\
& \text{s.t} \\
& \sum_{j=1}^n c_{ij}^m t_{ij} + s_i^- = c_{AI}^m && i = 1, \ldots, m
\end{aligned}
$$

204 Chapter 10 Finding efficient solutions of the multicriteria assignment problem

$$\sum_{j=1}^{n} p_{ij}^r t_{ij} - s_r^+ = p_{AI}^r \quad r = 1, \ldots, s$$

$$\sum_{j=1}^{n} t_{ij} = 1 \quad i = 1, \ldots, n$$

$$\sum_{i=1}^{n} t_{ij} = 1 \quad j = 1, \ldots, n \quad (10.12)$$

$$t_{ij} \in \{0, 1\} \quad i, j = 1, \ldots, n$$
$$s_i^- \geq 0 \quad i = 1, \ldots, m$$
$$s_r^+ \geq 0 \quad r = 1, \ldots, s$$

Model (10.12) can be obtained if one considers the anti-ideal point (C_{AI}, P_{AI}) as the DMU under evaluation, and we log noting that the mth cost-type function $\sum_{i=1}^{n} \sum_{j=1}^{n} c_{ij}^m t_{ij}$ and the rth profit-type function $\sum_{i=1}^{n} \sum_{j=1}^{n} p_{ij}^r t_{ij}$ correspond to the input and output values of the under evaluation DMU. It can be verified that t^* is an efficient assignment associated with the nondominated point $(C(t^*), P(t^*))$ if (t^*, s^{-*}, s^{+*}) is obtained as an optimal solution. Model (10.12) only provides us the opportunity to obtain an efficient assignment at an instance solution. However, to obtain the next distinct efficient assignment(s), we suggest an iterative approach that, once the optimization model is not infeasible, will ensure that the MCS of efficient assignments is obtained. We propose the following extension of the MILP:

$$\max \quad \sum_{m=1}^{k} s_m^- + \sum_{r=1}^{s} s_r^+$$

s.t.

(P_d)

$$\sum_{i=1}^{n} \sum_{j=1}^{n} c_{ij}^m t_{ij} + s_m^- = c_{AI}^m \quad m = 1, 2, \ldots, k$$

$$\sum_{i=1}^{n} \sum_{j=1}^{n} p_{ij}^r t_{ij} - s_r^+ = p_{AI}^r \quad r = 1, 2, \ldots, s$$

$$\sum_{j=1}^{n} t_{ij} = 1 \quad i = 1, \ldots, n$$

$$\sum_{i=1}^{n} t_{ij} = 1 \quad j = 1, \ldots, n \quad (10.13)$$

$$\sum_{i=1}^{n} \sum_{j=1}^{n} c_{ij}^m t_{ij} - c^m(t^J) + 1 \leq M\alpha_J^m \quad m = 1, 2, \ldots, k; \; J = 0, 1, \ldots, d-1$$

$$p^r(t^J) - \sum_{i=1}^{n} \sum_{j=1}^{n} p_{ij}^r t_{ij} + 1 \leq M\beta_J^r \quad r = 1, 2, \ldots, s; \; J = 0, 1, \ldots, d-1$$

$$\sum_{m=1}^{k} \alpha_J^m + \sum_{r=1}^{s} \beta_J^r \leq (k+s) - 1 \quad J = 0, 1, \ldots, d-1$$

$$t_{ij} \in \{0, 1\} \quad i, j = 1, \ldots, n$$
$$s_m^- \geq 0 \quad m = 1, 2, \ldots, k$$
$$s_r^+ \geq 0 \quad r = 1, 2, \ldots, s$$
$$\alpha_J^m \in \{0, 1\} \quad m = 1, 2, \ldots, k; \; J = 0, 1, \ldots, d-1$$
$$\beta_J^r \in \{0, 1\} \quad r = 1, 2, \ldots, s; \; J = 0, 1, \ldots, d-1$$

where the binary variable α_j^m, $J = 0, 1, \ldots, d-1$ is equal to 1 when the mth input of the obtained assignment is greater or equal to the mth input of the current assignment, t^J, else α_j^m could assume the value of zero. Put differently, $\sum_{i=1}^{n} \sum_{j=1}^{n} c_{ij}^m t_{ij} \geq c^m(t^J)$ and the constraint $\sum_{i=1}^{n} \sum_{j=1}^{n} c_{ij}^m t_{ij} - c^m(t^J) + 1 \leq M\alpha_j^m$ lead to $\alpha_j^m = 1$. Similarly, the binary variable $\beta_j^r = 1$ if $\sum_{i=1}^{n} \sum_{j=1}^{n} p_{ij}^r t_{ij} \leq p^r(t^J)$. In both the mth and rth constraints involve α_j^m and β_j^r, big-M is a sufficiently large positive number, it could be observed that a suitable value for the parameter M plays a key role in solving the model. A practical way to find this is to resort to ideal and anti-ideal points. We can select $M > \max\{c_{AI}^m - c_I^m, p_I^r - p_{AI}^r : m = 1, \ldots, k, r = 1, \ldots, s\}$, which are the upper bounds of the terms $\sum_{i=1}^{n} \sum_{j=1}^{n} c_{ij}^m t_{ij} - c^m(t^J)$ and $p^r(t^J) - \sum_{i=1}^{n} \sum_{j=1}^{n} p_{ij}^r t_{ij}$ respectively.

We now consider phase II of the algorithm. In this phase, nondominated efficient assignments obtained in Phase I are evaluated as a set of observed DMUs. To that end, the BCC model (10.7) is utilized. Referencing Theorem 1, the nondominated efficient assignments are supported only if they are BCC efficient, otherwise, they are nonsupported efficient assignments. In the DEA, the search consists of solving the BCC model n times, each comparing the current assignment to the all assignments prior to discriminating it as supported or nonsupported. The BCC model can be solved by well-known solvers such as the CPLEX, GUROBI, etc. Here, we use the optimization software, GAMS with the CPLEX solver to find the MCS of efficient assignments from the algorithm, and subsequently classify them.

10.4.2 The proposed algorithm

The following algorithm describes the steps involved in solving models in phases I and II and the iterative procedure to find all MCS of efficient solutions and its corresponding nondominated points in objective space.

Proposed algorithm

Phase I: Obtaining an MCS of efficient assignments

Input: Objective function coefficient matrices: $C_m = [c_{ij}^m]_{n \times n}$, $m = 1, 2, \ldots, k$, $P_r = [p_{ij}^r]_{n \times n}$, $r = 1, 2, \ldots, s$, anti-ideal point $(C_{AI}, P_{AI}) = (c_{AI}^1, c_{AI}^2, \ldots, c_{AI}^k, p_{AI}^1, p_{AI}^2, \ldots, p_{AI}^s)$.
Step 0: Let $\mathcal{X}_E = \phi$, $Z_N = \phi$ and $d = 0$.
Step 1: Solve the model (P_d), find an efficient assignment, $t^d \in \mathcal{X}$, and its associated objective vector $(C(t^d), P(t^d))$. Let $\mathcal{X}_E = \mathcal{X}_E \cup \{t^d\}$, $Z_N = Z_N \cup \{(C(t^d), P(t^d))\}$.
Step 2: Let $d = d + 1$. If the model (P_d) is infeasible, then stop.
Step 3: Solve the model (P_d) and find an optimal vector $(s^{+d}, s^{-d}, t^d, \alpha^d, \beta^d)$. Let $\mathcal{X}_E = \mathcal{X}_E \cup \{t^d\}$, $Z_N = Z_N \cup \{(C(t^d), P(t^d))\}$, then go to Step 2.
Output: An MCS of efficient assignments $\mathcal{X}_E$ and its related nondominated set Z_N in the objective space.

Phase II: Classifying the efficient assignments

Input: $\mathcal{X}_E$ and Z_N and d, obtained in phase I.
Step 0: Let $(\mathbf{x}_J, \mathbf{y}_J) = (C(t^J), P(t^J))$ as the input–output vector of DMU_J, $J = 0, \ldots, d-1$. Let $o = 0$
Step 1: Evaluate DMU_o by solving the BCC model (10.7) and find its BCC-efficiency score θ_o^*.
Step 2: If $\theta_o^* = 1$, then let $\mathcal{X}_{SE} = \mathcal{X}_{SE} \cup \{t^o\}$ and $Z_{SN} = Z_{SN} \cup \{(C(t^o), P(t^o))\}$.
Step 3: Let $o = o + 1$. If $o \leq d - 1$ go to Step 1.

Step 4: Let $\mathcal{X}_{NSE} = \mathcal{X}_E - \mathcal{X}_{SE}$ and $Z_{NSN} = Z_N - Z_{SN}$.
Output: Sets of supported and nonsupported efficient assignments and their associated nondominated points in the objective space, that is $\mathcal{X}_{SE}$, $\mathcal{X}_{NSE}$, $\mathcal{X}_{SN}$, and Z_{NSN}.

10.5 Numerical examples

Let us consider a three-criterion AP with the following cost and profit matrices:

$$C^1 = \begin{bmatrix} 9 & 3 & 9 & 5 & 2 \\ 6 & 5 & 4 & 4 & 9 \\ 7 & 2 & 5 & 7 & 12 \\ 6 & 4 & 7 & 9 & 10 \\ 4 & 3 & 6 & 4 & 5 \end{bmatrix},$$

$$P^1 = \begin{bmatrix} 6 & 9 & 8 & 4 & 4 \\ 5 & 7 & 6 & 9 & 5 \\ 9 & 2 & 8 & 7 & 6 \\ 4 & 7 & 5 & 8 & 4 \\ 3 & 8 & 3 & 6 & 3 \end{bmatrix},$$

$$P^2 = \begin{bmatrix} 6 & 7 & 6 & 4 & 7 \\ 7 & 11 & 4 & 6 & 3 \\ 3 & 9 & 7 & 7 & 9 \\ 4 & 5 & 10 & 3 & 7 \\ 6 & 2 & 11 & 5 & 5 \end{bmatrix}.$$

Clearly, in this problem, we have $n=5$, $m=1$, and $s=2$. The anti-ideal point is $(c_{AI}^1, p_{AI}^1, p_{AI}^2) = (41, 18, 17)$ which has been obtained by solving the models (10.11).

Phase I

Step 0: Let $\mathcal{X}_E = \phi$, $Z_N = \phi$ and $d = 0$.
Step 1: Solve the following model (P_0):

$$(P_0) \quad \begin{aligned} \max \quad & s_1^- + s_1^+ + s_2^+ \\ \text{s.t.} \quad & \sum_{i=1}^{5}\sum_{j=1}^{5} c_{ij}^1 t_{ij} + s_1^- = 41 \\ & \sum_{i=1}^{5}\sum_{j=1}^{5} p_{ij}^1 t_{ij} - s_1^+ = 18 \\ & \sum_{i=1}^{5}\sum_{j=1}^{5} p_{ij}^2 t_{ij} - s_2^+ = 17 \\ & \sum_{j=1}^{5} t_{ij} = 1 \qquad i = 1,\ldots 5 \\ & \sum_{i=1}^{5} t_{ij} = 1 \qquad j = 1,\ldots,5 \\ & t_{ij} \in \{0, 1\} \qquad i, j = 1,\ldots,5 \\ & s_1^- \geq 0,\ s_1^+ \geq 0,\ s_2^+ \geq 0 \end{aligned}$$

we obtain $t^0_{15} = t^0_{24} = t^0_{32} = t^0_{43} = t^0_{51} = 1$ (all other t^0_{ij}-values are zero). Its associated objective vector is $(C(t^0), P(t^0)) = (19, 23, 38)$, which is a nondominated point. Let $\mathcal{X}_E = \{t^0\}$ and $Z_N = \{(19, 23, 38)\}$.

Step 2: Let $d = 1$. The following model (P_1) is feasible:

$$\max \quad s^-_1 + s^+_1 + s^+_2$$

$$\text{s.t.} \quad \sum_{i=1}^{5} \sum_{j=1}^{5} c^1_{ij} t_{ij} + s^-_1 = 41$$

$$\sum_{i=1}^{5} \sum_{j=1}^{5} p^1_{ij} t_{ij} - s^+_1 = 18$$

$$\sum_{i=1}^{5} \sum_{j=1}^{5} p^2_{ij} t_{ij} - s^+_2 = 17$$

$$\sum_{j=1}^{5} t_{ij} = 1, \quad i = 1, \ldots, 5,$$

(P_1)
$$\sum_{i=1}^{5} t_{ij} = 1, \quad j = 1, \ldots, 5,$$

$$\sum_{i=1}^{5} \sum_{j=1}^{5} c^1_{ij} t_{ij} - 19 + 1 \leq M\alpha^1_0$$

$$23 - \sum_{i=1}^{5} \sum_{j=1}^{5} p^1_{ij} t_{ij} + 1 \leq M\beta^1_0$$

$$38 - \sum_{i=1}^{5} \sum_{j=1}^{5} p^2_{ij} t_{ij} + 1 \leq M\beta^2_0$$

$$\alpha^1_0 + \beta^1_0 + \beta^2_0 \leq 3 - 1$$
$$t_{ij} \in \{0, 1\}, \quad i, j = 1, \ldots, 5,$$
$$s^-_1 \geq 0, \ s^+_1 \geq 0, \ s^+_2 \geq 0$$

Step 3: Solve the problem (P_1). The optimal solution is $s^{-1}_1 = 21$, $s^{+1}_1 = 13$, $s^{+1}_2 = 14$, $t^1_{15} = t^1_{24} = t^1_{33} = t^1_{42} = t^1_{51} = 1$, $\alpha^{11}_0 = 1$, $\beta^{11}_0 = 0$, $\beta^{21}_0 = 1$. Hence, its associated point in the objective function is $(C(t^1), P(t^1)) = (20, 31, 31)$. Let $\mathcal{X}_E = \mathcal{X}_E \cup \{t^1\}$ and $Z_N = Z_N \cup \{(20, 31, 31)\}$. We summarize only the final results in Table 10.1 and ignore the details of further steps for the reader. The second and last columns of Table 10.1 list the MCS of efficient solutions and corresponding nondominated points in the objective space.

Phase II

To classify the efficient solutions and their associated nondominated points obtained in phase I, we utilize the BCC model to evaluate the nondominated points. Table 10.2 exhibits the details of this evaluation and resulted classification.

As is seen in Table 10.2, all nondominated points except $(C(t^4), P(t^4))$, $(C(t^5), P(t^5))$, $(C(t^6), P(t^6))$, $(C(t^9), P(t^9))$, $(C(t^{13}), P(t^{13}))$, and $(C(t^{17}), P(t^{17}))$ are BCC efficient in the objective space, and so their corresponding assignments are supported. Consequently, t^4, t^5, t^6, t^9, t^{13}, and t^{17}, are nonsupported efficient assignments.

Table 10.1 Results of phase I of the proposed algorithm.

d	t^d	s_1^{-d}	s_1^{+d}	s_2^{+d}	$(C(t^d), P(t^d))$
0	$t_{15}^{0}=1, t_{24}^{0}=1, t_{32}^{0}=1, t_{43}^{0}=1, t_{51}^{0}=1$	21	13	14	(19,23,38)
1	$t_{15}^{1}=1, t_{24}^{1}=1, t_{33}^{1}=1, t_{42}^{1}=1, t_{51}^{1}=1$	22	5	21	(20,31,31)
2	$t_{15}^{2}=1, t_{22}^{2}=1, t_{34}^{2}=1, t_{43}^{2}=1, t_{51}^{2}=1$	16	8	24	(25,26,41)
3	$t_{15}^{3}=1, t_{22}^{3}=1, t_{31}^{3}=1, t_{43}^{3}=1, t_{54}^{3}=1$	16	13	19	(25,31,36)
4	$t_{15}^{4}=1, t_{22}^{4}=1, t_{33}^{4}=1, t_{41}^{4}=1, t_{54}^{4}=1$	19	11	17	(22,29,34)
5	$t_{12}^{5}=1, t_{24}^{5}=1, t_{33}^{5}=1, t_{45}^{5}=1, t_{51}^{5}=1$	15	15	16	(26,33,33)
6	$t_{15}^{6}=1, t_{24}^{6}=1, t_{31}^{6}=1, t_{42}^{6}=1, t_{53}^{6}=1$	17	14	15	(24,32,32)
7	$t_{12}^{7}=1, t_{24}^{7}=1, t_{31}^{7}=1, t_{43}^{7}=1, t_{55}^{7}=1$	15	17	14	(26,35,31)
8	$t_{12}^{8}=1, t_{24}^{8}=1, t_{35}^{8}=1, t_{43}^{8}=1, t_{51}^{8}=1$	11	14	21	(30,32,38)
9	$t_{12}^{9}=1, t_{24}^{9}=1, t_{33}^{9}=1, t_{41}^{9}=1, t_{55}^{9}=1$	18	15	12	(23,33,29)
10	$t_{15}^{10}=1, t_{24}^{10}=1, t_{31}^{10}=1, t_{43}^{10}=1, t_{52}^{10}=1$	17	17	11	(24,35,28)
11	$t_{12}^{11}=1, t_{24}^{11}=1, t_{31}^{11}=1, t_{45}^{11}=1, t_{53}^{11}=1$	11	16	17	(30,34,34)
12	$t_{15}^{12}=1, t_{24}^{12}=1, t_{33}^{12}=1, t_{41}^{12}=1, t_{52}^{12}=1$	20	15	9	(21,33,26)
13	$t_{12}^{13}=1, t_{21}^{13}=1, t_{34}^{13}=1, t_{45}^{13}=1, t_{53}^{13}=1$	9	10	22	(32,28,39)
14	$t_{11}^{14}=1, t_{22}^{14}=1, t_{35}^{14}=1, t_{43}^{14}=1, t_{54}^{14}=1$	4	12	24	(37,30,41)
15	$t_{15}^{15}=1, t_{23}^{15}=1, t_{32}^{15}=1, t_{41}^{15}=1, t_{54}^{15}=1$	23	4	12	(18,22,29)
16	$t_{11}^{16}=1, t_{22}^{16}=1, t_{34}^{16}=1, t_{45}^{16}=1, t_{53}^{16}=1$	4	9	25	(37,27,42)
17	$t_{13}^{17}=1, t_{24}^{17}=1, t_{31}^{17}=1, t_{42}^{17}=1, t_{55}^{17}=1$	11	18	8	(30,36,25)
18	$t_{13}^{18}=1, t_{24}^{18}=1, t_{31}^{18}=1, t_{45}^{18}=1, t_{52}^{18}=1$	7	20	7	(34,38,24)
19	The model (P_{19}) is infeasible				

Table 10.2 Results of BCC evaluations for phase II.

Efficient assignment	$(C(t^*), P(t^*))$	BCC-efficiency score	Type of efficient solution and corresponding nondominated point
t^0	(19,23,38)	1	Supported
t^1	(20,31,31)	1	Supported
t^2	(25,26,41)	1	Supported
t^3	(25,31,36)	1	Supported
t^4	(22,29,34)	0.95	Nonsupported
t^5	(26,33,33)	0.96	Nonsupported
t^6	(24,32,32)	0.94	Nonsupported
t^7	(26,35,31)	1	Supported
t^8	(30,32,38)	1	Supported
t^9	(23,33,29)	0.95	Nonsupported
t^{10}	(24,35,28)	1	Supported
t^{11}	(30,34,34)	1	Supported
t^{12}	(21,33,26)	1	Supported
t^{13}	(32,28,39)	0.78	Nonsupported
t^{14}	(37,30,41)	1	Supported
t^{15}	(18,22,29)	1	Supported
t^{16}	(37,27,42)	1	Supported
t^{17}	(30,36,25)	0.91	Nonsupported
t^{18}	(34,38,24)	1	Supported

10.6 Conclusion

In this chapter, a two-phase algorithm based on the BCC and FDH models of DEA was proposed for solving the MCAP. In the first phase, MILP based on the ADD-FDH model is formulated to find an MCS of efficient assignments. Finding supported efficient assignments for multicriteria problems can be straightforward; however, classifying all efficient assignments as supported and nonsupported is a complex combinatorial problem and controversial in the literature. Thus, the result of the second phase of the algorithm is encouraging. In the second phase, the BCC model is utilized to classify all efficient assignments obtained in the first phase into supported or nonsupported efficient assignments. A great contribution initiated in this chapter to studies on MCAPs has been the ability to find all efficient assignments without enumerating them and more importantly classifying them. Further research in this direction could be to utilize the approach here for other combinatorial optimization problems.

Acknowledgments

Dr. Mehdi Toloo and Dr. Esmaeil Keshavarz are grateful for the partial support they received from the Czech Science Foundation (GACR 19-13946S) for this research

References

[1] C. Adiche, M. Aïder, A hybrid method for solving the multi-objective assignment problem, J. Math. Model. Algorithms 9 (2) (2010) 149–164.
[2] R.V.K. Ahuja, T.L. Magnanti, J.B Orlin, Network Flows: Theory, Algorithms, and Applications (1st ed), Prentice Hall, New Jersey, 1993.
[3] R.D. Banker, A. Charnes, W.W. Cooper, Some models for estimating technical and scale inefficiencies in data envelopment analysis, Manage. Sci. 30 (9) (1984) 1078–1092.
[4] M.S. Bazaraa, J.J. Jarvis, H.D. Sherali, Linear Programming and Network Flows, 4th ed., John Wiley & Sons,, New York, 2010.
[5] L. Belhoul, L. Galand, D. Vanderpooten, An efficient procedure for finding best compromise solutions to the multi-objective assignment problem, Comput. Oper. Res. 49 (2014) 97–106.
[6] A. Charnes, W.W. Cooper, E. Rhodes, Measuring the efficiency of decision making units, Eur. J. Oper. Res. 2 (1978) 429–444.
[7] M. Dell'Amico, P. Toth, Algorithms and codes for dense assignment problems: the state of the art, Discret. Appl. Math. 100 (1–2) (2000) 17–48.
[8] D. Deprins, L. Simar, H. Tulkens, Measuring labor efficiency in post offices, in: The Perfonnance of Public Enterprises: Concepts and Measurement, M. Marchand, P. Pestieau and H. Tulkens (eds.), North Holland, Amsterdam, 1984, pp. 243–267.
[9] Matthias. Ehrgott, A discussion of scalarization techniques for multiple objective integer programming, Ann. Oper. Res. 147 (1) (2006) 343–360.
[10] Mattias. Ehrgott, Multicriteria Optimization, 2nd ed, Springer Science & Business Media, Berlin/Heidelberg, 2006.
[11] A. Hatami-Marbini, M. Toloo, An extended multiple criteria data envelopment analysis model, Exp. Syst. Appl. 73 (2017) 201–219.

[12] T. Joro, P. Korhonen, J. Wallenius, Structural comparison of data envelopment analysis and multiple objective linear programming, Manage. Sci. 44 (7) (1998) 962–970.
[13] Esmaiel Keshavarz, M. Toloo, Finding efficient assignments: an innovative DEA approach, Meas.: J. Int. Meas. Confed. 58 (2014) 448–458.
[14] Esmail Keshavarz, M. Toloo, Efficiency status of a feasible solution in the multi-objective integer linear programming problems: a DEA methodology, Appl. Math. Model. 39 (12) (2015) 3236–3247.
[15] H.W. Kuhn, The Hungarian method for the assignment problem, Nav. Res. Logist. Q. 2 (1–2) (1955) 83–97.
[16] X.B. Li, G.R. Reeves, Multiple criteria approach to data envelopment analysis, Eur. J. Oper. Res. 115 (3) (1999) 507–517.
[17] F.H.F. Liu, C.C. Huang, Y.L Yen, Using DEA to obtain efficient solutions for multi-objective 0-1 linear programs, Eur. J. Oper. Res. 126 (1) (2000) 51–68.
[18] R. Malhotra, H.L. Bhatia, M.C. Puri, Bicriteria assignment problem, Oper. Res. 19 (2) (1982) 84–96.
[19] D.W. Pentico, Assignment problems: a golden anniversary survey, Eur. J. Oper. Res. 176 (2) (2007) 774–793.
[20] A. Przybylski, X. Gandibleux, M. Ehrgott, Two phase algorithms for the bi-objective assignment problem, Eur. J. Oper. Res. 185 (2) (2008) 509–533.
[21] J. Sarkis, Comparative analysis of DEA as a discrete alternative multiple criteria decision tool, Eur. J. Oper. Res. 123 (3) (2000) 543–557.
[22] Sosnowska, D. (2000). Optimization of a simplified fleet assignment problem with metaheuristics: simulated annealing and GRASP: In: Pardalos P.M. (Ed.) Approximation and Complexity in Numerical Optimization. Nonconvex Optimization and Its Applications, Springer , Boston, MA pp. 477–488.
[23] T. Stewart, Relationships between data envelopment analysis and multicriteria decision analysis, J. Oper. Res. Soc. 47 (5) (1996) 654–665.
[24] M. Toloo, E.K. Mensah, Robust optimization with nonnegative decision variables: a DEA approach, Comput. Ind. Eng. 127 (2019) 313–325.
[25] E.L. Ulungu, J. Teghem (1992). Multicriteria assignment problem – a new method. Technical report, Faculte Polytechnique de Mons, Belgium.
[26] E.L. Ulungu, J. Teghem, The two phases method: an efficient procedure to solve bi-objective combinatorial optimization problems, Found. Comput. Decis. Sci. 20 (2) (1995) 149–165.
[27] W.W. Cooper, L.M. Seiford, K. Tone, Data Envelopment Analysis: A Comprehensive Text with Models, Applications, References and DEASolver Software, second ed., Springer, 2007.

CHAPTER 11

Application of multiobjective optimization in thermal design and analysis of complex energy systems

A. Baghernejad[a] and E. Aslanzadeh[b]

[a]*Department of Mechanical and Aerospace Engineering, College of Engineering, Garmsar Branch, Islamic Azad University, Garmsar, Iran,* [b]*Department of Basic Sciences, Garmsar Branch, Islamic Azad University, Garmsar, Iran*

11.1 Introduction

Optimization is a required tool for many engineering designs. Using optimization modeling, one can find an optimum design, for instance, without the need to examine all possible cases and how each case affects the desired optimum design. That is, optimization will ensure an optimum case and reduce simulation time. To carry out optimization, some elements of optimization formulation need to be explained. These principles include system boundaries, optimization criteria, variables, mathematical model, and suboptimization.

11.1.1 System boundaries

The first step in an optimization study is to define clearly the boundaries of the systems to be optimized. These boundaries are real or imaginary surfaces that isolate the system from the surrounding. In the case of a complex system, the system could be divided into subsystems. The optimization could be performed on each subsystem independently and the optimization of these subsystems is called suboptimization.

11.1.2 Optimization criteria

The selection of criteria on the basis of which the system design will be evaluated and optimized is a key principle in formulating an optimization problem. Optimization criteria may be economic (total capital investment, total annual levelized cost, annual levelized net profit, return on investment or any of the profitability evaluation criteria), technological (thermodynamic efficiency, production time, production rate, reliability, total weight, etc.), and environmental (e.g., rate of emitted pollutants). An optimized design is determined by a minimum or maximum value, as appropriate, for each selected criterion. In practice, it is usually desirable to develop a design that is "best" with respect to more than one criterion. As these criteria usually compete with each other, it is impossible to find a solution that, for instance, simultaneously minimizes costs and environmental impact while maximizing efficiency and reliability. Advanced techniques for solving certain optimization problems with multiple competing criteria are brought up in the literature [1-3].

11.1.3 Variables

Another vital element in formulating an optimization problem is the selection of the independent variables that adequately specify the possible design options in selecting these variables, it is required to (1) include all the important variables that affect the performance and cost effectiveness of the system, (2) not include variables of minor importance, and (3) distinguish among independent variables whose values are inclined to change: decision variables and the parameters whose values are fixed by the particular application. In optimization studies, only the decision variables may be varied; the parameters are independent variables that are each given one specific and unchanging value in any particular model statement. The variables whose values are obtained from independent variables using the mathematical model are the dependent variables.

In some thermal systems, there are design variables that affect the product cost without affecting the thermodynamic performance (efficiency) of the system. As a simplification in practice, these variables often may be considered separately and optimized after the decision variables affecting both cost and efficiency have been optimized. Also, in selecting these variables, the practical range where these variables are operated needs to be identified.

11.1.4 The mathematical model

A mathematical model is a description in terms of mathematical relations, invariably involving some idealization of the functions of a physical system. The mathematical model describes the manner in which all problem variables are related and the way in which the independent variables affect the performance criterion. The mathematical model for an optimization problem consists of the following:

- Optimization criteria: objective function(s) to be minimized or maximized; and
- Equality and inequality constraints.

The objective function expresses the optimization criterion as a function of the dependent and independent variables.

The equality and inequality constraints are provided by suitable thermodynamic and cost models as well as by proper boundary conditions. These models generally include material and energy balance equations for each component, relations associated with the engineering design such as maximum or minimum values for temperature and pressure, the performance and the cost of each component as well as physical and chemical properties of substances involved. The models also contain equations and inequalities that specify the allowable operating ranges, the maximum or minimum performance requirements, and the bounds on the availability of resources.

The development of mathematical model requires a good comprehension of the system and is normally a time-consuming activity.

11.1.5 Suboptimization

Suboptimization is usually used for complex thermal systems, specifically when the optimization of the whole system may not be feasible owning to complexity. Suboptimization is the optimization of one part of a problem or of a subsystem, ignoring some variables that influence the objective function or other subsystems. Suboptimization is useful when neither the problem formulation nor the available optimization techniques allow a solution to the entire problem. In practice, suboptimization of a system may also be important because of economic and practical considerations such as limitation on time or manpower. However, suboptimization of all subsystems separately does not necessarily ensure optimization of the overall system.

11.2 Types of optimization problems

A function optimization problem may be of different types, depending on the desired aim of the optimization task. The optimization problem may have only one objective function (known as a single-objective (SO) optimization problem), or it may have multiple conflicting objective functions (known as a multiobjective optimization problem).

11.2.1 Single-objective optimization

Many real-world decision-making problems need to achieve several objectives: minimize risks, maximize reliability, minimize deviations from desired levels, minimize cost, etc. The main aim of SO optimization is to find the "best" solution, which corresponds to the minimum or maximum value of an SO function that lumps all different objectives into one. This type of optimization is useful as a tool that should provide decision makers with insights into the nature of the problem, but usually cannot provide a set of alternative solutions that trade different objectives against each other.

The use of an SO function, which is usually a weighted combination of several objectives, does not facilitate the judgment of the decision maker as it is often difficult to interrelate several objectives of different natures properly. Moreover, it is not often proper to claim that there is only one optimal solution because the tradeoffs between the conflicting objectives are important. Without tradeoff so, certain optimal solution may be lost as they may never be explored. On the contrary, using the multiobjective techniques, a set of several nondominated optimal solutions will be obtained which known as a Pareto-optimal set. As a result, the solution is not a single mathematically figure but a set of efficient solutions that can be examined using a judgment of the tradeoffs involved, giving the decision maker more flexibility.

11.2.2 Multiobjective optimization

A multiobjective optimization problem requires the simultaneous satisfaction of a number of different and often conflicting objectives. These objectives are characterized by specified measures of performance that may be (in) dependent and/or incommensurable. For thermal system design, for instance, the energetic efficiency and CO_2 emission or productions unit cost may have little dependence on each other. A global optimal solution to a multiobjective optimization problem is unlikely to exist: this means that there is no combination of decision variable values that minimizes all the conflicting objectives simultaneously. Multiobjective optimization problems generally demonstrate a possibly uncountable set of solutions, whose evaluated vectors represent the best possible tradeoffs in the objective function space. Pareto optimality [4] is the key concept to establish a hierarchy among the solutions of a multiobjective optimization problem to determine whether or not a solution is really one of the best possible tradesoff. The set of Pareto-optimal solutions is also called the Pareto-optimal set and the set of the corresponding evaluated vectors in the objective function space is called the Pareto frontier. The search space of the optimization is defined by the decision variables and their bounds. The solution of such a multiobjective optimization problem is a set of points in the decision variables space that expresses the possible tradeoff between the objectives. In the domain of the objective functions, this tradeoff is represented by the Pareto frontier that shows the set of nondominated solutions, which delimits the unfeasible domain from the feasible but suboptimal one. In the case of multiple objectives optimization, it does not necessarily exist a solution that is best with respect to all objectives because of differentiation between objectives. A solution may be best in one objective but worst in another. Therefore, there exists usually a set of solutions for the multiple-objective case, which cannot simply be compared with each other. For such solutions, called Pareto-optimal solutions or nondominated solutions, no improvement is possible in any objective function without sacrificing at least one of the other objective functions. The optimal tradeoff solutions of certain conflicting objective criteria are valuable for the decision maker to choose the best solution suited to his/her needs [3].

11.2.2.1 Multiobjective evolutionary algorithms

The two issues in multiobjective optimization are: (1) to find solutions close to the true Pareto-optimal set and (2) to find solutions that are widely different from each other to cover the entire Pareto-optimal set as well as not to introduce bias toward any particular objective. Classical search and optimization methods are not efficient in following the Pareto approach to multiobjective optimization for three main reasons [5]:

1. Most of them are unable to find multiple solutions in a single run, so they have to be applied as many times as the number of Pareto-optimal solutions required.
2. The solutions found through the repeated application of these methods are not guaranteed to be widely different from each other.
3. Most of them are unable to handle problems having multiple optimal solutions; for example, difficulties arise using the linear objective function aggregation technique, in which an objective function is built by weighting the original objective functions through priority coefficients, or searching the optimal solutions by treating all the objective functions, except one, as constraints.

The class of search algorithms that implement the Pareto approach for multiobjective optimization in the most straightforward way is the class of multiobjective evolutionary algorithms (MOEAs). MOEAs have been developed over the past decade [6]; several tests on complex mathematical problems and on real-world engineering problems have shown that they can eliminate the above-cited difficulties of classical methods. An evolutionary algorithm is an optimization tool that imitates the natural evolution of biological organisms, that is, a randomly initialized population of individuals (a set of points in the search space) evolves following the Darwinian principle of survival of the fittest. New individuals are created using some simulated evolutionary operators, such as crossover and mutation, and the probability of survival for these newly generated solutions depends on their fitness, that is on how well they perform with respect to the objective(s) of the optimization problem.

11.2.2.2 Multiobjective Data Envelopment Analysis

Data Envelopment Analysis (DEA) is a mathematical programming technique useful for evaluating the relative efficiency of a homogeneous set of decision-making units (DMUs) in a production system with multiple inputs and outputs. Subsequent developments have proved DEA as a useful tool for performance evaluation in a wide number of fields, with interesting applications in engineering, manufacturing, education, banking, etc. In the DEA methodology, for each DMU an efficiency score is computed as a ratio of a weighted sum of outputs to a weighted sum of inputs, with such set of weights found to guarantee the most favorable result for the DMU under assessing. According to this score, every DMU is either found to perform efficiently or deemed inefficient, in which case DEA can find a corresponding set of efficient units to be used as a benchmark for improvement. The problem tackled by DEA highly resembles the one studied within Multicriteria Decision Making, in which a number of alternatives have to be evaluated and compared in terms of several conflicting criteria to achieve a ranking of the alternatives and/or select the best option [7]. In fact, ranking a set of alternatives in real-world applications often turns into a rather overwhelming problem for many decision makers that may require expertise and computational support, particularly when the number of alternatives and criteria grow, and so the problem has been extensively studied. However, when using standard DEA techniques with a ranking purpose some difficulties may arise. First, DEA efficiency scores do not always allow a complete ordering of the alternatives as many of the DMUs are usually classified as efficient. The lack of discrimination in DEA applications is well documented, particularly when the number of inputs and outputs is too high relative to the number of DMUs being evaluated [7].

11.3 Optimization of energy systems

Engineers involved thermal system design is required to answer question such as: what process or equipment item should be selected and how should they be arranged? What is the preferred size of a component or group of components? What is the best temperature, pressure, flow rate, and chemical composition of each stream in the system? To answer these questions, engineers need to formulate an appropriate optimization problem. Appropriate problem formulation is usually the most important and sometimes the most difficult step of a successful optimization study.

11.3.1 Thermodynamic optimization and economic optimization

Two types of optimization are considered in design of energy systems: thermodynamic and economic. The objective of thermodynamic optimization is to maximize exergetic efficiency of the system. In the economic optimization, the objective is to minimize the levelized costs of the product streams (electricity, heat, and cold) generated by the combined cooling, heating and power (CCHP) system. Thermodynamic optimization is based on thermodynamic model, whereas the economic optimization employs both the thermodynamic model and the cost model. Before proceeding with the thermodynamic or economic optimization of an overall system, the design engineer may use the thermodynamic model to study the effect on the overall exergetic efficiency of a change in some of the decision variable, while keeping all other variables constant.

11.3.2 Thermoeconomic optimization

Complex thermal systems cannot always be optimized using the mathematical techniques. The reasons include incomplete models, system complexity, and structural changes:

- Some of the input data and functions required for the thermodynamic and, particularly, the economic model might not be available or might not be in the required form. For example, it is not always

possible to express the purchased-equipment costs as a function of the appropriate thermodynamic decision variables.
- Even if all the required information is available, the complexity of the system might not allow a satisfactory mathematical model to be formulated and solved in a reasonable time.
- The analytical and numerical optimization techniques are applied to a specific structure of the thermal system. However, the significant decrease in the product costs may be achieved through changes in the structure of the system. It is not always practical to develop a mathematical model for every promising design configuration of a system. More importantly, analytical and numerical optimization techniques can not suggest structural changes that have the potential of improving the cost effectiveness.

As the application of the thermoeconomic evaluation techniques improves the engineer understanding of the interactions among the system variables, and generally reveals opportunities for design improvements that might not be detected by other methods, these techniques are always recommended regardless of the optimization approaches, however.

Thermodynamic optimization is minimizing the thermodynamic inefficiencies in the system [8]. The thermodynamic inefficiencies are the exergy destruction and exergy loss. On the other hand, thermoeconomic optimization is minimizing the costs, including the cost of the thermodynamic inefficiencies. Thermoeconomics can be considered as exergy-aided cost minimization. The optimal design of a system is characterized by a maximum or minimum value of one or more selected criteria. The other criteria (the nonselected criteria) are considered as problem constraints [8]. Different criteria can be selected for optimization. For example, when the overall thermal exergy efficiency of a plant is selected as an optimization criterion, the optimization will provide the best operating conditions within the simulation constraint(s) that result in the highest exergy efficiency. However, this optimization criterion does not consider the cost effect and it could result in high operating cost conditions. Similarly, finding the best operating condition(s) for the lowest operating cost could result in low exergy efficiency. Therefore, there is a need to find a parameter that takes into consideration both the cost and exergy of the plant considered for optimization. This parameter can be obtained through thermoeconomic analysis [9]. As mentioned in [9], the cost per exergy unit of the final product is an important parameter to optimize. Therefore, it is considered as the thermoeconomic optimization objective.

11.3.2.1 Single-objective thermoeconomic optimization

In general, a thermal system requires two conflicting objectives: one being increase in exergetic efficiency and the other is decrease in product cost, to be satisfied simultaneously. The first objective is governed by thermodynamic requirements and the second by economic constraints. Therefore, objective function should be defined in such a way that the optimization satisfies both requirements. For that, the optimization problem should be formulated as a minimization or maximization problem. The exergoeconomic analysis gives a clear picture about the costs related to the exergy destruction, exergy losses, etc. Thus, the objective function becomes a minimization problem. Baghernejad and Yaghoubi [10,11] demonstrated the use of this technique for optimizing an integrated system using genetic algorithm. In the same manner, the optimization of the system requires the solution of a minimization problem.

11.3.2.2 Multiobjective thermoeconomic optimization

A multiobjective optimization problem requires the simultaneous satisfaction of a number of different and often conflicting objectives. The cost of the final system is always an important factor for the design of plants. Other important factors are often related to the efficiency and impact of the system on the environment. Hence, while trying to minimize the cost of the plant, the engineer will try to maximize the exergy efficiency and minimize the unit cost of products and emissions of the plant [12-14].

11.4 Literature survey on the optimization of complex energy systems

Economic performance, energy savings, and emission reduction can be formulated by using indices that compare the benefits of trigeneration with energy systems by conventional means. It is obvious that the design of such a system is associated with the above conflicting objectives. For a detailed overview, the role of optimization modeling techniques for power generation is reviewed in [15]. However, most of these optimization models only included a single economic objective function, completed with environmental and energetic targets as constraints, rather than multiobjective optimization. Shabani and Sowlati [16] developed a mathematical model to optimize the supply chain of a forest biomass byproducts power plant by using a mixed integer nonlinear model. Other researchers developed an optimization model for the integration of power generation and heating plants to the cooling systems based on temperature intervals [17]. A SO optimal design of trigeneration plants and effect of various operation parameters on the investment cost is also studied in [18]. Some limitations related to the simultaneous consideration of the economic evaluation and the CO_2 emissions assessment which may appear by available optimization methods are developed in [19].

11.5 Thermodynamic modeling of energy systems

11.5.1 Mass balance

The conservation of mass principle is a fundamental principle in analyzing any thermodynamic system. This principle is defined for a control volume as

$$\sum_k \dot{m}_i - \sum_k \dot{m}_e = \frac{dm_{cv}}{dt} \tag{11.1}$$

where m and $\dot{m}$ are the mass and mass flow rate, respectively, and the subscripts i and e refer to the inlet and exit of the control volume, respectively. The subscript cv indicates the control volume.

11.5.2 Energy balance

The energy principle of a control volume deals with all the energy components of a selected control volume. The conservation of energy principle, which is known as the first law of thermodynamics, is defined as

$$\dot{Q} - \dot{W} + \sum_i \dot{m}_i \left(h_i + \frac{v_i^2}{2} + gz_i \right) - \sum_e \dot{m}_e \left(h_e + \frac{V_e^2}{2} + gz_e \right) = \frac{dE_{cv}}{dt} \tag{11.2}$$

where E, $\dot{Q}$, $\dot{W}$, and t are the energy, heat transfer rate, power and time, respectively. The other symbols, h, V, g, and z, stand for specific enthalpy, velocity, gravity, and elevation, respectively.

11.5.3 Entropy balance

Entropy generation is associated with the losses in the system. The entropy generated within a process is called entropy generation and it is denoted by $\dot{S}_{gen}$. The entropy generation rate for a control volume is defined as

$$\dot{S}_{gen} = \sum_e \dot{m}_e s_e - \sum_i \dot{m}_i s_i - \sum_k \frac{\dot{Q}_k}{T_k} + \frac{dS_{cv}}{dt} \tag{11.3}$$

where S is the entropy and s is the specific entropy.

11.5.4 Exergy balance

Unlike energy, exergy is not conserved. It is defined as the maximum work that could be obtained from a system at a given state. To understand the exergy, reversible work should be defined first. Reversible work is the maximum useful work that can be obtained as a system goes through a process between two defined states. Another exergy terminology is the exergy destruction. It is defined as the potential work lost due to irreversibility. The exergy balance of a control volume system is defined as

$$\frac{dEx_{cv}}{dt} = \sum_j \dot{Q}_j \left(1 - \frac{T_0}{T_j}\right) - \left(W_{cv} - p_0 \frac{dV_{cv}}{dt}\right) + \sum_i \dot{m}_i ex_i - \sum_e \dot{m}_e ex_e - \dot{E}x_d \tag{11.4}$$

where T, p, V, ex, and $\dot{E}x_d$ are temperature, pressure, volume, specific exergy, and rate of exergy destruction, respectively. The subscript j is the property value at state j and the subscript 0 is the value of a property at the surrounding. The physical exergy, $exph$, at a given state is defined as

$$ex^{ph} = (h - h_0) - T_0(s - s_0) + \frac{v_2 - v_0^2}{2} + g(z - z_0) \tag{11.5}$$

The chemical exergy of an ideal gas is defined as

$$ex_j^{ch} = x_j \overline{ex}_j^{ch} + RT_0 x \ln(x_j) \tag{11.6}$$

where $\overline{ex}_j^{ch}$ is the standard chemical exergy value of species j.

11.5.5 Energy efficiency

The energy efficiency is a measure of the useful energy from a system to the input energy for this system. The energy efficiencies of different systems are defined in the following text. The thermal efficiency of a thermal cycle is defined as

$$\eta_{cycle} = \frac{W_{cycle}}{Q_i} = 1 - \frac{Q_e}{Q_i} \tag{11.7}$$

The isentropic thermal efficiency of work-producing devices is defined as

$$\eta_{is} = \frac{W_{ac}}{W_{is}} \tag{11.8}$$

On the other hand, the isentropic thermal efficiency of work-consuming devices is defined as

$$\eta_{is} = \frac{W_{is}}{W_{ac}} \tag{11.9}$$

The performance of refrigerators is known as the coefficient of performance (COP). It is defined as

$$COP_R = \frac{\text{desired output}}{\text{required output}} = \frac{Q_L}{Q_H - Q_L} = \frac{1}{\frac{Q_H}{Q_L} - 1} \tag{11.10}$$

where the subscripts *is, ac, R, H,* and *L* indicate isentropic, actual, refrigerator, high temperature reservoir, and low temperature reservoir, respectively.

11.5.6 Exergy efficiency

The exergy efficiency is defined as the ratio of the actual thermal efficiency to the maximum reversible thermal efficiency when both are under the same conditions. In general, the exergy efficiency can be defined as

$$\eta_{ex} = \frac{\text{Exergy recovered}}{\text{Exergy supplied}} = 1 - \frac{\text{Exergy destroyed}}{\text{Exergy supplied}} \tag{11.11}$$

The exergetic efficiencies of different systems are defined as follows. For heat engines the exergetic efficiency is defined as

$$\eta_{ex} = \frac{\eta_{th}}{\eta_{rev}} \tag{11.12}$$

The exergetic efficiency of work-producing devices is defined as

$$\eta_{ex} = \frac{W_{cv}}{W_{rev}} = \frac{W_{cv}}{Exi_i - Ex_e} \tag{11.13}$$

For work-consuming devices, the exergetic efficiency is defined as

$$\eta_{ex} = \frac{W_{rev}}{W_{cv}} = \frac{Exi_i - Ex_e}{W_{cv}} \tag{11.14}$$

For heat exchanger, the exergetic efficiency is defined as

$$\eta_{ex} = \frac{Ex_{cold.e} - Ex_{cold.i}}{Ex_{hot.i} - Ex_{hot.e}} \tag{11.15}$$

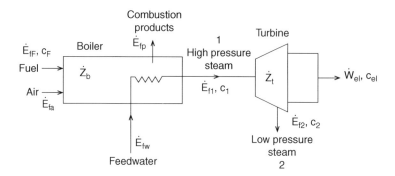

FIGURE 11.1

Simple cogeneration system.

11.6 Thermoeconomics methodology for optimization of energy systems

In the analysis and design of energy systems, the techniques such as thermoeconomic, which combine scientific disciplines with economic disciplines to achieve optimum design, are growing in the energy industries [20]. Exergoeconomic analysis is a powerful scheme that combines exergy analysis with economic studies. This method provides a technique to evaluate the cost of inefficiencies or cost of individual process streams, including intermediate and final products of any system. Also it can be utilized in optimization of thermodynamic systems, in which the task is usually focused on minimizing the unit cost of the system product. Exergy costing involves cost balances formulated for each system component separately. For a kth component receiving heat transfer, q, and generating power, w, the thermoeconomic balance equation of this component is [9]:

$$\sum_e (\dot{c}_{e,k}) + \dot{c}_{w,k} = \dot{c}_{q,k} + \sum_i (\dot{c}_{i,k}) + \dot{Z}_k \tag{11.16}$$

where $\dot{Z}_k$ is the levelized cost rate to own, maintain, and operate the kth component. Here $\dot{C}$ is the cost rate in $ per hour, for example. For exergy costing, $\dot{C}$ is defined as

$$\dot{C} = c\dot{E}x_j \tag{11.17}$$

where c is the cost per unit of exergy in $ per kW/h, or $ per GJ, for instance, and $\dot{E}x$ is the exergy transfer rate. In general, if there are Ne exergy streams exiting the component being considered, we have Ne unknowns and only one equation, the cost balance. Therefore, we need to formulate $Ne - 1$ auxiliary equations. This is accomplished with the aid of the F and P principles in the SPECO approach [21]. Thermoeconomics can be best explained through an example. The example on thermoeconomics from Moran and Shapiro [22] is adopted to explain it. Assume that a simple cogeneration system consists of a boiler and turbine, as shown in Fig. 11.1.

Apply the cost analysis on the boiler to get

$$\dot{C}_1 + \dot{C}_P = \dot{C}_F + \dot{C}_a + \dot{C}_w + \dot{Z}_b \tag{11.18}$$

11.6 Thermoeconomics methodology for optimization of energy systems

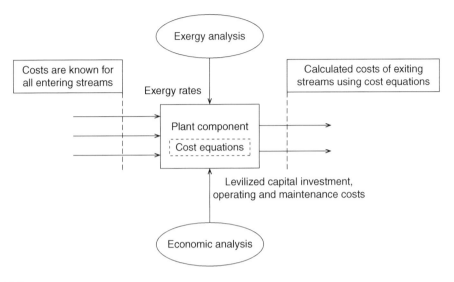

FIGURE 11.2

SPECO method [21].

where $\dot{C}$ is the cost rate and $\dot{Z}$ is the cost rate associated with owning and operating the boiler each in $ per hour, for instance. The subscripts 1, p, F, a, w, and b indicate stream at the exit of the boiler, combustion products, fuel, air, feed-water, and boiler, respectively. For simplicity, the exergy and cost of the feedwater and combustion air that enter the boiler are assumed to be negligible. Using this assumption, substitute Equation (11.17) into Equation (11.18) to obtain

$$c_1 \dot{E}x_{f1} = c_F \dot{E}_{fF} + \dot{Z}_b \tag{11.19}$$

Similarly, apply cost analysis on the turbine to obtain

$$\dot{C}_{el} + \dot{C}_2 = \dot{C}_1 + \dot{Z}_t \tag{11.20}$$

where the subscripts *el* and 2 refer to electricity and low-pressure steam exit, respectively. The variable $\dot{Z}_t$ refers to the cost rate of the operating and owing the turbine. Now, redefine Equation (11.20) in terms of cost and exergy to obtain

$$c_{el}\dot{W}_{el} + c_2 \dot{E}x_{f2} = c_1 \dot{E}x_{f1} + \dot{Z}_t \tag{11.21}$$

11.6.1 The SPECO method

The SPECO method refers to specific exergy costing. In this study, the SPECO method is used to carry out the thermoeconomic analysis (see Fig. 11.2). The SPECO method is selected in this study as it is the most widely accepted thermoeconomic method in the literature. An overview of the SPECO method was presented in [21]. The discussion of the SPECO method in this section is based on this reference. This method consists of the following three steps.

1. *Identification of exergy streams*:
   In this step, all exergy streams associated with the entering and exiting exergy streams of each component are identified and calculated. These exergy streams are in the power unit and not in power per mass flow rate.
2. *Definition of fuel and product costs*:
   The product cost is defined to be equal to the sum of
   - All the exergy values at the exit, plus
   - All the exergy increases between inlet and exit. Likewise, the fuel cost is defined to be equal to
   - All the exergy values at the inlet, plus
   - All the exergy decreases between inlet and exit, minus
   - All the exergy increases between the inlet and exit of a component that are not related to the component purpose.
3. *Identification of cost equations*:
   Cost equations are the cost rate and cost per exergy unit equations, as defined above at the beginning of this section and through Eqs. (11.16)–(11.21).

11.6.2 The F (fuel) and P (product) rules

When carrying thermoeconomic analysis using the SPECO method, usually further auxiliary equations are needed. These equations are obtained using the F and P rules, according to [21]. The F rule, for a considered component, refers to the removal of exergy from an exergy stream within the component when, for this stream, the exergy difference between inlet and outlet is considered in the definition of the fuel. The F rule states that the specific cost (cost per exergy unit) associated with this removal of exergy from a fuel stream must be equal to the average specific cost at which the removed exergy was supplied to the same stream in upstream components. The P rule for a considered component refers to the supply of exergy to an exergy stream within the component. The P rule states that each exergy unit is supplied to any stream associated with the product at the same average cost.

11.7 Sensitivity analysis of energy systems

Most energy systems are evaluated using sensitivity scenarios. A sensitivity analysis has been performed on the Pareto-optimal solutions to study the effect of some physical and economic parameters such as ambient temperature, interest rate, fuel cost, construction period, and operation period on the unit cost of products and efficiency.

11.8 Example of application (case study)

11.8.1 Integrated biomass trigeneration system

In the state of increasing the share of renewable energies to mitigate global warming and with respect to the global issue of sustainable energy development, biomass byproduct increasingly focused on as a potential source of renewable energy. There is a wide range of biomass resources potentially available to electricity or other products. This includes biodegradable fraction of products, waste, and residues from agriculture (including vegetable and animal substances), forestry, and related industries, as well as the biodegradable fraction of industrial and municipal waste. Presently, there is a diverse array of technologies to convert biomass resources into higher value products such as liquid and gaseous fuels or chemical products via thermochemical, biochemical, or chemical means [23]. However, most of these technologies, although feasible, are not yet cost competitive [24]. Currently, biomass resources are mainly used in the production of heat and electricity [25,26]. Biomass byproducts can be used as a fuel for a biomass combustor. One of the most common waste wood products is pine sawdust. Pine trees grow widely throughout the world and, thus, they are widely used for wood-based products. Pine sawdust is produced as a result of pine wood processing. This wasted sawdust is commonly used for biomass combustion. The trigeneration system consists of a biomass combustor, a combined cycle (gas and steam cycles) for electrical power production, a single-effect absorption chiller for cooling, and a heat exchanger for heating as shown in Fig. 11.3. The heat produced from the biomass combustor is used as a heat input to boil preheated water of economizer in heat recovery steam generator. In turn, the waste heat from the combined cycle is utilized to produce steam in the heating process using the heat exchanger and to produce cooling using the single-effect absorption chiller.

In the present study, the thermodynamic modeling, exergoeconomic analysis, and multiobjective optimization of the integrated biomass trigeneration plant is conducted. Two complete objective functions including the exergy efficiency (to be maximized) and the unit cost of products (to be minimized) are considered. Moreover, the sensitivity analysis of the changes in both objective functions with variations of important economic parameters for optimal solutions is carried out in details. To carry out the modeling and analysis of the system considered, several assumptions are used. It is assumed that the system is at a steady state. In addition, it is assumed that the combustion is complete and, hence, the combustion exhaust has only CO_2 as a greenhouse gas. The performance analysis is applied to a single-effect absorption chiller, which is similar to that used by Herold et al. [27]. More assumptions used in the single-effect absorption chiller are as follows:

- Pure water is used as a refrigerant (states 39–42).
- The liquid phases at states 33, 36, and 40 are considered saturated liquid.
- The water at state 42 is saturated vapor.
- The pressures in the generator and condenser are equivalent.
- The pressures in the evaporator and the absorber are equivalent.

11.8.1.1 Thermodynamic analysis

In this study, numerical results are based on a site design condition with ambient temperature of 25 °C. The biomass fuel selected for the trigeneration system is pine sawdust. The chemical compound of a biomass fuel is $C_{ZC}H_{ZH}O_{ZO}S_{ZS}$. The elements C, H, O, and S refer to carbon, hydrogen, oxygen, and sulfur, respectively. The subscripts of these components represent the percentage of these elements in the fuel compound. The fixed thermodynamic and economic input data for trigeneration system are listed in Table 11.1.

In this model, the variables selected for the optimization are as follows:

- Biomass share percent in the trigeneration system $\left(\% \text{ Biomass share} = \frac{\dot{H}_{30} - \dot{H}_{31}}{\dot{W}_{tot}}\right)$

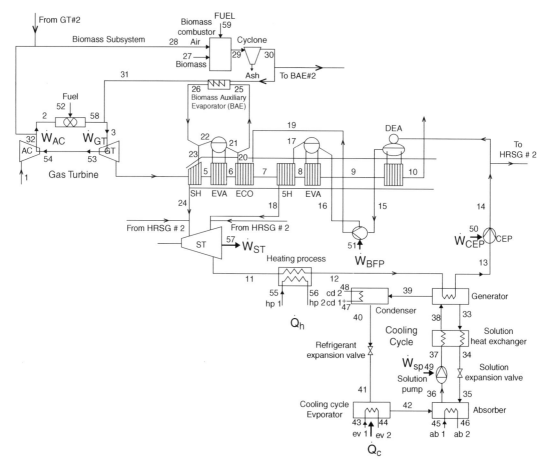

FIGURE 11.3
Schematic diagram of the integrated biomass trigeneration system.

Table 11.1 Thermodynamic and economic input data to analyze and optimization of the integrated biomass trigeneration system.

Thermodynamic input data		Economic input data	
Total power (MW) (power, heat and cold)	600	Maintenance factor (φ)	1.06
Effectiveness of solution heat exchanger	0.7	Interest rate (in),%	10
Evaporator temperature (°C)	6.8	Operation period (H), hour	2000
Condenser temperature (°C)	30.2	Construction period, year	3
Generator temperature (°C)	95	Fuel cost, ($/MWh)	120

- Ratio of produced electricity to heating $\left(r_{el/h} = \frac{\dot{W}_{tot}}{\dot{Q}_{heating}}\right)$
- Ratio of produced electricity to cooling $\left(r_{el/c} = \frac{\dot{W}_{tot}}{\dot{Q}_{cooling}}\right)$
- Isentropic efficiency of the compressor (η_{AC})
- Pressure ratio of the compressor $\left(P_r = \frac{P_1}{P_2}\right)$
- Temperature of the combustion in the gas turbine and biomass byproducts subsystem (T_3, T_{29})
- Isentropic efficiency of gas turbine (GT) and steam turbine (ST) (η_{GT}, η_{ST})
- Temperature and pressure of the steams leaving the heat recovery steam generator (T_{24}, P_{24}, P_{18})
- Moisture content in the biomass byproducts (%ω).

The present model with the specification given in Table 11.1 was treated as a base case and the following nominal values of the decision variables are selected for the optimization based on the operation program of the system.

$$\begin{aligned}
&\% \text{ Biomass share} = 20 \quad r_{el/h} = 4 \quad r_{el/c} = 5 \quad P_{18} = 9\,\text{bar} \\
&T_c = T_3 = T_{29} = 1400\text{K} \quad T_{24} = 780.15\text{K} \quad P_{24} = 84\,\text{bar} \quad \%\omega = 10 \\
&\eta_T = \eta_{GT} = \eta_{ST} = 0.85 \quad P_r = 11 \quad \eta_{AC} = 0.75
\end{aligned}$$

In this case study, the total power of the trigeneration system is 600 MW and is fed with natural gas and biomass (pine sawdust). The heat and cold produced are about 100 and 80 MW and it can be used for district heating and cooling, heat and cold storage systems, etc.

11.8.1.2 Exergoeconomic principles

The term thermoeconomics is defined usually as the methodologies combing thermodynamic properties and economics parameters to obtain a better design and operation of a thermal system. As thermoeconomics is based on exergy and cost, it is sometimes called exergoeconomics.

The system showed in Fig. 11.3 consists of 23 components and has 59 streams (53 for mass and 6 for work). Therefore 36 boundary conditions and auxiliary equations are necessary. For example in this system, cost balance (Eq. 11.16) and auxiliary costing equations (according to P and F rules) for ST and biomass heat exchanger (BHE) are formulated as follows [9]:

ST:

$$\dot{C}_{11} + \dot{C}_{57} = 2(\dot{C}_{18} + \dot{C}_{24}) + \dot{Z}_{ST} \tag{11.22}$$

$$c_{18} + c_{24} = c_{11} \text{ or } \frac{\dot{C}_{18} + \dot{C}_{24}}{\dot{E}_{18} + \dot{E}_{24}} = \frac{\dot{C}_{11}}{\dot{E}_{11}} \tag{11.23}$$

BHE:

$$\dot{C}_{26} + \dot{C}_{31} = \dot{C}_{25} + \dot{C}_{30} + \dot{Z}_{BHE} \tag{11.24}$$

$$c_{30} = c_{31} \text{ or } \frac{\dot{C}_{30}}{\dot{E}_{30}} = \frac{\dot{C}_{31}}{\dot{E}_{31}} \tag{11.25}$$

In the same way, developing Equation (11.16) for other element of plant along with auxiliary costing equations (according to P and F rules) leads to a system of 59 equations [21]. By solving the system of 59 equations and 59 unknowns, the cost of unknown streams of the system will be obtained. More details of the cost balance equations and exergoeconomic analysis are completely discussed in references [28].

11.8.1.3 Multiobjective exergoeconomic optimization

The objective functions of the multicriteria optimization for the present study are exergetic efficiency of the trigeneration system (to be maximized), the unit cost of productions of system (to be minimized), and CO_2 emissions cost (to be minimized). It should be noted that the environmental impact as the total cost rate of pollution damage (\$/hour) due to CO_2 emissions is determined by multiplying their respective flow rates by their corresponding unit damage costs. The unit damage costs due to CO_2 emission (C_{CO2}) is equal to 0.024 \$/kg [28]. The mathematical formulation of objective functions is as follows:

- Exergetic efficiency of the trigeneration system (to be maximized)

$$\eta_{ex,Trigeneration} = \frac{\dot{W}_{net} + \dot{Q}_h\left(1 - \frac{T_0}{T_h}\right) + \dot{Q}_c\left(1 - \frac{T_0}{T_c}\right)}{\dot{E}x_{f,tot}} \qquad (11.26)$$

- Unit cost of productions of trigeneration system (to be minimized)

$$^c products = {^c electricity} + {^c heating} + {^c cooling} \qquad (11.27)$$

- Total cost rate of pollution damage due to CO_2 emissions (to be minimized)

$$\dot{C}_{CO_2} = \dot{m}_{CO_2} c_{CO_2} \qquad (11.28)$$

To minimize the environmental impacts, the primary target is to increase the efficiency of energy conversion processes, and as a result, decreasing the amount of fuel and the related overall environmental impacts, especially the release of carbon dioxide, which is one of the main components of greenhouse gases. Therefore, in the present case study, the cost of pollution damage will be minimized simultaneously in multiobjective optimization process that maximizes exergy efficiency and minimizes unit cost of products in trigeneration system.

11.8.2 Results and discussion

The state properties and exergy rates of each line of the base case integrated biomass trigeneration system according to Fig. 11.3 are determined with a thermodynamic analysis and given in Table 11.2. In optimization procedure, although the decision variables may be varied, but each decision variable is normally required to be within a given practical range of operation as follows [29]:

$$10 < \% \text{ Biomass share} < 60 \qquad 80 \leq P_{24} \leq 100 \text{ bar}$$

$$2 < r_{el/h} < 10 \qquad 723.15 \leq T_{24} \leq 823.15 \text{ K}$$

$$2 < r_{el/c} < 10 \qquad 7 \leq P_{18} \leq 10 \text{ bar}$$

$$10 < \%\omega < 60 \qquad 0.7 \leq \eta_{AC} \leq 0.85$$

$$10 \leq pr \leq 15 \qquad 0.75 \leq \eta_T = \eta_{ST} = \eta_{GT} \leq 0.9$$

$$1200 \leq T_C = T_3 = T_{29} \leq 1500 \text{ K}$$

Table 11.2 State properties of integrated biomass trigeneration system corresponding to Fig. 11.3 in the base case design.

State	$\dot{m}$(kg/s)	T (K)	h (kJ/kg)	$\dot{E}$ (MW)	State	$\dot{m}$ kg/s	T (K)	h (kJ/kg)	$\dot{E}$(MW)
1	396.97	298.15	298.61	0	31	87.07	1074.8	876	51.9
2	373.46	689.32	405.96	56.87	32	396.97	689.32	574.71	60.45
3	467.14	1400	1179	417.65	33	458.46	368.15	239.43	42.75
4	467.14	835.26	574.71	116.44	34	458.46	322.79	137.68	28.43
5	467.14	736.27	468.79	86.54	35	458.46	302.79	137.68	43.88
6	467.14	543.52	262.55	34.25	36	493.13	303.35	94.08	23.33
7	467.14	482.18	196.91	21.43	37	493.13	303.35	94.08	23.33
8	467.14	479.52	194.06	20.12	38	493.13	345.52	188.67	33.51
9	467.14	434.19	145.57	12.26	39	34.67	368.15	2667.6	15.99
10	467.14	386.15	94.16	5.72	40	34.67	303.35	126.58	0.0031
11	167.01	320.83	2305.1	25.42	41	34.67	279.95	126.58	−0.14
12	167.01	320.83	1685.6	18.1	42	34.67	279.95	2513.4	−5.52
13	167.01	320.83	199.65	0.55	43	439.86	343.15	293.07	5.68
14	167.01	321.13	203.07	0.99	44	439.86	298.15	104.92	0
15	83.5	390.06	490.66	4.14	45	552.11	298.15	104.92	0
16	9.92	390.19	491.71	0.5	46	552.11	343.15	293.07	7.13
17	9.92	449.44	2773.9	8	47	468.28	298.15	104.92	0
18	9.92	505.43	2907.9	8.47	48	468.28	343.15	293.07	6.04
19	73.58	391.98	507.08	4.63	49	–	–	–	0.00095
20	73.58	488.15	923.78	14.45	50	–	–	–	0.57
21	53.05	488.15	923.78	10.42	51	–	–	–	1.37
22	53.05	578.04	2739.6	55.96	52	6.6	298.15	104.92	343.2
23	73.58	578.04	2739.6	77.61	53	–	–	–	282.28
24	73.58	780.15	3412	103.93	54	–	–	–	161.51
25	20.52	488.15	923.78	4.03	55	549.82	298.15	104.92	0
26	20.52	578.04	2739.6	21.64	56	549.82	343.15	293.07	7.1
27	127.13	298.15	198.98	2.61	57	–	–	–	174.85
28	47.01	689.32	405.96	7.16	58	380.07	1442	1223.9	353.3
29	174.14	1400	1304	160.66	59	3.8	298.15	104.92	197.97
30	174.14	1400	1304	160.66					

To optimize the trigeneration system, the enthalpy and entropy equations for each species are programmed as user-defined functions into MATLAB. Subroutines for each component as well as cost functions are then written in MATLAB's programming language. These subroutines iteratively determine the outlet conditions (temperature and composition) of each component given the inlet conditions. When inlet conditions are unknown, they are initially guessed and iteratively determined. All subroutines are combined into one plant algorithm that is used to determine all state properties, and hence optimize the trigeneration system. The structure of the MOEA and DEA method used in the present work are illustrated in [30,7]. Also, the tuning of MOEA is performed according to the values

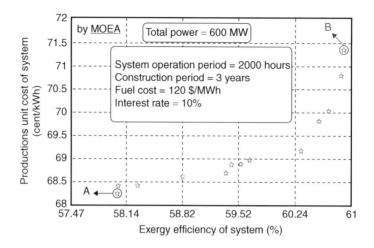

FIGURE 11.4

Pareto-optimal solutions obtained by MOEA and selecting two limited optimum points for the integrated biomass trigeneration system according to the objective functions (exergetic efficiency and unit cost of products).

indicated in [12]. The Pareto approach to multiobjective optimization is the key concept to establish optimal set of design variables, as the concepts of Pareto dominance and optimality are straightforward tools for determining the best tradeoff solutions among conflicting objectives. Figs. 11.4 and 11.5 present the Pareto optimum solutions for the integrated biomass trigeneration system based on MOEA and DEA method with the objective function indicated in Eqs. (11.26) and (11.27). As shown in these figures, the minimum unit cost of products occurs at design point A as the exergy efficiency is the lowest at this point. Also, the maximum exergy efficiency exists at design point B as unit cost of products is the greatest at this point. Design point A is the optimal situation when unit cost of products is the only objective function, while design point B is the optimum condition when exergy efficiency of trigeneration systems is the sole objective function. In multiobjective optimization, a process of decision making for selection of the final optimal solution from the available solutions is required. The process of decision making is usually performed with the aid of a hypothetical point in Figs. 11.6 and 11.7 (based on MOEA and DEA method, respectively) named as equilibrium point that both objectives have their optimal values independent to the other objective. It is clear that it is impossible to have both objectives at their optimum point, simultaneously and as shown in Figs. 11.6 and 11.7, the equilibrium point is not a solution located on the Pareto frontier. The closest point of Pareto frontier to the equilibrium point might be considered as a desirable final solution. In selection of the final optimum point, it is desired to achieve the better magnitude for each objective than its initial value of the base case problem. A final optimum solution with 60.3% and 60.21% exergetic efficiency and unit cost of product equal to 69.2 and 69.24 cent/kWh as indicated in Figs. 11.6 and 11.7 are selected. It should be noted that the selection of an optimum solution depends on the preferences and criteria of each decision-maker. Therefore, each decision-maker may select a different point as an optimum solution to better suit his/her desires.

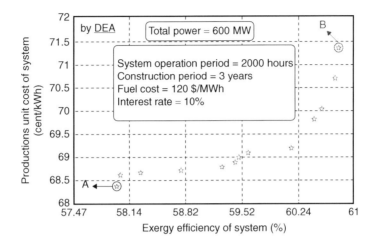

FIGURE 11.5

Pareto-optimal solutions obtained by DEA methodology and selecting two limited optimum points for the integrated biomass trigeneration system according to the objective functions (exergetic efficiency and unit cost of products).

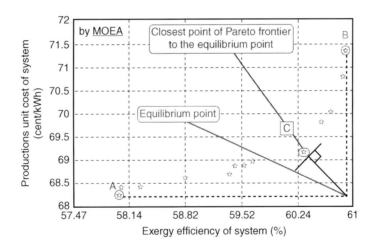

FIGURE 11.6

Pareto Frontier obtained by MOEA: best tradeoff values for the objective functions (total exergetic efficiency and unit cost of products) and selecting the optimal solution from Pareto Frontier.

The related values of decision variables in the optimum case with MOEA and DEA method are given in Table 11.3. These new parameters obtained in the optimized cases will help the designer to select components, that is turbines, compressor, as close to the optimum configuration. The comparative results of the base case and the optimum case are presented in Table 11.4. This table indicates that the multiobjective evolutionary algorithm (MOEA) leads to 22.8% increase in the exergetic efficiency

Chapter 11 Application of multiobjective optimization in thermal design

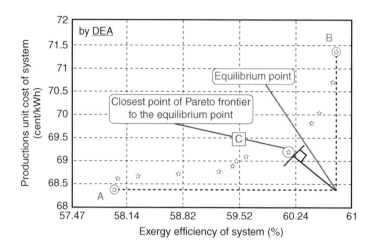

FIGURE 11.7

Pareto Frontier obtained by DEA methodology: best tradeoff values for the objective functions (total exergetic efficiency and unit cost of products) and selecting the optimal solution from Pareto Frontier.

Table 11.3 Comparison of the decisions variables in the integrated biomass trigeneration system shown in Fig. 11.3 for the base case and optimum case.

Decision variables	Base case	Optimum case (MOEA)	Optimum case (DEA)
Biomass share (%)	20	48.89	48.81
Ratio of electricity to heat	4	4.42	4.35
Ratio of electricity to cool	5	2.23	2.28
Compressor isentropic efficiency (%)	75	81.9	81.2
Compressor pressure ratio	11	11.84	11.5
Combustion temperature (K)	1400	1348.1	1346.4
Turbines isentropic efficiency (%)	85	88	87.3
Inlet pressure to the high pressure steam turbine (bar)	84	94.08	93.9
Inlet temperature to the high pressure steam turbine (K)	780.15	789	788.6
Inlet pressure to the low pressure steam turbine (bar)	9	11.68	11.6
Moisture content (%)	10	25.17	24.98

and 7.8% decrease for unit cost of products. Therefore, improvement for all objectives has been achieved using optimization process. According to this table; optimization process improves the total performance of the system in a way that the rate of fuel cost is decreased by 24.15% in the optimum case. Moreover exergy destructions is reduced about 26.49% and the related cost of the system inefficiencies is decreased about 36.95%, although the total owning and operation cost in the optimum cases is increased by 34.29%.

11.8 Example of application (case study)

Table 11.4 Comparative results of the base and optimum case in the integrated biomass trigeneration system shown in Fig. 11.3.

Multiobjective optimization results		Base case	Optimum case (MOEA)	Optimum case (DEA)
Objective functions	Unit cost of products ($c_{electricity}$ + $c_{heating}$ + $c_{cooling}$)(cent/kWh)	75.12	69.2	60.3
	Total exergy efficiency (%)	49.09	69.24	60.21
	Fuel exergy (MW)	886.99	673.14	672.92
	Exergy destruction (MW)	443.55	326.04	325.1
	Fuel cost ($/h)	107,000	81,153	81,151.7
Exergoeconomic parameters	Exergy destruction cost ($/h)	94,914	59,839	59,832
	Capital investment cost ($/h)	10,574	14,200	14,194
	CO_2 emission (kg/MWh)	639.29	571.3	570
	CO_2 emission cost ($/h)	9163.8	6954.5	6953.7

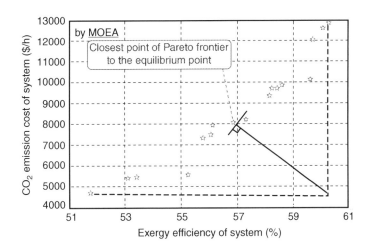

FIGURE 11.8

Pareto-optimal solutions of integrated biomass trigeneration system for the objective functions (exergetic efficiency and CO_2 emission cost).

11.8.2.1 Exergy efficiency and CO_2 emission cost

An additional run of the optimization algorithm is performed on the trigeneration system with MOEA because of the reduction in the CO_2 emission cost and increase in the exergy efficiency of the system, simultaneously [13,14]. The results of multiobjective optimization with given specification are shown in Fig. 11.8 for the objective functions accounting trigeneration exergy efficiency, (Eq. 11.26) and the

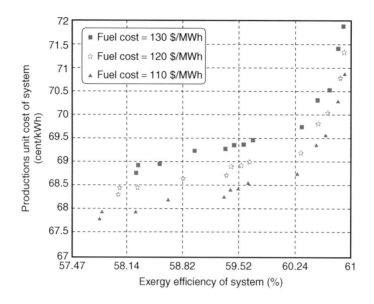

FIGURE 11.9

Sensitivity of Pareto optimum solutions to the specific fuel cost.

CO_2 emission cost, (Eq. 11.28). From this figure one can consider a desired point from Pareto frontier line and analyze the system for that optimum point.

11.8.3 Sensitivity analysis

The influences of physical parameters as well as economic parameters on the Pareto-optimal solution are evaluated using the developed program code. The results of run of the program showed that the impact of economic parameters is much more than physical parameters on the Pareto solutions according to objectives of exergy efficiency and trigeneration unit cost. Therefore due to the ineffectiveness of physical parameters on the Pareto solution, additional runs of the optimization algorithm were performed on the system to investigate the influence of economic parameters such as the unit cost of fuel, the construction period, the interest rate, and operation period on the Pareto optimal set of solutions obtained from MOEA. Fig. 11.9 shows the sensitivity of the Pareto-optimal Frontier to the variation of specific fuel cost. A comparison of the Pareto frontiers for the three optimizations shows that the economic minimum at higher unit costs of fuel is shifted upwards as expected but it is also shifted toward higher exergetic efficiencies. Similar behavior is observed for sensitivity of Pareto-optimal solution to the construction period and interest rate in Figs. 11.10 and 11.11. In addition, the sensitivity of the Pareto-optimal Frontier to the variation of operation period is illustrated in Fig. 11.12. This figure shows that the Pareto frontier not only shifts downward as the operation period increased but it is also shifted toward higher exergetic efficiencies.

11.8 Example of application (case study)

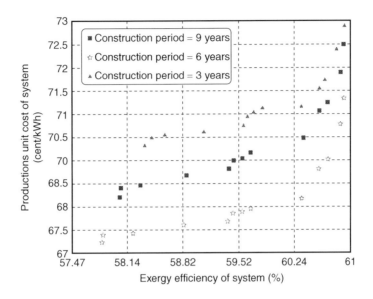

FIGURE 11.10

Sensitivity of Pareto optimum solutions to the construction period.

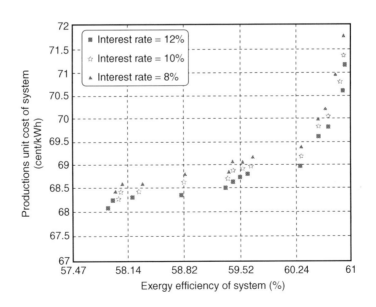

FIGURE 11.11

Sensitivity of Pareto optimum solutions to the interest rate.

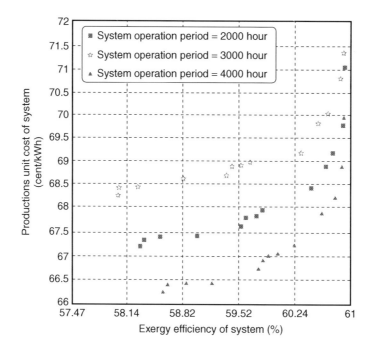

FIGURE 11.12

Sensitivity of Pareto optimum solutions to the system operation period.

11.9 Conclusions

The present chapter provides a comprehensive introduction to thermal system design and optimization from a contemporary prospective. Throughout this chapter, a methodology, based on the second law of thermodynamics, to evaluate the exergoeconomic potential and multiobjective exergoeconomic optimization of an integrated biomass trigeneration system as a case study is considered. A framework for multiobjective simulation optimization is proposed so that the power of MOEA, which can effectively search very large design spaces, compares with DEA used to evaluate the simulation results and guide the search process. In the multiobjective optimization, an example of decision-making process for selection of the final optimal solution from the Pareto frontier, which better suits with desires of decision maker, is presented. The final optimum solution is compared with the base case design and the following remarks can be extracted from this study:

1. Multiobjective optimization leads to 22.8% increase in the exergetic efficiency and 7.8% decrease of the unit cost of products.
2. Optimization process leads to the 24.1% reduction on the fuel exergy, 26.49% reduction in the total exergy destruction, and also 24.15% and 36.95% reductions in the rate of fuel cost and cost rate relating to the exergy destruction, respectively.

3. A sensitivity analysis is performed to examine the impact of the obtained Pareto solutions from MOEA to the fuel cost, construction period, system operation period, and interest rate.
4. The present algorithm can easily be used for energy, exergy, exergoeconomic analysis, and optimization of complex energy systems.
5. The analysis presented in this chapter demonstrates the strong potential of exergoeconomic analysis to increase thermal performance and decrease the unit costs of products.

References

[1] M.T.M. Emmerich, A.H. Deutz, Tutorial on multi-objective optimization: fundamentals and evolutionary methods, Nat. Comput. 17 (2018) 585–609.
[2] Z. Yang, K. Yang, Y. Wang, L. Su, H. Hu, The improved multi-criteria decision-making model for multi-objective operation in a complex reservoir system, J. Hydro Inform. 21 (2019) 851–874.
[3] M. Zare, C. Pahl, H. Rahnama, Multi-criteria decision making approach in E-learning: a systematic review and classification, Appl. Soft Comput. 45 (2016) 108–128.
[4] V. Pareto, Cours d'economie Politique, Rouge: Lausanne, Switzerland, 1896.
[5] W. Zhang, D. Yang, G. Zhang, M. Gen, Hybrid multi objective evolutionary algorithm with fast sampling strategy-based global search and route sequence difference-based local search for VRPTW, Exp. Syst. Appl. 1451 (2020) Article 113151.
[6] S. Jose, C. Vijayalakshmi, Design and analysis of multi-objective optimization problem using evolutionary algorithm, Procedia Comput. Sci. 172 (2020) 896–899.
[7] M. Carrillo, J.M. Jorgeb, A multiobjective DEA approach to ranking alternatives, Exp. Syst. Appl. 50 (2016) 130–139.
[8] G.B. de Campos, C. Bringhenti, A. Traverso, J.T. Tomita, Thermoeconomic optimization of organic Rankine bottoming cycles for micro gas turbines, Appl. Therm. Eng. 1645 (2020) 114477.
[9] A. Bejan, G. Tsatsaronis, M. Moran, Thermal Design and OPTIMIZATION, John Wiley and Sons, Inc., New York, 1996.
[10] A. Baghernejad, M. Yaghoubi, Exergoeconomic analysis and optimization of an integrated solar combined cycle system (ISCCS) using genetic algorithm, Energy Convers. Manage. 52 (2011) 2193–2203.
[11] A. Baghernejad, M. Yaghoubi, Thermoeconomic methodology for analysis and optimization of a hybrid solar thermal power plant, Int. J. Green Energy 10 (2013) 588–609.
[12] A. Baghernejad, M. Yaghoubi, Multi-objective exergoeconomic optimization of an integrated solar combined cycle system using evolutionary algorithms, Int. J. Energy Res. 35 (2010) 601–615.
[13] A. Baghernejad, M. Yaghoubi, K. Jafarpur, Optimum power performance of a new integrated SOFC-trigeneration system by multi-objective exergoeconomic optimization, Int. J. Electric. Power Energy Syst. 73 (2015) 899–912.
[14] A. Baghernejad, M. Yaghoubi, K. Jafarpur, Exergoeconomic optimization and environmental analysis of a novel solar-trigeneration system for heating, cooling and power production purpose, Solar Energy 134 (2016) 165–179.
[15] A.A. Bazmi, G. Zahedi, Sustainable energy systems: role of optimization modeling techniques in power generation and supply-a review, Renew. Sustain. Energy Rev. 15 (2011) 3480–3500.
[16] N. Shabani, T. Sowlati, A mixed integer non-linear programming model for tactical value chain optimization of a wood biomass power plant, Appl. Energy 104 (2013) 353–361.
[17] J. Soderman, P. Ahtila, Optimization model for integration of cooling and heating systems in large industrial plants, Appl. Therm. Eng. 30 (2010) 15–22.

[18] K. Kavvadias, A. Tosios, Z. Maroulis, Design of a combined heating, cooling and power system: sizing, operation strategy selection and parametric analysis, Energy Convers. Manage. 51 (2010) 833–845.
[19] A. Piacentino, F. Cardona, An original multi-objective criterion for the design of small-scale polygeneration systems based on realistic operating conditions, Appl. Therm. Eng. 28 (2008) 2391–2404.
[20] E.A. Pina, M.A. Lozano, L.M. Serra, Thermoeconomic cost allocation in simple trigeneration systems including thermal energy storage, Energy 15315 (2018) 170–184.
[21] A. Lazzaretto, G. Tsatsaronis, SPECO: a systematic and general methodology for calculating efficiencies and costs in thermal systems, Energy 31 (2006) 1257–1289.
[22] M. Moran, H. Shapiro, Fundamentals of Engineering Thermodynamics, 5th edn, John Wiley and Sons, Inc., New York, 2004.
[23] M. Saghaei, H. Ghaderi, H. Soleimani, Design and optimization of biomass electricity supply chain with uncertainty in material quality, availability and market demand, Energy 19715 (2020) Article 117165.
[24] R. Moradi, V. Marcantonio, L. Cioccolanti, E. Bocci, Integrating biomass gasification with a steam-injected micro gas turbine and an Organic Rankine Cycle unit for combined heat and power production, Energy Convers. Manage. 2051 (2020) Article 112464.
[25] Y. Zhang, P. Wang, Impact of a cold-village merger plan on the investment cost and energy utilization ratio of biomass combined heat and power plants, J. Clean. Prod. 25510 (2020) Article 120346.
[26] A.E. Broughel, Impact of state policies on generating capacity for production of electricity and combined heat and power from forest biomass in the United States, Renew. Energy 134 (2019) 1163–1172.
[27] K.E. Herold, R. Radermacher, S.A. Klein, Absorption Chillers and Heat Pumps, CRC Press, FL, 1996.
[28] P. Ahmadi, I. Dincer, Thermodynamic analysis and thermoeconomic optimization of a dual pressure combined cycle power plant with a supplementary firing unit, Energy Conversion and Management 52 (2011) 2296–2308.
[29] A. Baghernejad, M. Yaghoubi, Thermoeconomic methodology for analysis and optimization of a hybrid solar thermal power plant, Int. J. Green Energy 10 (2013) 588–609.
[30] H.P. Schwefel, Evolution and Optimum Seeking, John Wiley and Sons, New York, 1995.

CHAPTER 12

A multiobjective nonlinear combinatorial model for improved planning of tour visits using a novel binary gaining-sharing knowledge-based optimization algorithm

Said Ali Hassan[a], Prachi Agrawal[b], Talari Ganesh[b] and Ali Wagdy Mohamed[c,d]

[a]*Department of Operations Research and Decision Support, Faculty of Computers and Artificial Intelligence, Cairo University, Giza, Egypt,* [b]*Department of Mathematics and Scientific Computing, National Institute of Technology Hamirpur, Himachal Pradesh, India,* [c]*Operations Research Department, Faculty of Graduate Studies for Statistical Research, Cairo University, Giza, Egypt,* [d]*Wireless Intelligent Networks Center (WINC), School of Engineering and Applied Sciences, Nile University, Giza, Egypt*

12.1 Introduction

Managers of tourism companies always face the recurring problem of choosing the best tourist destinations during their tourist tours. This problem is evident in countries with a large number of famous tourist places at the global level, such as Egypt. Tour operators plan to try to differentiate between visiting the largest number of these tourist places and at the same time try to please tourists in visiting specific places they have heard about from other tourists or read about.

Section 12.2 is devoted to demonstrating the importance of tourism in Egypt as it is distinguished by the many and diverse tourist destinations and cities that can be visited. Among these famous cities is Cairo which is the capital characterized by diversity, which includes an infinite number of mosques and archaeological churches that date back to many centuries starting with the era of the Fatimids and ended with the Ottoman era.

Section 12.3 gives a brief review of the well-known travelling salesman problem (TSP) and the 0–1 Knapsack Problems (0–1 KP) and their variations. It reveals also the main differences between the new definition and formulation of the current presented problem and both TSP and 0–1 KP.

The mathematical model for the proposed multiobjective improved planning of tour visits is designed in Section 12.4 including the definition of problem variables, constraints, and the objective functions. The proposed model is a Multiobjective Nonlinear Binary Model with a dimension depending on the number of candidate places, and the steps of the solution procedure are also explained.

A real practical example is given in Section 12.5 including the best comprise choice between famous five touristic places in Cairo, the Capital of Egypt.

In Section 12.6, a novel binary version of a recently developed gaining-sharing knowledge-based optimization algorithm (GSK) is introduced to solve the Planning of Tour-visits Problem (PTP). GSK cannot solve the problem with binary space; therefore, binary-gaining-sharing knowledge-based optimization algorithm (BGSK) is proposed with two new binary junior and senior stages. These stages allow BGSK to explore and exploit the search space of the problem efficiently.

Section 12.7 represents the experimental results of PTP obtained by BGSK, and Section 12.8 gives the conclusions and the suggested points for future researches.

12.2 Tourism in Egypt: an overview
12.2.1 Tourism in Egypt

Egypt has been known throughout its history as a destination for many tourism lovers, but the discovery of the effects of the Pharaohs since the beginning of the last century has added a special charm to it. This is in addition to its unique religious and cultural monuments, along with its geographical location in the middle of the world, its suitable climate in both summer and winter, and its coasts of beautiful stretched beaches full of rare coral treasures. Egypt also overlooks the Red Sea from the east, which includes clean water, many fish, and meandering beaches, which led to the emergence of many natural ports, which has several sports that attract sports tourists, such as diving, golf, and others.

Tourism in Egypt depends on many touristic monuments that resulted from the diversity of civilizations and their differences over the ages such as Pharaonic, Coptic, Roman, and Islamic. Egypt also has a variety spread of mosques, temples, museums, monuments, historical and artistic buildings, and vast gardens on its lands. It also possesses an infrastructure serving the tourism sector, including hotel rooms, tourist resorts, and airline offices. Egypt is also distinguished by many and diverse tourist destinations and cities that can be visited, such as:

- Wadi El-Nile region includes cities such as Cairo, Minya, Luxor, Aswan, in addition to Abu Simbel temples.
- The Red Sea region includes the following places and cities: Dahab, El Gouna, Hurghada, Makadi Bay, Marsa Alam, Nuweiba, Safaga, Sahl Hasheesh, Sharm El Sheikh, Soma Bay, and the golf resort in Taba Heights.
- The Mediterranean region includes the following places and cities: Alexandria, Battlefield of El Alamein, Marsa Matruh, and the Northern Coast.
- The Western Desert region, which includes the following oases: Dakhla, Farafra, Big Gulf, Kharga, Siwa, and Bahariya Oasis.

12.2.2 Tourism in Cairo

Cairo is an ancient city and has kept pace with the development until it has also become a modern capital. It is considered one of the largest cities in the Middle East with traffic and noise problems, but as long as you are not looking for isolation, Cairo is a great place to explore Egyptian history and civilization. Cairo is the largest Egyptian city, as it is inhabited by about 20 million people out of the total 100 million population of Egypt and was called in the era of the Pharaohs as "Men Khar" which means the beautiful city.

Tourism in Cairo is characterized by diversity, starting from where Cairo started, from Fatimid and Old Coptic Cairo, which includes an infinite number of mosques and archaeological churches that date back to many centuries starting with the era of the Fatimids and ended with the Ottoman era. Cairo is the most important city in Egypt, and it is a wonderful place where many tourist places, including recreational, cultural, and historical places, meet. It is also famous for its popular atmosphere and nightlife, with cafes and a corniche crowded with visitors coming to enjoy watching the Nile in the evening. The River Nile is considered one of the longest rivers in the world, and attracts a lot of tourists, it provides tourist and entertainment services through riding boats and ships, and fishing. Moreover, several annual festivals are held in Cairo, such as the Cairo Film Festival, the Drum Festival, the Jazz Festival, and others.

Cairo is also distinguished by the presence of historical monuments, so the tourist finds museums, such as the Egyptian Museum, which contains the largest archaeological collection of monuments of the Pharaohs, the Coptic Museum, the Museum of Ceramics, the Museum of Railways, and the Museum of Agriculture.

As for the Islamic monuments, the tourist can find many of them, such as Amr Ibn Al-Aas Mosque, which is the first mosque built in Egypt and Africa, where Amr Ibn Al-Aas built it in the city of Fustat, Al-Azhar Al-Sharif, the Mosque of Ibn Tulun, and Salahuddin Castle, which is considered one of the largest castles warships built in the Middle Ages.

12.2.3 Planning of tour visits

Suppose that one of the tourism institutions organizes tourist tours for visitors in various cities of the country. The tours are divided into several days and include day and evening visits, in addition to some free and purchased tours.

A famous touristic city may include dozens of tourist and shopping places, and the visit of tourists to this city takes only a few days that are not enough to cover all these places. The management of the tourism establishment is trying to achieve the largest number of tourist visits in addition to meeting the desire of tourists to visit specific places that have priority from their point of view, so the problem has two opposing objectives that the planning authority would like to achieve.

This new improved PTP is defined on a graph G with a set of n nodes V representing the candidate places to be visits and an additional node denotes the Place to start the tour (the hotel for example), and a set of arcs representing the transportation times between two distinct places, Pinter [48]. The time of visit in a place and the transportation time between each two places are specified.

While the tour starts at a specific departure location, it ends at the last visited place. At the surroundings of this last visited place, normally a dinner is offered in a suitable chosen restaurant.

The PTP is then defined as follows:

- Each place is visited only once by the tourists.
- The tour starts at a predetermined starting location and terminates at the latest visited place.
- The problem has two competing objectives to be achieved which are to maximize the total number of visited places and to maximize the guest satisfaction measured by their relative weights given to different candidate touristic places under the time limit constraint.

The definition of the PTP is somewhat like the famous TSP and the KP. Such a proximity will be useful for creating the mathematical model for the new proposed problem (PTP). Meanwhile the PTP differs from the famous and well-known TSP and KP in many basic points that will be revealed later in Section 12.6.

12.3 PTP versus both the TSP and KP

This section will give a concise overview for the TSP and KP, also it will state the main distinct differences between the PTP and both TSP and KP. Table 12.3 summarizes also these itemized differences.

12.3.1 The Traveling Salesman Problem and its variations

The TSP is one of the problems excessively considered in studying of networks, it possesses broad real-life applications [5]. TSP is summarized as a salesman visiting a number of cities, he begins in his hometown and next wants to visit each place on a collection of places only once. Finally, he returns into hometown. He wants to do this so that the overall distance covered is a minimum. The problem sounds very simple, yet the solution is much more dificult. For almost 100 years mathematicians have tried to solve it. The problem has $\frac{(n-1)!}{2}$ possible solutions: finite, but enumeration intractable [30].

Droste [20] stated that the number of different tours is very large, so one might not think to solve the problem by simply calculating the length of each possible tour. This is only possible for small instances: for only 14 cities there are more than 3 billion possible tours. Therefore, we need suitable algorithms to solve these situations.

TSP is among the most intensively considered problems in combinatorial optimization and it is proven to be nondeterministic polynomial-time hardness (NP-hard). TSP in the nineteenth century was already known problem. In the 1930s, mathematicians became interested in it. Since then, better and better algorithms were developed. In 1954, the largest solved instance consisted of 49 cities in the United States of America. This was a great achievement at that time. In 2004, the shortest tour passing through all 24,978 cities in Sweden was found. The current record is an instance of 85,900 locations on a computer chip was solved by Appelgate et al. [6].

Numerous variations of the TSP are considered in the publications and formative programming formulations are frequently representing foundations for a great portion of these variations [53]. The main variations of the TSP are summarized in Table 12.1 which include authors (year), area of research, and the main contributions.

12.3.2 Multiobjective 0–1 KP

The *KP* involves the maximization or minimization of a value, such as profits or costs. A knapsack can only hold a certain weight or volume that can accommodate different types of items but with limitation in total volume, weights, or both. The question is to decide the number of each item to pick-up so that to minimize the total charge or to maximize the total gain [52]. In combinatorial optimization problems like that of KP, exact methods are impractical in finding an optimal solution because the run time is exponentially increased as the problem size increases [10]. Therefore, interest in the application of the metaheuristic algorithms has become a necessary to solve these problems and obtain the results in a reasonable time[11],[23],[58],[59].

Table 12.1 Summary of the main variations of the Traveling Salesman Problem.

Author (year)	Area of research	Main contributions
Pureza et al. [50]	TSP with priority prizes.	Maximizing all the customer visits profi.t taking the involved costs and prizes in the tour.
Mosheiov [43]	TSP with Pickup and Delivery (TSPPD), or TSP with Deliveries and Collections (TSPDC).	Suggested a formulation of TSPDC, in plus he introduced heuristic models supported with extended formulations of TSP.
Anily and Mosheiov [4]		A novel approach derived from the algorithm of Shortest Spanning Trees.
Gendreau et al. [29]		Suggested two approaches for TSPDC.
Baldacci et al. [9]		Investigate a formulation with integer variables when adding mixed deliveries and collections.
O'Neil, R.J. and Hoffman, K. (2018)		Exact Methods for Solving TSPPD in Real Time
Dumitrescu et al. [21]		Modeled a mixed-integer programming model and its polyhedral framework for the TSPPD.
Halse [33]	Generalization of the TSPDC.	A heuristic approach and a mathematical algorithm relying on a Lagrangian relaxation.
Gendreau et al. [28] and Aramgiatisiris [7]	TSP with backhauls (TSPB).	Consider the case when all distribution clients must be visited before the collection clients.
Gendreau et al. [28]		Heuristics for the TSPB.
Süral et al. [56]		An evolutionary approach for the TSPB.
Kara and Bektas [36]	Generalized TSP (GTSP).	Presented a mixed integer programming formulation of the GTSP.
Pop [49]		Described six linear programming (LP) formulations and the relationships between the polytopes related to the linear relaxations.
Bektas [12]	Multiple Traveling Salesman Problem (mTSP).	Names some variations of the mTSP: Multiple depots, specifications on the number of salesmen, Fixed charges and Time windows (mTSPTW).
Oberlin et al. [46].	(mTSP) with Multiple depots.	mTSP applied in robots with land and air vehicles.
Demiral and Şen [19]	Double Traveling Salesman Problem (dTSP).	Used simulated annealing to optimize the tours of salespersons
Silva et al. [55]	Traveling Repairman Problem (TRP).	A simple and effective metaheuristic for the TRP.
Onder et al. [47]	Multiple Traveling Repairman Problem (mTRP).	New integer programming formulation for mTRP.

Table 12.2 Some main solution methods and variations of the Knapsack Problem.

Author (year)	Area of research	Main contributions
Alaya et al. [2]	Multiple KP (mKP)	Ant algorithm for the (mKP).
Fidanova [25]		Ant colony optimization for mKP.
Fidanova [26]		Probabilistic Ant Colony Optimization for mKP.
Ji et al. [35]	Multiple 0–1 Knapsack Problem (m0–1 KP)	An ant colony optimization algorithm for solving m0–1 KP.
Ke et al. [38]		An ant colony optimization approach for the multidimensional knapsack problem
Iqbal et al. [54]		Solving m0–1 KP with the help of ants.
Captivo et al. [13]	Bicriteria {0–1} KP	Solving bicriteria 0–1 KP using a labeling algorithm.
da Silva et al. [16]		A scatter search method for bicriteria {0, 1} KP.
Gandibleux and Freville [27]		Tabu search for solving the bicriteria {0–1} KP.
da Silva et al. [17]		Integrating partial optimization with scatter search for solving bicriteria {0, 1} KP.
Erlebach et al. [24]	Multiobjective Knapsack Problem (MOKP)	Approximating multiobjective knapsack problems.
Groşan et al. [[31] and b]		Evolutionary algorithms for solving MOKP.
Zitzler et al. [60]		Performance assessment of multiobjective optimizers
Chabane et al. [14]		Real-world application case studies.

In the Integer type KP, items can have arbitrary integer values, whereas the 0–1 *KP* (0–1 *KP*) limits the number of each type to 0 or 1. The 0–1 KP does not allow the user to put multiple copies of the same items in their knapsack. It is a special case in which each input can be loaded or not into the knapsack [1]. The integer KP and the 0–1 KP can be tackled by some designed dynamic programming algorithms by deriving a recurrence equation expressing a solution to an instance in terms of solutions to its small-scale instances [57]. Dynamic Programming follows the principle of optimality to reach the final solution [3]. The 0–1 KP and its versions are famous NP-hard problems, dynamic programming technique can solve such problems in pseudo-polynomial time [[40],[39]]. The 0–1 KP can be solved also by greedy genetic algorithm, genetic algorithm, rough set theory, and ant weight-lifting algorithm [51].

In *Multiobjective KP (MOKP)*, a major challenge is to produce efficient solutions. A bibliography about multiobjective combinatorial optimization is presented in Ehrgott and Gandibleux [22].

Other numerous solution methods and variations of the KP are considered in the publications, some basic methods and variations are summarized in Table 12.2 which include authors (year), area of research, and the main contributions.

12.3.3 Basic differences between PTP and both the TSP and KP

The main basic differences between PTP and TSP are listed as follows:

1. In the TSP, the time is open until completing the visits of all customers, while in the PTP the available time is limited by the day visiting hours.
2. In the TSP, the salesman will reach all customers, while in the PTP the tour-guide will determine a tour containing some or all the places which improve the utilization of the available predetermined time limit.
3. In the TSP, no time is consumed in customer places or it is immaterial, while in the PTP the visiting time is a basic factor influencing the time limit constraint and hence, it directly affects the selection of the optimum solution of the problem.
4. In the TSP, the aim is the minimization of the travelling times, while the PTP is aiming at maximizing the utilization of daily available time calculated as maximizing the total number of visited places and maximizing the satisfaction of tourists.
5. The main applications of TSP are sales, mask plotting in printed circuit board production, vehicle routing, order-picking problem in warehouses, X-ray crystallography, circuit boards, overhauling engines, and computer wiring. The principal applications of PTP are the advising, counseling, and visiting matters in most practical fields such as tourism, industry, pollution, agriculture, business, education, telecommunications, banking, quality assurance, social and community services, tourism, sales, advertising, medical, sports, arts, cooking, and others.

The main basic differences between PTP and 0–1 KP are listed as follows:

1. In the 0–1 KP, the order of picking-up the items is immaterial, the only importance matter to be considered is only taking or leaving each one; while in the PTP the ordering of visiting different places is very important as this ordering will affect significantly the transportation times and then the choice of the optimum solution.
2. In the 0–1 KP, the problem is considered as an event that takes place in a distinct instant of time, while in the PTP the available time is limited by the day visiting hours.
3. In the 0–1 KP, some or all the items are chosen within the capacity limits measured in most cases as volume or weight, while in the PTP some or all places are chosen within the available predetermined time limit.
4. In the 0–1 KP, the choice of items will be constrained by their characteristics s volume or weight or both, while in the PTP, the choice of places will be constrained by their characteristics based on the consumed visiting time and the related transportation time.
5. In the 0–1 KP, the aim is generally to maximize the total profit from the carried items, while the PTP is aiming at maximizing the utilization of daily available time calculated as the total number of visited places and maximizing the satisfaction of tourists.
6. Some applications of 0–1 KP are home energy management, cognitive radio networks, resource management in software, a mining application, relay selection in secure cooperative wireless communication, power allocation management, selecting adverts garden city radio, production planning problem, 5G mobile edge computing, selection of renovation actions, sensor selection in distributed multiple-radar, architectures for localization, appliance scheduling optimization for demand response, adaptive variable density sampling, secure cooperative wireless communication,

Table 12.3 Main differences between PTP and both TSP and KP.

Item	PTP	TSP	KP
Order of picking-up items or visiting places	Effectively considered	Effectively considered	Immaterial
Completion time for problem	Limited	Open	Instant
Served customers	Places	Customers	Items
Number of served customers	Some or all	All	Some or all
Selection constraints	Transportation and counseling times	Travelling times	Weight or volume of items
Objective function	Maximize number of visited places and satisfaction of tourists	Minimize total travelling time	Maximize profit

optimizing power allocation to electrical appliances, computation offloading in wireless multi-hop networks, tour conducting, plastic bags waste management, workflow mapping, content delivery networks and network selection for mobile nodes. The fields of applications for PTP were stated in the previous part.

The main basic differences between PTP and both TSP and 0–1 KP can be summed up as indicated in Table 12.3.

12.4 Mathematical model for planning of tour visits

Decision variables:
Let:
$$\begin{cases} 1, \text{ if place is visited by the tourist group on position } m \text{ of their tour, and } m = 1, \\ 2, \ldots, n. \\ 0, \text{ otherwise.} \end{cases}$$

Constraints:
1. *Positions constraints*:
Each position m in the tourist tour has at most one place:

$$\sum_{i=1}^{n} x_i^m \leq 1, m = 1, 2, \ldots, n. \tag{12.1}$$

2. *Places constraints*:
Each place i can be in one position of the tourist tour or not visited:

$$\sum_{m=1}^{n} x_i^m \leq 1, i = 1, 2, \ldots, n. \tag{12.2}$$

3. Consecutive positions constraints

A position $(m + 1)$ cannot exist in the tourist tour unless the preceding position m exists, this is achieved by the following set of constraints:

$$\sum_{i=1}^{n} x_i^{m+1} \leq \sum_{i=1}^{n} x_i^m, \quad m = 1, 2, \ldots, n - 1 \tag{12.3}$$

If $\sum_{i=1}^{n} x_i^{m+1} = 1$, then $\sum_{i=1}^{n} x_i^m = 1$, $m = 1, 2, \ldots, n - 1$,

If $\sum_{i=1}^{n} x_i^{m+1} = 0$, then there is no restriction on the value of $\sum_{i=1}^{n} x_i^m$, $m = 1, 2, \ldots, n - 1$.

4. Tour time constraints

The total time spent by the tourist group in transporting and site visits should be within the maximum tour $time = T$ hours.

$$\sum_{i=1}^{n} t_{0i} x_i^1 + \sum_{m=1}^{n}\sum_{i=1}^{n}(t_i x_i^m) + \sum_{i=1}^{n} \sum_{\substack{j=1 \\ j \neq i}}^{n} t_{ij} \cdot \left(\sum_{m=1}^{n-1} x_i^m . x_j^{m+1}\right) \leq T. \tag{12.4}$$

where

t_{0i} = Transportation time between the hotel and place i, $i = 1, 2, \ldots, n$,
t_{ij} = Transportation time between the two adjacent places $i = 1, 2, \ldots, n$ and j, $i, j = 1, 2, \ldots, n$
t_i = Visiting time for place i, $i = 1, 2, \ldots, n$

This is a quadratic inequality in two variables, the first part is for transportation from the hotel to the first position in the tour, the second part is the visiting times at places, and the third part is the transportation time between different positions in the tour.

5. Binary constraints

All the decision variables are 0–1.

$$x_i^m = 0 \text{ or } 1, \; i, \; m = 1, 2, \ldots, n. \tag{12.5}$$

6. The objective functions

The tourism experts decide on two main competing objectives, the first is to maximize the number of visited touristic places, and the second one is to give relative weights depending on the satisfaction level of tourists evaluated from previous questionnaires.

The problem is then a multiobjective one with two objectives. The motivation in Multiobjective Optimization (MOO) is that it searches for a tradeoff on some contradictory matters. There is no single optimum solution, but many several solutions.

The weighted sum or scalarization method is one of the classic MOO methods, it puts a set of objectives into one by adding each objective premultiplied by a user-supplied weight in proportion to the relative importance of the objective. In the scalarization method, answer is a set of solutions that define the best tradeoff between competing objectives that form in its entirety the nondominated Pareto-optimal set for the problem [32].

This is simple, but it is not easy to put the weight vectors to obtain a Pareto-optimal solution in a desired region in the objective space and it cannot obtain certain solutions in the nonconvex objective space [41].

The weighted sum approach treats the MOO as a composite objective function [34]. This function is expressed as follows:

$$Z = \text{Maximize} \sum_{i=1}^{q} w_i \cdot f_i(x)$$

where w_i is the positive weight values, $f_i(x)$ is one of the objective functions and q is the number of objective functions. As the objective in this paper is to find a compromise between maximizing of total visited places (TVP) and maximizing of total relative satisfaction (TRS), the following composite objective functions is formulated as

$$Z = w_1 f_1(x) + w_2 f_2(x) \tag{12a}$$

$$\text{Maximize } f_1(x) = (\text{TVP}) = \sum_{i=1}^{n} \sum_{m=1}^{n} x_i^m \text{ and} \tag{12b}$$

$$\text{Maximize } f_2(x) = (\text{TRS}) = \sum_{i=1}^{n} s_i \sum_{m=1}^{n} x_i^m \tag{12c}$$

The weights w_1 and w_2 are related based on the following expression:
$w_2 = 1 - w_1$. w_1 is chosen is in the range of [0–1]. where s_i is the relative satisfaction level of place no i.

From (12a, 12b, and 12c), the composite objective function will be

$$\text{Maximize } Z = w_2 \left(\sum_{i=1}^{n} \sum_{m=1}^{n} x_i^m \right) + w_1 \left(\sum_{i=1}^{n} s_i \sum_{m=1}^{n} x_i^m \right) \tag{12.6}$$

where s_i = relative weight of touristic place number i based on relative satisfaction level,

To specify proper weights in Eq. (12.6), an evaluation is done by varying w_1 from 1 to 0 with an increment of 0.1 [45].

Finally, we have a suggested model that contains (n^2) binary variables and $(3n)$ constraints.

Any final chosen efficient solution will produce two distinct situations:

1. If $\sum_{m=1}^{n} x_i^m = 1 \; \forall i$, then all the touristic places are visited during the planned time, and the problem is completed.

2. If $\sum_{m=1}^{n} x_i^m = 0$ for any i, then the corresponding place i is not visited. In this case, add this place to the updated new list for places to be visited in the next tour and repeat the procedure for another time available slot.

The solution procedure is presented in Fig. 12.1.

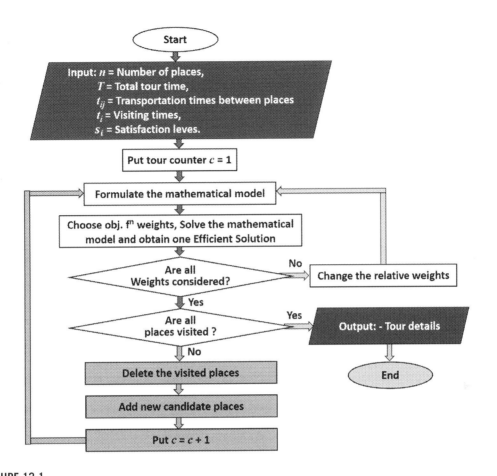

FIGURE 12.1

Steps of the solution procedure of the problem.

12.5 A real application case study

In this case, we assume as an illustrative real example for the mathematical model application that a tourist group resides in the Ramses Hilton hotel in Cairo, the Capital of Egypt. In one of the morning tours that last 9 h, five tourist places are identified to choose between them, these places are defined by serial numbers (1, 2,..., 5), they are located at different positions in Great Cairo Governorate with the data given in Table 12.4, where numbers inside the cells (i, j) represent the traveling times t_{ij}.

A brief information about the residence hotel from which the tour will start and each of these tourist places is given in the next paragraphs.

Table 12.4 Data for the case study example.

s_i (from 5)	t_i	ij	1	2	3	4	5
		0	1.30	0.25	0.167	0.75	1.0
5	2.0	1		1.5	1.5	1.65	1.7
1	1.5	2			0.167	0.7	0.6
5	2.0	3				0.65	0.5
3.5	1.5	4					0.7
2	1.0	5					

12.5.1 Ramses Hilton Hotel

Ramses Hilton Hotel is 24 km from Cairo International Airport, where an airport shuttle service is provided to the hotel. The hotel overlooks the Nile and offers a terrace, pool, casino, a billiards room, and gym. Guest rooms feature balconies with panoramic views of the city or the Nile. The hotel also features massage services, several saunas, and a hot tub are available at the spa for guests looking to relax. A range of international cuisine is available at the hotel, from Mediterranean specialties to traditional English cuisine and a specialized restaurant provides authentic and delicious Indian cuisine. An Oriental Café offers daily live entertainment in the evening for guests.

The hotel also features an on-site shopping exhibition with 250 stores and galleries. The adjacent and surrounding area of the hotel has many tourist and shopping places that can be reached on foot, such as Ramses Hilton Hotel Shopping Mall, the Egyptian Museum, Ministry of Foreign Affairs, Tahrir Square, Arab League HQ, Andalus Park, the Nile Palace Bridge, Cairo Tower, and the Egyptian Opera House.

1. *Giza Pyramids and Sphinx*

The pyramids are one of the most important places of tourism in Egypt. They are located on Giza Plateau in the Giza Governorate, that is, on the West Bank of the Nile River. It includes three pyramids of Khufu, Khafre, and Menkaure, next to the Sphinx which were built around 2500 BC. The pyramids are royal tombs, each bearing the name of the king who built it and where he was buried.

The three pyramids of Egypt are characterized as one of the most important Seven Wonders of the World and one of the greatest mysteries of ancient and modern history. The tourist can also enjoy a camel trek and take a tour around one of the three pyramids.

The Great Sphinx is also one of the most famous historical monuments in Egypt due to its unique appearance and its distinctive location near the pyramids of Giza. Many historians and archaeologists note that this great statue dates back specifically to the period of the Fourth Dynasty that ruled Egypt at that time, and some of them dated the ownership of this statue to King Khafre, one of the ancient Egyptian kings.

2. *Cairo Tower*

The tower is one of the most important tourist places in Cairo that attracts the attention of tourists due to the magnificence of its design. The design takes the form of the famous Egyptian lotus flower, in addition to its majestic height that reaches 187 m. The tower building offers a wonderful panoramic view of Cairo and the Nile. The visitor can enjoy this wonderful view while eating his meal in the round tower restaurant on the fourteenth floor or in the cafeteria or through the telescopic observatory above the top of the tower.

At night, the tower building launches wonderful gradient lights where photos of the tower building can be taken from the outside.

3. *The Egyptian Museum*

It is one of the most important landmarks of Cairo and, in general, Egypt; it is one of its largest museums, space, and collectibles. The ancient museum contains about 150,000 pieces dealing with the history of Egypt, Pharaonic, Romanian, and Greek. The museum is divided into two floors, one of which contains light holdings such as small sculptures, manuscripts and pictures, and the other includes heavy pieces such as coffins, mummies, and colossal statues.

It may require the use of a special tourist guide to explain the development of the Egyptian historical periods according to their historical arrangement, or it is possible for the tourist to resort to the guided paintings and posters set alongside the exhibits. From the outside, the museum has a magnificent red clad architectural design, along with its dome and distinctive windows, which make it very similar to the architecture of Islamic buildings.

4. *Saladin Citadel*

The castle is considered one of the most tourist sites in Cairo, it attracts the attention of tourists from everywhere in the world. It is of Islamic design distinguished by the minarets and huge silver domes dating back to the era of the Ayyubid state in Egypt. It has a magnificent view that offers a panoramic image of entire Cairo as it is building on the highest hill in a mountain Mokattam.

The ancient castle was established by the great leader Salah Al-Din Al-Ayoubi, which dates back to the fourteenth century. It includes three main mosques: Al-Nasir Muhammad ibn Qalawun Mosque dating from the history of the castle, Sulayman Pasha Mosque which dates back to the sixteenth century and topped by Muhammad Ali Mosque with distinctive Ottoman style, which dates back to the nineteenth century. It also includes several palaces, the most famous of which is Al-Gawhara Palace, and museums such as the Police Museum and the Military Museum.

5. *Baron Palace*

Baron Palace is located in Heliopolis district on the airport road, and it is one of the most important and beautiful buildings in Cairo. Its foundation dates back to the nineteenth century by order of a Belgian millionaire who spent most of his life in India. So, the designs of this majestic palace are inspired by the unique styles of Indian architecture visible in the designs of its balconies and windows that lean on statues of elephants and Buddhist gods.

The palace consists of two floors, and includes seven rooms rich with precious holdings, along with a unique clock in calculating the timings where one likes to except in London. Its floors are completely covered with European marble and alabaster. Adjacent to the palace is a mighty four-story tower, whose staircases are decorated with marble, bronze pieces, and finely carved Indian statues, and is decorated inside and outside with delicate sculptures.

Fig 12.2 shows Cairo location in Egypt, and Fig 12.3 shows the locations of the hotel where the tour starts and the five tourist places. Table 12.2 shows the time required to travel from each location i to

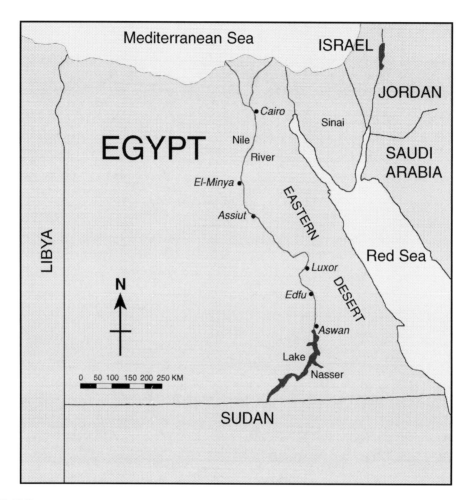

FIGURE 12.2

Cairo location in Egypt.

another one j where the hotel is indicated by (0), it also represents the visiting time t_i and the satisfaction level s_i for each tourist place i.

The mathematical model for the problem is formulated completely including the design of decision variables, constraints, and the two objective functions. The description of the solution method is introduced in the following section.

12.6 Proposed methodology

Metaheuristic algorithms have been developed to solve the complex optimization problem with continuous variables. Mohamed et al. [42] recently proposed a novel GSK, which is based on the ideology

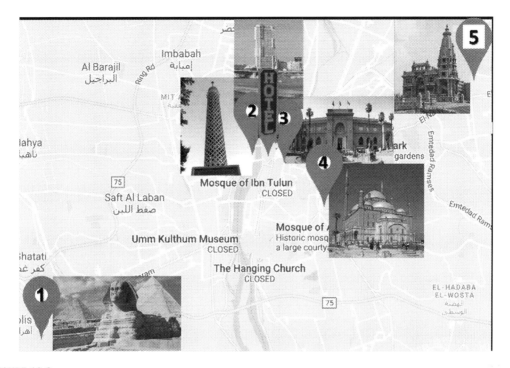

FIGURE 12.3

Candidate touristic places for the example tour.

of acquiring knowledge and share it with others throughout their lifetime. The original GSK solves optimization problems over continuous space, but it cannot solve the problem with binary space. So, a new variant of GSK is introduced to solve the proposed PTP. A novel BGSK is proposed over discrete binary space with new binary junior and senior gaining and sharing stages.

On the other hand, there are many constraint handling techniques in the literature [[18],[15]],[44]. In this chapter, the augmented Lagrangian method is used to handle the constraints, in which a constrained optimization problem is converted into an unconstrained optimization problem [[37],[8]]. The proposed methodology is described below.

12.6.1 Gaining Sharing Knowledge-based optimization algorithm (GSK)

A constrained optimization problem can be formulated mathematically as

$$\text{Min } f(X); \; X = [x_1, x_2, \ldots, x_{Dim}]$$

$$\text{subject to } g_i(X) \leq 0; \; i = 1, 2, \ldots, m$$

$$X \in [\alpha_p, \beta_p]; \; p = 1, 2, \ldots, Dim$$

where f denotes the objective function; $X = [x_1, x_2, \ldots, x_{Dim}]$ are the decision variables; $g_i(X)$ are the inequality constraints and α_p, β_p are the lower and upper bounds of decision variables, respectively, and Dim represents the dimension of individuals. If the problem is in maximization form, then consider minimization $= -$ maximization.

The human-based algorithm GSK is of two-stages: junior and senior gaining and sharing stage. All persons acquire knowledge and share their views with others. The people from early-stage gain knowledge from their small networks such as family members, relatives, neighbors, etc. and want to share their opinions with the others who might not be from their networks, due to curiosity of exploring others. These may not have the experience to categorize the people. In the same way, the people from the middle or later age enhance their knowledge by interacting with friends, colleagues, social media friends, etc. and share their views with the most suitable person, so that they can improve their knowledge. These people have the experience to judge other people and can categorize them (good or bad). The process mentioned above can be formulated mathematically in the following steps:

Step 1: To obtain the starting solution of the optimization problem, the initial population must be obtained. The initial population is created randomly within the boundary constraints as

$$x_{tp}^0 = \alpha_p + rand_p(\beta_p - \alpha_p) \tag{12.7}$$

where t is for the number of populations; $rand_p$ denotes uniformly distributed random number between 0 and 1.

Step 2: At the beginning, the dimensions of the junior and senior stage should be computed through the following formula:

$$Dim_J = Dim \times \left(\frac{Gen^{max} - G}{Gen^{max}}\right)^k \tag{12.8}$$

$$Dim_S = Dim - Dim_J \tag{12.9}$$

where $k \, (> 0)$ denotes the knowledge rate, that controls the experience rate. Dim_J and Dim_S represent the dimension for the junior and senior stage, respectively. Gen^{max} is the maximum number of generations, and G denotes the generation number.

Step 3: Junior gaining sharing knowledge stage: In this stage, the early aged people gain knowledge from their small networks and share their views with the other people who may or may not belong to their group. Thus, individuals are updated through as

i. According to the objective function values, the individuals are arranged in an ascending order. For every x_t ($t = 1, 2, \ldots, NP$), select the nearest best (x_{t-1}) and worst (x_{t+1}) to gain knowledge, also choose randomly (x_r) to share knowledge. Therefore, to update the individuals, the pseudocode is presented in Fig. 12.4.

Where $k_f \, (> 0)$ is the knowledge factor.

Step 4: Senior gaining sharing knowledge stage: This stage comprises the impact and effect of other people (good or bad) on the individual. The updated individual can be determined as follows:

i. The individuals are classified into three categories (best, middle, and worst) after sorting individuals into an ascending order (based on the objective function values).

```
for t=1:NP
    for p=1:Dim
        if rand≤ k_r (knowledge ratio)
            if f(x_t) > f(x_r)
                x_{tp}^{new} = (x_t + k_f * ((x_{t-1} − x_{t+1}) + (x_r − x_t)))
            else
                x_{tp}^{new} = (x_t + k_f * ((x_{t-1} − x_{t+1}) + (x_t − x_r)))
            end (if)
        else  x_{tp}^{new} = x_{tp}^{old}
        end (if)
    end (for p)
end (for t)
```

FIGURE 12.4

Pseudocode for Junior gaining sharing knowledge stage.

```
for t=1:NP
    for p=1:Dim
        if rand≤ k_r (knowledge ratio)
            if f(x_t) > f(x_{middle})
                x_{tp}^{new} = (x_t + k_f * ((x_{best} − x_{worst}) + (x_{middle} − x_t)))
            else
                x_{tp}^{new} = (x_t + k_f * ((x_{best} − x_{worst}) + (x_t − x_{middle})))
            end (if)
        else  x_{tp}^{new} = x_{tp}^{old}
        end (if)
    end (for p)
end (for t)
```

FIGURE 12.5

Pseudocode for Senior gaining sharing knowledge stage.

Best individual = 100 $p\%$ (x_{best}), middle individual = $Dim - 2 \times 100p\%$ (x_{middle}), worst individual = 100 $p\%$ (x_{worst})

For every individual x_t, choose two random vectors of the top and bottom 100 $p\%$ individual for gaining part and the third one (middle individual) is chosen for the sharing part. Therefore, the new individual is updated through the following pseudocode presented in Fig. 12.5.

Where $p \in [0, 1]$ is the percentage of best and worst classes.

Table 12.5 Results of the binary junior gaining and sharing stage of Case 12.1 with $k_f = 1$.

	x_{t-1}	x_{t+1}	x_r	Results	Modified results
Subcase (a)	0	0	0	0	0
	0	0	1	1	1
	1	1	0	0	0
	1	1	1	1	1
Subcase (b)	1	0	0	1	1
	1	0	1	2	1
	0	1	0	−1	0
	0	1	1	0	0

12.6.2 Binary Gaining Sharing Knowledge-based optimization algorithm (BGSK)

To solve problems in discrete binary space, a novel BGSK proposed. In BGSK, the new initialization and the working mechanism of both stages (junior and senior gaining sharing stages) are introduced over binary space, and the remaining algorithms remain the same as the previous one. The working mechanism of BGSK is presented in the following subsections:

Binary initialization

The initial population is obtained in GSK using Eq. (12.7) and it must be updated using the following equation for binary population:

$$x_{tp}^0 = \text{round}(\text{rand}(0, 1)) \qquad (12.10)$$

Where the round operator is used to convert the decimal number into the nearest binary number.

Binary junior gaining and sharing stage

The binary junior gaining and sharing stage is based on the original GSK with $k_f = 1$. The individuals are updated in original GSK using the pseudocode (Fig. 12.4) which contains two cases. These two cases are defined for binary stage as follows

Case 12.1. When $f(x_r) < f(x_t)$: There are three different vectors (x_{t-1}, x_{t+1}, and x_r), which can take only two values (0 and 1). Therefore, a total of 2^3 combinations are possible, which are listed in Table 12.3. Furthermore, these eight combinations can be categorized into two different subcases [(a) and (b)] and each subcase has four combinations. The results of each possible combination are presented in Table 12.5.

Subcase (a): If x_{t-1} is equal to x_{t+1}, the result is equal to x_r.

Subcase (b): When x_{t-1} is not equal to x_{t+1}, then the result is the same as x_{t-1} by taking −1 as 0 and 2 as 1.

The mathematical formulation of Case 12.1 is as follows:

$$x_{tp}^{new} = \begin{cases} x_r; & \text{if } x_{t-1} = x_{t+1} \\ x_{t-1}; & \text{if } x_{t-1} \neq x_{t+1} \end{cases}$$

Case 12.2. When $f(x_r) \geq f(x_t)$: There are four different vectors (x_{t-1}, x_t, x_{t+1}, and x_r), that consider only two values (0 and 1). Thus, there are 2^4 possible combinations that are presented in Table 12.6.

Table 12.6 Results of the binary junior gaining and sharing stage of Case 12.2 with $k_f = 1$.

	x_{t-1}	x_t	x_{t+1}	x_r	Results	Modified results
Subcase (c)	1	1	0	0	3	1
	1	0	0	0	1	1
	0	1	1	1	0	0
	0	0	1	1	-2	0
Subcase (d)	0	0	0	0	0	0
	0	1	0	0	2	1
	0	0	1	0	-1	0
	0	0	0	1	-1	0
	1	0	1	0	0	0
	1	0	0	1	0	0
	0	1	1	0	1	1
	0	1	0	1	1	1
	1	1	1	0	2	1
	1	0	1	1	-1	0
	1	1	0	1	2	1
	1	1	1	1	1	1

Moreover, the 16 combinations can be divided into two subcases [(c) and (d)] in which (c) and (d) has 4 and 12 combinations, respectively.

Subcase (c): If x_{t-1} is not equal to x_{t+1}, but x_{t+1} is equal to x_r, the result is equal to x_{t-1}.

Subcase (d): If any of the condition arises $x_{t-1} = x_{t+1} \neq x_r$ or $x_{t-1} \neq x_{t+1} \neq x_r$ or $x_{t-1} = x_{t+1} = x_r$, the result is equal to x_t by considering -1 and -2 as 0, and 2 and 3 as 1.

The mathematical formulation of Case 12.2 is as

$$x_{tp}^{new} = \begin{cases} x_{t-1}: & \text{if } x_{t-1} \neq x_{t+1} = x_r \\ x_t: & \text{Otherwise} \end{cases}$$

Binary senior gaining and sharing stage:

The working mechanism of binary senior gaining and sharing stage is the same as the binary junior gaining and sharing stage with value of $k_f = 1$. The individuals are updated in the original senior gaining sharing stage using pseudocode (Fig. 12.5) that contains two cases. The two cases further modified for binary senior gaining sharing stage in the following manner:

Case 12.1. $f(x_{middle}) < f(x_t)$: It contains three different vectors (x_{best}, x_{middle}, and x_{worst}), and they can assume only binary values (0 and 1), thus total eight combinations are possible to update the individuals. These total eight combinations can be classified into two subcases [(a) and (b)] and each subcase contains only four different combinations. The obtained results of this case are presented in Table 12.7.

Subcase (a): If x_{best} is equal to x_{worst} then the obtained results are equal to x_{middle}.

Subcase (b): On the other hand, if x_{best} is not equal to x_{worst} then the results are equal to x_{best} with assuming -1 or 2 equivalent to their nearest binary value (0 and 1, respectively).

Table 12.7 Results of binary senior gaining and sharing stage of Case 1 with $k_f = 1$.

	x_{best}	x_{worst}	x_{middle}	Results	Modified results
Subcase (a)	0	0	0	0	0
	0	0	1	1	1
	1	1	0	0	0
	1	1	1	1	1
Subcase (b)	1	0	0	1	1
	1	0	1	2	1
	0	1	0	-1	0
	0	1	1	0	0

Table 12.8 Results of binary senior gaining and sharing stage of Case 2 with $k_f = 1$.

	x_{best}	x_t	x_{worst}	x_{middle}	Results	Modified results
Subcase (c)	1	1	0	0	3	1
	1	0	0	0	1	1
	0	1	1	1	0	0
	0	0	1	1	-2	0
Subcase (d)	0	0	0	0	0	0
	0	1	0	0	2	1
	0	0	1	0	-1	0
	0	0	0	1	-1	0
	1	0	1	0	0	0
	1	0	0	1	0	0
	0	1	1	0	1	1
	0	1	0	1	1	1
	1	1	1	0	2	1
	1	0	1	1	-1	0
	1	1	0	1	2	1
	1	1	1	1	1	1

Case 12.1 can be mathematically formulated in the following way:

$$x_{tp}^{new} = \begin{cases} x_{middle}; & \text{if } x_{best} = x_{worst} \\ x_{best}; & \text{if } x_{best} \neq x_{worst} \end{cases}$$

Case 12.2. $f(x_{middle}) > f(x_t)$: It consists of four different binary vectors (x_{best}, x_{middle}, x_{worst}, and x_t), and with the values of each vector, a total of 16 combinations are presented. The 16 combinations are also divided into two subcases [(c) and (d)]. The subcases (c) and (d) further contain 4 and 12 combinations, respectively. The subcases are explained in detail in Table 12.8.

Subcase (c): When x_{best} is not equal to x_{worst} and x_{worst} is equal to x_{middle}, then the obtained results are equal to x_{best}.

Table 12.9 Numerical values of parameters.

Parameters of BGSK	Considered values
NP	200
K	10
k_r	0.9
P	0.1
k_f	1
Max number of iterations	200

Subcase (d): If any case arises other than (c), then the obtained results are equal to x_t by taking -2 and -1 as 0 and 2 and 3 as 1.

The mathematical formulation of Case 12.2 is given as

$$x_{tp}^{new} = \begin{cases} x_{best}; & if\ x_{best} \neq x_{worst} = x_{middle} \\ x_t; & Otherwise \end{cases}$$

The Pseudocode of BGSK is presented in Fig. 12.6:

1. **Begin**
2. **Initialize the value of parameters Gen^{max}, NP, k_r, k, p**
3. **Initialize the generation ($G = 0$)**
4. **Create binary population using equation (12.7)**
5. **Evaluate $f(x_t)$.**
6. **For $G = 1$ to Gen^{max}**
7. **Compute the dimensions of both stages (Binary junior and senior gaining sharing stage)**
8. **Apply Binary Junior gaining sharing stage**
9. **Apply Binary Senior gaining sharing stage**
10. **Update the population**
11. **Select the global best solution**
12. **End for NP**
13. **End for G**
14. **End for Begin**

FIGURE 12.6

Pseudocode for BGSK the equation number is Eq. (12.7).

12.7 Experimental results

The PTP is handled by using the proposed novel BGSK algorithm, the used parameters are presented in Table 12.9. BGSK runs over personal computer Intel Core i5-7200U CPU @ 2.50GHz and 4 GB RAM and coded on MATLAB R2015a. To get the compromise or effective solutions, 30 independent runs are complete, and the obtained statistics are provided in Table 12.10, including the best, median, average,

Table 12.10 Statistical results of PTP using BGSK.

Weights	$f_1(x), f_2(x)$	Bes	Median	Average	Worst	Standard deviation
$w_1 \in [0.7: 1]$	(4,11.5)	(4,11.5)	(4,11.5)	(4,11.5)	(4,11.5)	0.00
$w_1 \in [0: 0.6]$	(3,13.5)	(3,13.5)	(3,13.5)	(3,13.5)	(3,13.5)	0.00

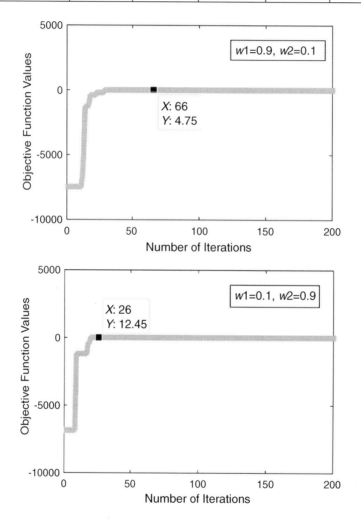

FIGURE 12.7

Convergence graph of BGSK with first and second group of the weights.

worst solutions, and the standard deviations at the three groups of weights. Moreover, Fig 12.7 shows the convergence graph of the solutions of PTP using weights of (0.9,0.1) and (0.1,0.9), respectively. From Fig. 12.7, it can be observed that after the 66th and 26th iteration, respectively, it converges to the two efficient solutions 4.75 and 12.45, which shows a high convergence speed of BGSK.

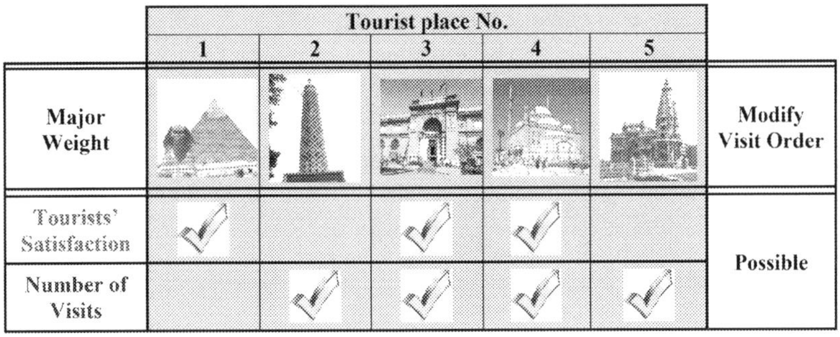

	Tourist place No.					
Major Weight	1	2	3	4	5	Modify Visit Order
Tourists' Satisfaction	✓		✓	✓		Possible
Number of Visits		✓	✓	✓	✓	

FIGURE 12.8

Efficient solutions of the illustrated case study.

Fig 12.8 represents the efficient solutions of the illustrated case study. The efficient solutions for the problem for different weights of the two objective functions are presented in Table 12.11. Columns 1 and 2 are the relative weights for the objective functions, Column 3 contains the obtained efficient solutions, Column 4 represents the remaining time from the total tour time, and the last two columns are for the values of the two objective functions. The efficient solution to visit places 1, 3, and 4 is obtained when the relative weights are in favor of the satisfaction of the tourists and efficient solution to visit places 2, 3, 4, and 5 is obtained when the relative weights are in favor of the number of visited places.

The same efficient solution appears many times with different sequence of the visits, that is, different arrangements of places. This gives the decision maker more room for scheduling the tour by choosing several possibilities for the tour sequence. This will be helpful to adjust the sequence of places according to factors such as the time window, the weather conditions, remainder time of the tour, and other factors.

12.8 Conclusions and points for future studies

The main conclusions for this paper can be summarized as follows:

1. A new formulation for improved PTP is presented, where a tour manager is likely to select the most efficient tour for visiting some selected places among the candidate ones. The problem arises mainly in the tourism specialization.
2. The PTP looks like both the famous TSP and the 0–1 KP but with basic distinct differences. The main differences are explained in detail in Section 12.3.3.
3. A nonlinear binary mathematical model is formulated for the given problem. The binary decision variables represent the allocation of chosen places into the positions of the proposed tour, the nonlinearity appears in one of the constraints. The final aim is to maximize the total number of visited places and to maximize the satisfaction of tourists.
4. To obtain the solution of the proposed PTP Nonlinear Binary Programming Model, a novel BGSK is introduced which includes the two main binary junior and senior gaining sharing stages with the knowledge factor $k_f = 1$. The BGSK is the first variant of the developed GSK, and the proposed algorithm is applied with an augmented Lagrangian method to handle the constraints.

Table 12.11 Efficient solutions of the illustrated case study.

	w_2	X values	Remaining time (h)	$f_1(x)$	$f_2(x)$
0.9	0.1	$x_2^1, x_4^2, x_5^3, x_3^4 = 1$	0.85	11.5	4
		$x_4^1, x_3^2, x_5^3, x_2^4 = 1$	0.5		
		$x_2^1, x_4^2, x_3^3, x_5^4 = 1$	0.05		
		$x_2^1, x_3^2, x_4^3, x_5^4 = 1$	1.233		
		$x_5^1, x_4^2, x_3^3, x_2^4 = 1$	0.983		
0.8	0.2	$x_4^1, x_2^2, x_3^3, x_5^4 = 1$	0.883		
		$x_2^1, x_3^2, x_5^3, x_4^4 = 1$	1.383		
		$x_4^1, x_3^2, x_5^3, x_2^4 = 1$	0.5		
0.7	0.3	$x_5^1, x_2^2, x_3^3, x_4^4 = 1$	1.083		
		$x_3^1, x_4^2, x_5^3, x_2^4 = 1$	0.89		
		$x_2^1, x_5^2, x_3^3, x_4^4 = 1$	0.00		
0.6	0.4	$x_1^1, x_3^2, x_4^3 = 1$	0.05	13.5	3
		$x_3^1, x_4^2, x_1^3 = 1$	1.04		
		$x_4^1, x_3^2, x_1^3 = 1$	0.6		
0.5	0.5	$x_1^1, x_3^2, x_4^3 = 1$	0.05		
		$x_3^1, x_1^2, x_4^3 = 1$	0.19		
0.4	0.6	$x_3^1, x_4^2, x_1^3 = 1$	1.04		
		$x_3^1, x_1^2, x_4^3 = 1$	0.19		
		$x_4^1, x_3^2, x_1^3 = 1$	0.6		
0.3	0.7	$x_1^1, x_3^2, x_4^3 = 1$	0.05		
		$x_4^1, x_3^2, x_1^3 = 1$	0.6		
		$x_3^1, x_4^2, x_1^3 = 1$	1.04		
0.2	0.8	$x_1^1, x_3^2, x_4^3 = 1$	0.05		
		$x_3^1, x_4^2, x_1^3 = 1$	1.04		
0.1	0.9	$x_3^1, x_4^2, x_1^3 = 1$	1.04		
		$x_1^1, x_3^2, x_4^3 = 1$	0.05		
		$x_4^1, x_3^2, x_1^3 = 1$	0.6		

5. The nonlinear binary mathematical model and the solution method are used to solve a real application case study for planning a suitable tour visits in Cairo, Egypt. The detailed mathematical model and the solution procedure are exhaustively explained.
6. The obtained results by BGSK show its robustness and convergence and prove that it can find the efficient solutions of the PTP.

The points for future studies can be stated in the following points:

1. To propose other mathematical models' formulations for the same problem starting with the design of the decision variables and compare the effectiveness of computations for each model.

2. To augment the proposed PTP with its variations as: PTP with Time Window (PTPTW), Stochastic PTP (SPTP), Multiobjective SPTP (MOSPTP), and other variations.
3. To apply the same problem formulation to other similar advisory fields that can show up in many other domains such as industry, agriculture, business, education, telecommunications, investing, quality assurance, social and community services, pollution, medical, tourism, marketing, sales, advertising, sports, arts, cooking, and others.
4. To check the performance of the BGSK approach in solving different complex optimization problems, and further works can be investigated by the extension of BGSK with different kinds of constraint handling methods.

References

[1] M. Abdel-Basset, D. El-Shahat, I. El-Henawy, Solving 0–1 knapsack problem by binary flower pollination algorithm, Neural Comput. Appl. 2019 (31) (2019) 5477–5495. https://doi.org/10.1007/s00521-018-3375-7(0123456789.,-volV)(0123456789.,-volV).

[2] I. Alaya, C. Solnon, K. GHéDIRA, Ant algorithm for the multi-dimensional knapsack problem, in: Proc. International Conference on Bioinspired Optimization Methods and their Applications (BIOMA 2004), 2004.

[3] L. Anany, Introduction to the Design and Analysis of Algorithms, 3rd edn, Pearson, USA, 2012, ISBN 10: 0-13-231681-1 ISBN 13: 978-0-13-231681-1.

[4] S. Anily, G. Mosheiov, The traveling salesman problem with delivery and backhauls, Oper. Res. Lett. 16 (1) (1994) 11–18.

[5] D.L. Applegate, R.E. Bixby, V. Chvatal, W.J. Cook, The Traveling Salesman Problem: A Computational Study, Princeton university press, 2006.

[6] D.L. Applegate, R.E. Bixby, V. Chvátal, W. Cook, D.G. Espinoza, M. Goycoolea, K. Helsgaun, Certification of an optimal TSP tour through 85,900 cities, Oper. Res. Let. 37 (1) (2009) 11–15.

[7] T. Aramgiatisiris, An exact decomposition algorithm for the traveling salesman problem with backhauls, J. Res. Eng. Technol. 1 (2004) 151–164.

[8] A. Bahreininejad, Improving the performance of water cycle algorithm using augmented Lagrangian method, Adv. Eng. Softw. 132 (2019) 55–64.

[9] R. Baldacci, E. Hadjiconstantinou, A. Mingozzi, An exact algorithm for the traveling salesman problem with deliveries and collections, Networks: Int. J. 42 (1) (2003) 26–41.

[10] G.T. Basheer, Z.Y. Algamal, Nature-inspired optimization algorithms in knapsack problem: a review, Iraq. J. Stat. Sci. 16 (30) (2019) 55–72.

[11] Z. Beheshti, S.M. Shamsuddin, S.S. Yuhaniz, Binary accelerated particle swarm algorithm (BAPSA) for discrete optimization problems, J. Global Optimiz. 57 (2) (2013) 549–573.

[12] T. Bektas, The multiple traveling salesman problem: an overview of formulations and solution procedures, Omega 34 (3) (2006) 209–219.

[13] M.E. Captivo, J. Clímaco, J. Figueira, E. Martins, J.L. Santos, Solving bicriteria 0–1 knapsack problems using a labeling algorithm, Comput. Oper. Res. 30 (12) (2003) 1865–1886.

[14] B. Chabane, M. Basseur, J.K. Hao, A practical case of the multiobjective knapsack problem: design, modelling, tests and analysis, in: Proc. International Conference on Learning and Intelligent Optimization, Springer, Cham, 2015, pp. 249–255.

[15] C.A.C Coello, Theoretical and numerical constraint-handling techniques used with evolutionary algorithms: a survey of the state of the art, Comput. Methods Appl. Mech. Eng. 191 (11-12) (2002) 1245–1287.

[16] C.G. da Silva, J. Clímaco, J. Figueira, A scatter search method for bi-criteria {0, 1}-knapsack problems, Eur. J. Oper. Res. 169 (2) (2006) 373–391.

[17] C.G. da Silva, J. Figueira, J. Clímaco, Integrating partial optimization with scatter search for solving bi-criteria {0, 1}-knapsack problems, Eur. J. Oper. Res. 177 (3) (2007) 1656–1677.
[18] K. Deb, An efficient constraint handling method for genetic algorithms, Comput. Methods Appl. Mech. Eng. 186 (2-4) (2000) 311–338.
[19] M.F. Demiral, H. Şen, Integer programming model for two-centered double traveling salesman problem, Eur. J. Econ. Bus. Stud. 2 (2) (2016) 80–86.
[20] I.E.A.C Droste, Algorithms for the travelling salesman problem (Bachelor's thesis), Utrecht University, 2017.
[21] I. Dumitrescu, S. Ropke, J.F. Cordeau, G. Laporte, The traveling salesman problem with pickup and delivery: polyhedral results and a branch-and-cut algorithm, Math. Program. 121 (2) (2010) 269.
[22] M. Ehrgott, X. Gandibleux, A survey and annotated bibliography of multiobjective combinatorial optimization, OR-Spektrum 22 (4) (2000) 425–460.
[23] T. El-Ghazali, Metaheuristics: From Design to Implementation, 9, Jonh Wiley and Sons Inc., Chichester, 2009, pp. 10–11.
[24] T. Erlebach, H. Kellerer, U. Pferschy, Approximating multiobjective knapsack problems, Manage. Sci. 48 (12) (2002) 1603–1612.
[25] S. Fidanova, Ant colony optimization for multiple knapsack problem and model bias, in: Proc. International Conference on Numerical Analysis and Its Applications, Berlin, Heidelberg, Springer, 2004, pp. 280–287.
[26] S. Fidanova, Probabilistic model of ant colony optimization for multiple knapsack problem, in: Proc. International Conference on Large-Scale Scientific Computing, Berlin, Heidelberg, Springer, 2007, pp. 545–552.
[27] X. Gandibleux, A. Freville, Tabu search based procedure for solving the 0-1 multiobjective knapsack problem: the two objectives case, J. Heuristics 6 (3) (2000) 361–383.
[28] M. Gendreau, A. Hertz, G. Laporte, The traveling salesman problem with backhauls, Comput. Oper. Res. 23 (5) (1996) 501–508.
[29] M. Gendreau, G. Laporte, D. Vigo, Heuristics for the traveling salesman problem with pickup and delivery, Comput. Oper. Res. 26 (7) (1999) 699–714.
[30] A.M. Gleixner, Introduction to constraint integer programming, in: Proc. 5th. Porto Meeting on Mathematics for Industry, Porto, Zuse Institute Berlin, MATHEON, Berlin Mathematical School, 2014 April 10-11.
[31] C. Groşan, M. Oltean, D Dumitrescu, A new evolutionary algorithm for the multiobjective 0/1 knapsack problem, in: Proc. the International Conference on Theory and Applications of Mathematics and Informatics – ICTAMI 2003, Alba Iulia, 2003a.
[32] N. Gunantara, A review of multi-objective optimization: Methods and its applications, Cogent Eng., 5, 2018.
[33] K. Halse, Modeling and Solving Complex Vehicle Routing Problems (Doctoral dissertation), Technical University of Denmark, 1992.
[34] S. Hemamalini, S.P. Simon, Economic/emission load dispatch using artificial bee colony algorithm, Int. Conf. Cont. Comm. Power Eng. 1 (2010), pp.338–343.
[35] J. Ji, Z. Huang, C. Liu, X. Liu, N. Zhong, An ant colony optimization algorithm for solving the multidimensional knapsack problems, in: Proc. 2007 IEEE/WIC/ACM International Conference on Intelligent Agent Technology (IAT'07), IEEE, 2007, pp. 10–16.
[36] I. Kara, T. Bektas, Integer linear programming formulation of the generalized vehicle routing problem, in: Proc. EURO/INFORMS Joint International Meeting, Istanbul, 2003, pp. 06–10. *July*.
[37] W. Long, X. Liang, Y. Huang, Y. Chen, A hybrid differential evolution augmented Lagrangian method for constrained numerical and engineering optimization, Computer-Aided Des. 4 (12) (2013) 1562–1574.
[38] L. Ke, Z. Feng, Z. Ren, X. Wei, An ant colony optimization approach for the multidimensional knapsack problem, J. Heuristics 16 (1) (2010) 65–83.
[39] H. Kellerer, U. Pferschy, D. Pisinger, Multidimensional knapsack problems, in: H. Kellerer, U. Pferschy, D. Pisinger (Eds.), Knapsack Problems, Springer, 2004, pp. 235–283.

[40] K. Klamroth, M.M. Wiecek, Dynamic programming approaches to the multiple criteria knapsack problem, Naval Res. Logist. 47 (1) (2000) 57–76.

[41] R.T. Marler, J.S. Arora, Survey of multi-objective optimization methods for engineering, Struct. Multidiscip. Optim. 26 (6) (2004) 369–395.

[42] A.W. Mohamed, A.A. Hadi, A.K Mohamed, Gaining-sharing knowledge based algorithm for solving optimization problems: a novel nature-inspired algorithm, Int. J. Mach. Learn. Cybern. 11 (2020) 1501–1529.

[43] G. Mosheiov, The travelling salesman problem with pick-up and delivery, Eur. J. Oper. Res. 79 (2) (1994) 299–310.

[44] N. Muangkote, L. Photong, A. Sukprasert, Effectiveness of Constrained Handling Techniques of Improved Constrained Differential Evolution Algorithm Applied to Constrained Optimization Problems in Mechanical Engineering, 3rd Technology Innovation Management and Engineering Science International Conference (TIMES-iCON), December 2018, https://doi.org/10.1109/TIMES-iCON.2018.8621654.

[45] K. Naidu, H. Mokhlis, A.A. Bakar, Multiobjective optimization using weighted sum artificial bee colony algorithm for load frequency control, Int. J. Electric. Power Energy Syst. 55 (2014) 657–667.

[46] P. Oberlin, S. Rathinam, S. Darbha, A transformation for a heterogeneous, multiple depot, multiple traveling salesman problem, in: Proc. 2009 American Control Conference, IEEE, 2009, pp. 1292–1297.

[47] G. Onder, I. Kara, T. Derya, New integer programming formulation for multiple traveling repairmen problem, Transp. Res. Proc. 22 (2017) 355–361.

[48] C.C. Pinter, A Book of Set Theory, Courier Corporation,, North Chelmsford, MA, 2014.

[49] P.C. Pop, New integer programming formulations of the generalized traveling salesman problem, Am. J. Appl. Sci. 4 (11) (2007) 932–937.

[50] V. Pureza, R. Morabito, H.P. Luna, Modeling and solving the traveling salesman problem with priority prizes, Pesqui. Oper. 38 (3) (2018) 499–522.

[51] N. Raj, J. Vitthalpura, Literature review on implementing binary knapsack problem, IJARIIE, 3, 2017, pp. 2395–4396.

[52] B. Render, R.M. Stair Jr, Quantitative Analysis for Management, 12e, Pearson Education India, 2016.

[53] J.F.M Sarubbi, H.P.L. Luna, The multi-commodity traveling salesman problem, in: In: Proc. International Network Optimization Conference, April 2007, 2007.

[54] S. Iqbal, M.F. Bari, M.S. Rahman, Solving the multi-dimensional multi-choice knapsack problem with the help of ants, in: International Conference on Swarm Intelligence, Springer, Berlin, Heidelberg, 2010, pp. 312–323.

[55] M.M. Silva, A. Subramanian, T. Vidal, L.S. Ochi, A simple and effective metaheuristic for the minimum latency problem, Eur. J. Oper. Res. 221 (3) (2012) 513–520.

[56] H. Süral, N.E. Özdemirel, I. Önder, M.S. Turan, An evolutionary approach for the TSP and the TSP with Backhauls, in: Y. Tenne, C-K. Goh (Eds.), Computational Intelligence in Expensive Optimization Problems, Springer-Verlag, Berlin Heidelberg, 2010, pp. 371–396, doi:10.1007/978-3-642-10701-6_15.

[57] P. Vickram, A.S. Krishna, V., S. Srinivas, A survey on design paradigms to solve 0/1 knapsack problem, Int. J. Sci. Eng. Res. 7 (11) (2016) 112–117, ISSN 2229-5518.

[58] X. S. Yang, (Ed.), Cuckoo Search and Firefly Algorithm: Theory and Applications (Vol. 516). Springer, Cham. (2013).

[59] X.S. Yang (Ed.), Recent Advances in Swarm Intelligence and Evolutionary Computation, Springer International Publishing, New York City, 2015.

[60] E. Zitzler, L. Thiele, M. Laumanns, C.M. Fonseca, V.G. Da Fonseca, Performance assessment of multiobjective optimizers: an analysis and review, IEEE Trans. Evol. Comput. 7 (2) (2003) 117–132.

CHAPTER 13

Variables clustering method to enable planning of large supply chains

Emilio Bertolotti
Fast.square, Milan, Italy

13.1 Introduction

As supply chains (SCs) are expanding their boundaries and size, Supply Chain Planning (SCP) practices have to cope with increasingly complex and challenging decision processes [1]. The resulting overwhelming number of potential SCP decisions is causing the inadequate quality of decision making, which is affecting the financial bottom line more than ever. As a result, the availability of automated procedures that could effectively support human planners is critical and increasingly essential.

Any SCP instance with its distinctive manufacturing and distribution constraints can be represented as a Multiobjectives Combinatorial Optimization model (MOCO). Recently, many efforts have been made to improve the scalability of MOCO solution algorithms [8]. Researchers in the field are continuously trying to find smarter and more efficient techniques to resolve the ever-growing challenge of planning large and complex SCs [6]. However, despite these valuable contributions, complex and large-scale SCP instances remain a challenge for the development of effective automated planning tools.

This chapter outlines a new planning approach that reduces the search space for large SCP instances using the "Variables Clustering Method"(VCM). A general SCP framework description is first introduced in Section 13.2. Section 13.3 shows how an SCP problem with all its specific constraints derived from the description of manufacturing and distribution operations can be precisely mapped into a MOCO model. Section 13.4 discusses how the idea of clustering decision variables has been already applied a few times in some specific car manufacturing domains. Section 13.5 introduces the general VCM that can be applied to a more universal SCP framework. From this description, it becomes clear how VCM, other than significantly reducing the number of decision variables, can naturally combine two planning approaches such as SCP and DDMRP (Demand Driven Material Requirements Planning) that are sometimes considered competing alternatives [15].

13.2 SCP at a glance

A Physical SC consists of a set of infrastructures and resources that enable sourcing, manufacturing, distribution, transportation, purchasing, and selling of goods and/or services. A Physical SC consists of the set of all locations and resources that support the flows of thousands of different products from the most upstream sources of supply to the most downstream sales and distribution channels [1]. SCP

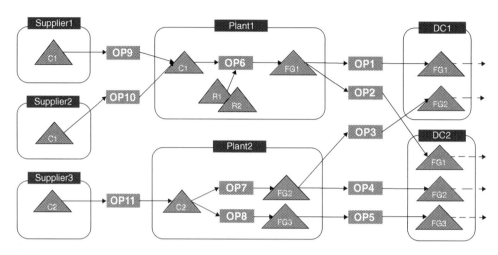

FIGURE 13.1

Representation of physical supply chain.

is the decision process that plans these resources and coordinates all physical flows within the SC [2]. Fig. 13.1 illustrates a simple example of a SC model. Materials flow from the left (supply) to the right (fulfillment of final demand) through a series of stages such as suppliers, plants, and distribution centers that involve different resources such as machines, people, vehicles.

In summary, Fig. 13.1 represents the following elements:

- SC locations are represented by rounded frames:
- Distribution centers (DC_x), manufacturing plants ($Plant_y$), external suppliers ($Supplier_z$)
- SC buffers are represented by triangles:
- "FG_i and C_j triangles" represent buffers holding finished goods (FG_i) and components (C_j) that can be stored/stocked before being transferred to the next stage.
- "R_l triangles" represent "resource-buffers," resource capacity (R_l).
- SC flows (material/capacity) represented by directed arrows:
- Material/capacity "in-flows" entering a buffer (e.g., flow entering the FG_1 buffer)
- Material/capacity "out-flows" exiting a buffer (e.g., flow exiting the C_1 buffer)
- Material "special out-flows": the dotted lines flowing out from DC_1/DC_2 represent deliveries that fulfill "external SC demand" (SC-independent demand)
- "OP_l rectangles" represent "transportation, manufacturing, and purchasing operations" that can manipulate flows (time-shifting of flows, flow quantity multipliers, merging of different flows).

The SCP process must then decide, within the entire planning horizon, the timing and quantities of flows to be produced for each buffer to satisfy the *SC-independent demand* while observing a set of SC constraints (*lead times, resource capacity,* etc.) and optimization objectives (*minimize lateness, minimize stock carrying costs,* etc.). For instance, to plan the delivery of "100 pieces" of FG_1 from DC_1 on "day 10," SCP has to do the following:

- Plan a delivery flow of 100 pieces of FG_1 from DC_1 to the customer's delivery location on "day 10." Being subject to conflicting goals → Minimize Lateness, Minimize Carrier Costs.
- Plan a distribution flow of 100 pieces of FG_1 from $Plant_1$ to DC_1 on "day 8" (two days transfer time). Being subject to conflicting goals → Minimize Lateness, Optimize Load.
- Plan a production flow of 100 pieces of FG_1 through operation OP_1 in $Plant_1$ on time (no later than day 6), choosing between one of the two alternate assembly machines R_1/R_2. Subject to conflicting goals → Minimize Lateness, Optimize Production Efficiency.
- Plan a purchase flow of component C_1 choosing one of the two alternative suppliers. Subject to conflicting goals → Chose Lowest Purchase Price, Minimize Supply Lead Time.
- Plan a transfer flow of C_1 from Supplier to $Plant_1$ combining additional components to fill the truck. Subject to conflicting goals → Optimize Loads, Minimize Lateness.

With large SCs the number of buffers, operations, resources, and flows increases accordingly (#Buffers $\approx 10^5$, #Operations $\approx 10^4$, #Resources $\approx 10^3$, #Flows $\approx 10^5$). This means that the SCP process owners should be able to scrutinize masses of "interconnected decisions." In reality, despite the considerable number of human planners and the variety of supporting software solutions available, at present, there is no way to adequately (exactly) evaluate the large number of possibilities presented by these complex, large SCs. As a consequence, suboptimal decisions are taken every day that significantly affect the SC operational bottom line. To reduce this harmful "bottleneck," researchers and engineers within the SC management community are constantly looking for new techniques and new software solutions that could support SCP analysts more effectively. The next sections illustrate how any SCP instance can be described as a MOCO model and introduce the "VCM" developed to support the kind of large SCP instances often prevalent in the industrial environments.

In our definition of the SCP process, we do not include short-term planning (continuous-time scheduling, sequencing), which is a supplementary planning step. Therefore, we consider it to be a part of the execution process. Scheduling is done only after all critical supply material and capacity flows have been coordinated and synchronized by the end-to-end Bucketed SCP process [2]. Nevertheless, our definition of SCP should not be mistaken for the outdated definition of the high-level "Master Production Scheduling"(MPS) [1]. Unlike MPS, SCP is synchronizing all stages of the SC: distribution, manufacturing, and purchasing. Moreover, the SCP process offers a detailed, in-depth representation of all SC flows, observes all critical SC constraints, and generates feasible and executable plans providing all the planned production orders and purchase orders that drive the SC "execution"[2].

13.3 SCP instances as MOCO models

This section will demonstrate how any SCP instance, with all its specific constraints derived from SC and products structures (BOM: Bill of Material/Distribution [5]), can be defined as a "Multiobjective Network Flow Optimization"(MNFO) problem [3] and categorized as a Multiobjective Combinatorial Optimization (MOCO) problem [4]. We are proposing a formalization of the SCP model that includes all its typical BOM constraints. As mentioned in the introduction, the size of these industrial SCP models can become prohibitive. Section 13.5 will define the comprehensive "VCM" that enables pruning of general SCP large instances without compromising the quality of planning results.

We can use the simple example illustrated in Fig 13.2 as the base model for our mapping of SCP into a Network Flow Optimization Model:

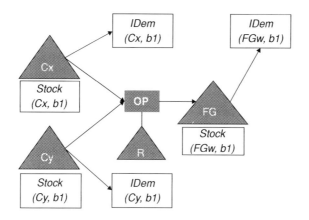

FIGURE 13.2

SCP base model.

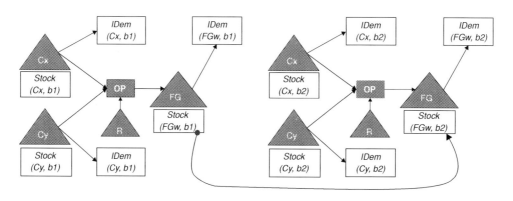

FIGURE 13.3

Modelling the SCP time dimension.

- As a first task, we must capture the "SCP Time Dimension" and model it into our MNFO/MOCO model.

  In SCP, the time dimension is often represented through a set of discrete disjoint time intervals called time-buckets. In our MOCO model, we will capture this time dimension replicating the whole SCP model for each bucket as shown in Fig 13.3. The index "b_k," represents a specific time bucket. The left side of the illustration represents the SC model/status for the bucket "b_1" while the right side represents the SC model/status in the bucket "b_2." The line that goes from FG_w, b_1 to FG_w, b_2 indicates the stock carryover (cumulated flow) across different buckets.

  We can then define B as a set of time buckets covering the entire planning horizon H:

$$B := \{b_1, b_2, \ldots b_h\} \quad B \text{ is a set of disjoint time intervals } b_k \in Z$$

$$H = \Sigma_k b_k \quad H \text{ represents the SCP planning horizon}$$

- Modeling "SCP Buffers" as "Network Nodes" of the MNFO/MOCO model.
  As already mentioned, the "FG_w" triangles represent buffers for the finished goods and the "C_x" triangles represent buffers for components (or semifinished goods). They can hold/store material flows (stocked material) that can be carried over across buckets. SCP resources are just a special type of buffers (R triangle in Fig 13.3) whose outflow consists of "resource capacities" (machine-hours, people-hours). Resource buffers cannot carryover "capacity quantities" across time buckets. In our MOCO model, we represent all SCP buffers as network nodes. As SCP material buffers can hold/store material stock, the associated nodes must reflect this property of accumulating flow quantities across time buckets. Nodes associated with resource buffers cannot accumulate flow quantities across time buckets. All these different behaviors will be captured by MOCO flows' equations defined in point d of the "SCP constraints to MOCO" description. The set of "n" nodes will then be replicated for all buckets "h":

$$N := \{no_{i,k}\}^{n \times h} \quad no_{i,k} \text{ represents the buffer } i \text{ in bucket } k$$

$$S := \{in_{i,k}\}^{n \times h} \quad in_{i,k} \text{ quantity of stock (inventory) at node } i \text{ in bucket } k$$

- Modeling "SCP Material and Capacity Flows" as "Network Flows" of the MNFO/MOCO model.
  Within our SCP model, any material/capacity flow (represented by arrows in Fig. 13.3) from a "source buffer" to a "destination buffer" with flow transformations in the middle is defined as "operations" (OP boxes in "Fig. 13.3"). In SCP, flows must be expressed as integers (pieces, cases, lots, machine-hours). In our MOCO model, flows will be represented by a set of variables:

$$X := \{x_{i,j,k}\} \quad \text{where } x_{i,j,k} \text{ is the flow quantity from } no_i \text{ to } no_j \text{ in bucket } k$$

$$x_{i,j,k} \in Z \quad \text{positive integers}$$

Valid flows are defined through a "Nodes Connection Matrix" $NM := \{1, 0\}^{(n \times n \times h)}$ where "n" is the number of nodes in one single bucket, "h" is the total number of buckets. To complete the MOCO modeling of the SCP flows, we must still define how operations OP transform flows and how we treat the two different types of flows: IDem and DDem.

a. An "operation" op_t belonging to a set of SCP "operations" $OP := \{op_t\}^p$ represents SC tasks such as "manufacturing-operation" or "transfer-operation" that manipulate flows: $op_t: x_{i,j,k} \to x'_{j,l,k}$. These operations are derived from the specific product/SC structures defined by the traditional "BOM"[5]. Each operation op_t transforms one or more "in-flows" $x_{i,j,k}$ into an "out-flow" $x'_{i,j,k}$ in one of the following modes:

$$x'_{i,j,k} = op_t(x_{i,j,k}) \quad \text{with } k' = k + s, \text{ flow time-shift of "s" buckets}$$

$$x'_{l,j,k} = op_t(\Sigma_i \, x_{i,j,k}) \quad \text{flows merge}$$

$$x'_{i,j,k} = op_t(m_{i,j} * x_{i,j,k}) \quad \text{flow multiplier}$$

For example: to produce $100 FG_w$ we must run a packaging operation which takes 2 days *(flow time-shift with $s = 2$ days)*, for each piece of FG_w two different components C_x and C_y are required *(flows merge)*, for each piece of FG_w two pieces of C_x *(flow multiplier with $m = 2$)* are needed. Any actual operation op_t can be a combination of one or more of these three transformations.

b. A buffer "outflow" $x_{i,k}$ is the sum of two different flows:
- IDem flows (e.g., outflow of FG_w)

  "IDem" flows are defined for material buffers only and represent materials flowing out of the SC to satisfy the "external" demand for finished goods sold and "external" demand for components (semifinished goods) that can also be sold. For resource buffers, outflows "IDem" is always zero.
- DDem flows (e.g., outflows of C_x, C_y to "operations" OP)

  "DDem" flows represent the dependent demand that any SC destination stage (e.g., FG_w) generates for all its SC source stages (e.g., C_x, C_y).

The buffer outflow at any given stage must satisfy independent-demand requirements (IDem) coming from external sales as well as dependent-demand requirements (DDem) generated by SC destination stages. In the example of Fig. 13.2, the outflow from the component C_x must satisfy IDem external demand as well as the DDem amount generated by FG_w. The total outflow $x_{i,l,k}$ of any node no_i in any bucket b_k is the sum of its IDem flows plus all its DDem flows to all l connected nodes.

$$x_{i,j,k} = IDem_{i,k} + \Sigma_l x_{i,l,k} \quad x_{i,l,k} \text{ is the DDem flow from node } no_i \text{ to node } no_l$$

- Mapping "SCP constraints" into "Network Flow constraints" of the MNFO/MOCO model

  The set of network constraints $g(x) := \{g_1(x), g_2(x),\ldots, g_q(x)\}$, where $x \in X$ of our MOCO model must include the following SCP constraints categories:

  a. Stock Constraints (with slack variables $\rho_{i,k}$ and $in_{i,k}$ being the stock at bucket k):

  $$\{in_{i,k} - MAXSTOCK_{i,k} + \rho_{i,k} = 0\} \quad \text{for all } x_{i,j,k} \in X \text{ (material flows only)}$$

  b. Capacity Constraints (with slack variables $\mu_{i,k}$):

  $$\{x_{i,j,k} - MAXCAP_{i,j,k} + \mu_{i,k} = 0\} \quad \text{for all } x_{i,j,k} \in X \text{ (capacity flows only)}$$

  c. Flow Conservation Constraints (with slack variables $\delta_{j,k}$ and stock carryover):

  $$\{in_{j,k-1} + \Sigma_i x_{i,j,k} - in_{j,k} - (IDem_{j,k} + \delta_{j,k}) = 0\} \quad \text{for all } x_{i,j,k} \text{ in } X$$

  d. Network Flow Integrity Constraints:

  $$\{x_{i,j,k} \in Z\} \quad \text{for all flows } x_{i,j,k} \text{ in } \in X$$

  "MAXSTOCK" and "MAXCAPACITY" are parameters derived from the SC/BOM information [5]

- Mapping "SCP Objectives" into "Net Flow Optimization Objectives" of the MNFO/MOCO model.

  The set of objectives $OB = \{ob_o(x)\}$ for typical SCP instances can be quite large and could include goals such as *minimize unsatisfied demand, minimize stock, minimize resources' idle time, optimize profit, minimize the use of alternates, saturate resources in short terms* (objectives can be time dependent and applied to a subrange of buckets). For the sake of simplicity, we will consider a simpler subset of representative objectives directed at minimizing "slacks" in our MNFO model:

  a. *Minimize unsatisfied independent demand (demand flow maximization):*

  $$\text{Min} \left(ob_1(x) = \Sigma_{i,k} \delta_{i,k} \times cd_i \right) \quad cd_i = \text{IDem priorities}$$

b. *Minimize overstock carrying costs*

$$\text{Min } ob_2(x) = \Sigma_{i,k} \left(\rho_i + s_{i,k}\right) \times cs_i \quad cs_i = \text{stock carrying costs}$$

c. *Minimize resources' idle time and overtime (applies to production resources)*

$$\text{Min } ob_3(x) = \Sigma_{i,k} \left(\mu c_i + x_{i,k}\right) \times cr_i \quad cr_i = \text{resource costs}$$

d. *Minimize logistic costs (applies to transportation resources)*

$$\text{Min } ob_4(x) = \Sigma_{i,k} \left(\mu l_i + x_{i,k}\right) \times cl_i \quad cl_i = \text{tranportation costs}$$

We are now ready to synthesize the mapping of SCP into the MNFO model with the tuple:

$$SCP := \langle B, N, NM, D, X, S, OP, g(x), ob(x) \rangle \tag{13.1}$$

where

$B := \{b_1, b_2, \ldots b_h\}$ set of disjoint time intervals (buckets), $b_k \in R$
$N := \{no_{i,k}\}^{n \times h}$ set of nodes, where $no_{i,k}$ node i is in bucket k (buffers)
$NM := \{1, 0\}^{(n \times n \times h)}$ where $NM_{i,j,k} = 1$ if n_i is connected to n_j in bucket b_k
$D := \{d_{j,k}\}^{n \times h}$ where $d_{j,k} :=$ independent demand for n_j in bucket k, $d_{j,k} \in Z$
$X := \{x_{i,j,k}\}^{(n \times n \times h)}$ where $x_{i,j,k} :=$ flow from n_i to n_j in bucket k, $x_{i,j,k} \in Z$
$S := \{in_{i,k}\}^{n \times k}$ where $in_{i,k} :=$ stock at buffer n_i in bucket k, $I_{i,k} \in Z$
$OP := \{op_t\}^p$ where op_t represent a BOM defined SC "operation"
$g(x) := \{g_1, g_2, \ldots, g_q\}$ $g(x): R^n \to R^q$ defining feasible solutions $x \in U \subset X$
$ob(x) := \{ob_1, ob_2, \ldots ob_p\}$ with $ob(x): R^n \to R^p$ and $x \in U \subset X$

Solving an SCP problem is equivalent to solving the following MNFO problem:

$$\text{Min } \{ob(x), \text{ with } x \text{ subject to } g(x) \leq 0 \text{ and } x \in Z^n\} \tag{13.2}$$

In a typical MOCO model [6] where "$ob(x)$" is the objectives vector, the optimization of "$ob(x)$" becomes a quite different task compared to single-objective optimizations. Several multiobjective optimization methods have been developed [8,9]. Most of these MOCO optimization methods fall into one of the following two categories (or a combination of the two):

a. *Pareto-analysis*: It builds the "Pareto-efficient frontier" in a multiobjective solution space.
b. *Scalarization*: It transforms multiple-objectives in multiple instances of single-objectives.

A complete analysis of these methods and their formal properties can be found in [6] and [8]. At this point, we want just to clarify why "scalarization" is the most suitable set of methods for large SCP instances. For this purpose, let us consider a simple SCP example with its small set of feasible solutions "U":

$$U := \{x1, x2, x3, x4\}$$

Each solution $xf \in U$ consists of a set of values $\{x_{i,j,k}\}$ defining how much material/capacity must flow from any node n_i to the node n_j during time bucket b_k for all i, j, k while observing all constraints

Table 13.1 Solutions space.

Feasible Solutions	x1	x2	x3	x4
$ob_1(s)$: minimize unsatisfied demand	80	60	40	20
$ob_2(s)$: minimize stock carrying costs	20	80	40	60

$\{g_1, g_2,..., g_p\}$. Let us also assume that each feasible solution xf is evaluated through the following two objective functions that are the components of the objective vector $ob(xf)$:

$$ob_1(xf) \rightarrow \text{minimize unsatisfied demand}$$

$$ob_2(xf) \rightarrow \text{minimize stock carrying costs.}$$

The evaluation of all the feasible solutions can be represented by the matrix of Table 13.1, which maps feasible solutions space into objectives space:

- x1 minimizes stock and makes demand satisfaction more problematic
- x4 minimizes unmet demand but requires higher stock availability

We will now see how the two solution approaches (paretization, scalarization) would work for our SCP example and why *scalarization* methods are more appropriate for large SCP instances.

a. Solving SCP problems through "pareto-analysis"
   The "Pareto-analysis" supports the evaluation of multiple-objectives by constructing a special solution set: the "Pareto Set." The "Pareto Set" consists of the feasible solutions that have at least one "objectives value" lower (better) than any other feasible solution. Fig. 13.4 illustrates this definition and shows that solution x2 cannot be part of the "Pareto Set" as $ob_1(x2)$ and $ob_2(x2)$ are both higher than $ob_1(x3)$, $ob_2(x3)$, and the solution x2 is thus "dominated" by the solution x3. For the same reason, x2 is also dominated by x4. On the contrary, solutions x1, x3, x4 belong to the "Pareto Set" as they are "nondominated" solutions. Any feasible solutions above the frontier (dotted line of Fig 13.4) can be disregarded because they are "dominated" by one or more solutions on the border.

   For certain industrial SCM (Supply Chain Management) applications, this method is suitable. An example can be found in [7] where "optimal coils cutting plans" must be generated considering multiple objectives such as "trim loss minimization" and "minimization of cutting patterns." In that case, given the nature of the planning problem, a small set of "nondominated" solutions can be allowed to spill over from a large set of feasible solutions, and "pareto-analysis" becomes effective. Unfortunately, this is not the case for large SCP instances where the number of "dominant" solutions would remain unmanageable.

b. Solving SCP problems through scalarization
   "Scalarization" methods aim to transform the "multiobjective" optimization problem into "single-objective" models (scalars), which can then be normally tackled through single-objective "Mixed-integer Programming" MIPs [13]. This category of solution methods is of great interest to SCP. Ehrgott [6,8] has extensively studied these algorithms and their properties. One primary outcome of their research consists of the so-called "elastic constraints method" that offers a reasonable compromise between completeness and computability. The "elastic constraints method" solves MOCO problems for one of the "p" objectives at a time while considering other "p-1" objectives to be

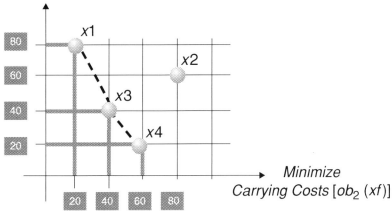

FIGURE 13.4

Pareto set and efficient frontier.

soft constraints and allowing their violations. The violations of the "p-1"objectives are penalized through costs within the overall objective function. Different variants of this technique have been implemented [10]. We just choose one of the simplest variants to complete our example:
* We must first prioritize all the different objectives:

$$ob_1(xf) - \text{Minimize unsatisfied demand} \rightarrow \text{highest priority}$$

$$ob_2(xf) - \text{Minimize stock carrying costs} \rightarrow \text{lowest priority}$$

* Then we define a degradation-range (δ) for the higher priority objective
  * By how much the higher priority objective can be "deteriorated"to optimize a lower priority objective. For instance, $\delta = 20$ (for the sake of simplicity we use an absolute number in our example, but δ is normally expressed as a percentage or in some form of normalized value).

The complete solution to the multiobjective optimization problem is then obtained by applying two consecutive optimization steps (Step 1 and Step 2).
* Step 1: Search for optimal solution xf' that minimizes ob_1

$$ob_1^*(xf') = \text{Min } \{ob_1(xf), xf \in U\}$$

the results of Step 1 applied to our example are

$$xf' = x4$$

$$ob_1^*(x4) = 20$$

$$ob_2(x4) = 60$$

- Step 2: Search for a new feasible solution xf'' that minimizes $ob_2(xf'')$ and limits degradation of $ob_1^*(xf')$

$$ob_2^*(xf'') = \text{Min}\left\{ob_2(xf),\ xf \in U,\ \left[ob_1^*(xf'') - ob_1^*(xf')\right] \leq \delta\right\}$$

the results of Step 2 are

$$xf'' = x3$$
$$ob_1^*(x3) = 40$$
$$ob_2^*(x3) = 40$$

The solution $x3$ is the final preferred solution and is the solution that better balances all different optimization objectives.

Scalarization methods, especially the "elastic constraints method," are effectively applied to solve real-life SCP instances. Nevertheless, with large SC instances, the considerable number of decision variables to be handled remains a significant challenge. In Section 13.5, we will define the Variable Clustering Method (VCM) that addresses this challenge pruning the size of a general SCP model without compromising plan accuracy. Before defining our general VCM applicable to the universal SCP paradigm, let us analyze in the next section) how this clustering principle has been already successfully applied in the automotive industry to solve very specific planning problems that could not be practically solved with traditional methods because of their size and complexity.

13.4 Orders clustering for mix-planning

A typical SC of a car manufacturer incorporates all classical ingredients of a traditional SC: distribution networks (dealers, central warehouses, etc.), multistage manufacturing facilities (chasses welding, painting, assembly), and a large number of supplied components (many "Just in Time" supplied components to be coordinated). Still, their SCP processes are very industry specific due to the distinct nature of the car manufacturing sector: high production volumes, extremely configurable products (vehicle options), a high value of every single product (car price) produced. They must synchronize the production/distribution of millions of these high-value products and therefore their end-to-end SCP consists of a complex set of highly specialized business-critical processes. Moreover, given the maturity of this industry sector, each company has developed its in-house planning landscape and terminology. Different articles and documents in the automotive literature often describe the same planning problem using different models and different terminologies. The terminology that we are adopting in this article is deliberately kept general without referring to any specific car manufacturer SCM landscape.

In this section, we will:

1. Define a universal structure for the "automotive planning workflow" with all its typical planning layers.
2. Define for the first time a "core automotive planning problem" that can be used to model all these different planning layers: "mix-planning" (within the automotive context the term "mix-planning" is used to identify just one of these specific planning layers, in this chapter the term "mix-planning" will

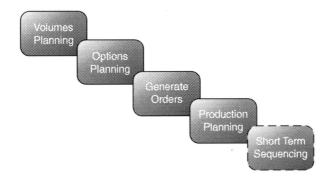

FIGURE 13.5

High-level planning framework for a car manufacturer.

refer to the formal description of a newly defined common planning problem that can be used to model all the main automotive planning layers).

3. Formalize "mix-planning" as an SCP and MOCO model. This SCP model remains specific to mix-planning and automotive context and is a subclass of the more comprehensive SCP model definition as clarified in Section 13.5.
4. Show how a clustering approach, "Vehicles Orders Clustering," enables the solution of larger mix-planning instances (SCP/MOCO model). The idea of clustering Vehicles Orders into Classes is not new and has already been introduced in some of the automotive literature, for instance [11]. A couple of independent research and implementation projects conducted by a software vendor for two large car manufacturers have independently used this clustering principle to solve two different planning problems: high-level vehicles volumes planning and low-level vehicles production planning. The specific algorithms and the applications developed for these two car-manufacturers are covered by patent [12], so we will not disclose their specific details in this chapter.

Let us start to describe the general high-level planning framework, with all its planning layers, applicable for a typical car manufacturer, as depicted in Fig. 13.5.

In the first box (Volumes Planning), the free forecasted demand (vehicles quantities per model, per market, and per time-bucket) are generated. In the second box, Options Planning, the percentages for each Vehicle Option, Market, and Time-Bucket are planned (e.g., for model SUV, market Germany, in week 28, the 40% of vehicles will require "alloy wheels"). These percentages are not unconstrained and must obey the so-called "engineering rules." The role of engineering rules/constraints is to ensure consistency of percentages across different options. For instance, if 100% of the vehicles for market "Germany" should be configured with Radio "$R1$" and the presence of Radio "$R1$" is excluding Screen type "$S1$," then the percentage of vehicles configured with "$S1$" in the market "Germany" cannot be > 0. In a nutshell, the volume of nonconfigured vehicles must be planned against market targets while options percentages (mix) must reflect market demand and observe engineering rules. The definition of this optimization problem has been generalized in such a way that becomes a common optimization "kernel" for the different planning layers of Fig 13.5. To do so we have defined a templatized "kernel" that can be described at a high level as: "the generation of the vehicle/options mix that satisfies market targets while observing engineering configuration constraints and vehicles/options production

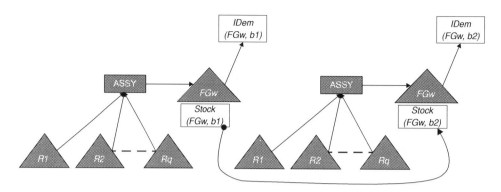

FIGURE 13.6

Order-based MIX planning: SCP model.

capacity constraints." In a nutshell, we have designed a template model, called "mix-planning" that can be configured to represent Engineering-configuration/Options-Vehicle-production constraints and Market-Volumes/Target objectives at different planning levels. "Mix-planning" can be seamlessly applied to represent the core optimization problem for all planning layers of Fig 13.5 (excluding the last layer "Short Term Sequencing" as articulated in the explanation of the "fifth box").

In the third box, planned vehicle volumes and option percentages are combined and translated into "forecasted configured vehicles" that should be produced for each market and each bucket. This represents the second application of the "mix-planning" core problem where a mix of configured vehicles must be generated optimizing satisfaction of projected demands for configured vehicles while observing high-level production constraints. Notice that thanks to the templatized "mix-planning" approach, the feasibility against engineering-rules, established at the upper planning layers, can be preserved in this third step which is rarely the case for more decoupled approaches.

In the fourth box, the mix of configured vehicles must be planned in weekly/daily buckets (slotting), taking into account both the assembly line constraints (assembly capacity and options capacities) and the market targets. This represents the third application of the "mix-planning" template. Again, with the unified "mix-planning" template it is easier to preserve feasibility and optimization achieved at the upper planning layers that could be otherwise lost.

The fifth box represents the short-term sequencing of production lines (welding, painting, assembly). As anticipated in Section 13.2, we are considering the short-term sequencing step as an execution step rather than being a part of the SCP.

We will now show how this templatized "mix-planning" model can be represented through an SCP model offering a general conceptual framework even if specific details could require ad hoc interpretations and implementations. Fig. 13.6 illustrates the SCP model representation of a mix-planning model whereby:

- The manufacturing operations are modeled through a single critical production stage (parallel assembly lines).
- Constraints related to the options (e.g., "sunroof") or subconfiguration (e.g., "diesel-engine and alloy-wheels") are represented as capacity constraints: $R_1, R_1, \ldots, R_q$.

- We do not need to represent component/material constraints as, within the automotive industry, material supply is planned separately (just in time call-offs).

As we have already done in Section 13.3 we can now map our SCP for mix-planning with its specific variables and parameters into a MOCO model. Then we will discuss the application of the "Vehicles Orders Clustering" to the SCP/MOCO for mix-planning and show how this approach can significantly reduce the huge size of mix-planning instances.

- Mix-planning parameters and variables:

  $B := \{b_1, b_2, \ldots b_h\}$ set of disjoint time intervals $b_k \in Z$
  $N := \{no_{i,k}\}^{n \times h}$ set of nodes, where $no_{i,k}$ node i is in bucket k (buffers)
  $O := \{o_i\}^d$ o_i represents single demand orders (customer or forecast)
  $PM := \{1,0\}^{d \times s \times h}$ validity of assembly line a_j for order o_i within a range of buckets $b_k \div b_{k+w}$
  $OPT := \{op_l\}^q$ set of all "q" possible options
  $OO := \{1,0\}^{d \times q}$ configuration matrix, $oo_{i,l} = 1$ if o_i is configured with option op_l
  $M := \{m_r\}^m$ set of "m" markets
  $MO := \{1,0\}^{d \times m}$ $mo_{i,r} = 1$ if o_i belongs to market m_r
  $DD_i \in Z$ due date of order o_i (day rank in the horizon)
  $A := \{a_1, a_2, \ldots, a_s\}$ set of parallel assembly lines
  $X := \{x_{i,j,k}\} = \{1,0\}^{d \times s \times h}$ $x_{i,j,k} = 1$ if order o_i is planned on assembly line a_j in bucket b_k
  $S := \{in_{i,k}\}^{n \times k}$ where $in_{i,k} :=$ stock at buffer n_i in bucket k, $I_{i,k} \in Z$

  We must also consider the following aspects:

- Values of $x_{i,j,k}$ are subject to the "producibility matrix": The selection of the assembly line a_j for a given order o_i can only be done within a range of buckets $b_k \div b_{k+w}$
- Each element op_i of the vector OPT is a Boolean function of one or more car options (for instance $op_{12} := ((diesel = "Y") \text{ AND } (sunroof = "Y"))$. Each function op_i can identify a specific subconfiguration other than a single option. These subconfigurations functions enable the planners to flexibly define "option capacity" constraints (e.g., in November and December we cannot plan to produce more than 50% of vehicles that have both a diesel engine and a sunroof on the same day, otherwise the cables assembly station will be overloaded, impacting the overall cadence of the line). These "options capacity" constraints are defined below (see point "d").
- Mix-planning constraints ($g(x)$):

  a. Values of $x_{i,j,k}$ are subject to the "producibility matrix" $PM := \{1,0\}^{n \times s \times h}$ constraining the selection of the assembly line a_j for the given order o_i (car model) within a range of buckets $b_k \div b_{k+w}$
  b. Market target constraints (with slack $\upsilon_{r,k}$):
  $\{(\Sigma_{i,j,k} \, x_{i,j,k} \times MO_{i,r}) - MAXM_{r,k} + \upsilon_{r,k} = 0\}$ for all m_r in M
  c. Order due dates constraints (with slack $\delta_{i,k}$):
  $\{x_{i,j,k} \times (b_k - DD_i - \delta_{j,k}) = 0\}$ for all o_i
  d. Options capacity constraints (with slack $\rho_{i,k}$):
  $\{(\Sigma_{i,k} \, x_{i,j,k} \times OO_{i,l}) - MAXO_{l,k} + \rho_{l,k} = 0\}$ for all op_l in OPT an all a_j in A
  e. Assembly capacity constraints (with slack $\mu_{j,k}$):

$\{(\Sigma_i\, x_{i,j,k}) - MAXC_{j,k} + \mu_{j,k} = 0\}$ for all a_j in AL and all k in B

  f. Flow conservation constraints (with slack variables $\delta_{j,k}$ and stock carry-over):

$\{in_{j,k-1} + \Sigma_i\, x_{i,j,k} - in_{j,k} - (IDem_{j,k} + \delta_{j,k}) = 0\}$ for all $x_{i,j,k}$ in X

  g. Network flow integrity constraints:

$\{x_{i,j,k} \in Z\}$ for all flows $x_{i,j,k}$ in $\in X$

- Mix-planning optimization objectives (a representative set of objectives selected from the whole model – $ob(x)$):

  a. Minimize deviations from market targets:
  $$\text{Min } ob_1(x) = \Sigma_{r,k}\, \upsilon_{r,k} \times cm_r \quad cm_r = \text{priority of market } h$$

  b. Minimize order lateness
  $$\text{Min } ob_2(x) = \Sigma_i\, \delta_i \times cl \quad cl = \text{cost of lateness}$$

  c. Minimize options unbalancing
  $$\text{Min } ob_3(x) = \Sigma_{l,k}\, \rho_l \times co \quad co = \text{options unbalacing cost}$$

  d. Minimize assembly line idle time and overtime
  $$\text{Min } ob_3(x) = \Sigma_{j,k}\, \mu c_j \times cr \quad cr = \text{resource costs}$$

  e. Mix planning can then be defined through the tuple

$$MIX - PLANNING = \langle B, N, O, PM, OPT, OO, M, MO, DD, A, X, S, g(x), ob(x) \rangle \quad (13.3)$$

Solving the mix-planning problem is equivalent to solving the following MOCO problem:

$$\text{Min } \{ob(x),\ \text{with } x \text{ subject to } g(x) \leq 0 \text{ and } x \in Z^n\} \quad (13.4)$$

Classical scalarization methods can be applied to solve this problem. But real-life data sets can easily cause scalability issues:

#Orders: up one million (customer-orders plus fictitious orders)
#Options constraints: in the order of hundreds
#Buckets: up to 200/300 buckets
#Parallel Assembly Lines: up to 3 (per car model)

The idea of clustering "vehicles orders" to reduce these huge search space has been already sucesfully applied for some specific automotive planning problem. These experiences are described in [11,12] as well as within some private documentation referring to industrial "car slotting" projects (JDA-software). The same clustering principle can also be applied to our general modelization of "mix planning" reducing thus the search space for all automotive planning layers.

Two main factors influencing the size of the mix-planning model: high cardinality of decision variables set (number of orders) and high cardinality of the possible orders configurations (with 100 options the number of possible configurations is 2^{100}). Still, when analyzing the sales history of any given car model, the real number of option configurations sold is much lower. It usually ranges from a few hundred to 3000, depending on the specific car model and market. We can then "cluster vehicles orders" generating a set of classes $\{Cl_c\}$ where each class represents one of the configurations sold. As a result, two orders o_s, o_t would belong to the same class Cl_c if they have the same configuration vector:

$$OO_s = OO_t$$

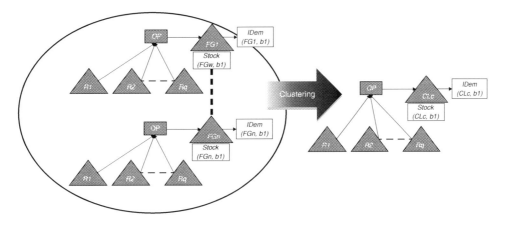

FIGURE 13.7

Variables clustering for MIX planning SCP model.

Beyond "orders configurations clustering," we had to consider other factors that generate additional subclasses:

- The effectivity dates (model Y can be produced on the assembly line a_l only after June 2021)
- The order due dates (required for the postprocessing of actual orders)
- The different markets (cannot mix configured vehicles quantities from different markets)

Nonetheless, even with these extra diversities that increase the number of classes, the drastic reduction of the solution space remained almost unaltered. Thanks to this "vehicles orders clustering" approach, we could transform the original mix-planning problem based on the $x_{i,j,k}$ decision variables into a much smaller problem based on $cl_{c,j,k}$ decision variables where:

$cl_{c,j,k}$ is the number of orders in class Cl_c to be produced on the assembly line j in the bucket k

with

$$\# cl_{c,j,k} \ll \# x_{i,j,k}$$

Through postprocessing (polynomial time), we could consistently allocate vehicle orders to optimally planned class quantities.

From the SCP modeling perspective, what we have done is to transform our original order-based planning model into a smarter Variables Clustered SCP model shown in Fig. 13.7.

Table 13.2 illustrates a comparison between the number of decision variables to be modeled with the classes-based mix-planning and with the orders-based mix-planning model. These values have been normalized and expressed as the "V0" unit of measure to remove the dependency on model details.

The quality of the solutions generated by the classes-based MIP is equivalent to the one produced by the orders-based MIP [13]. Fig. 13.8 illustrates the "less than linear" growth when the "vehicles orders clustering" method is applied.

Table 13.2 No of decision variables with and without clustering.

g(x): number of constraints	100	100	100	100	100
n: number of orders	0÷10,000	50,000	100,000	200,000	400,000
#Decision variables with classes	2.5 × V0	4 × V0	5.5 × V0	8 × V0	12 × V0
#Decision variables with orders	0.5 × V0	8 × V0	24 × V0	32 × V0	92 × V0

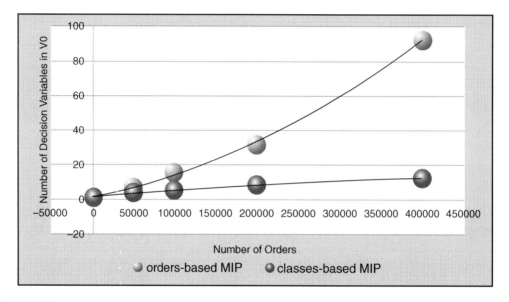

FIGURE 13.8

Number of decision variables with or without variables clustering.

In section 13.5, we define the "VCM," a clustering method that can be applied to the general SCP paradigm rather than to the more specific automotive context.

13.5 Variables clustering for the general SCP paradigm

To generalize the "vehicles orders variables" approach so that it can be systematically applied within a broader SCP context, we must go back to a more general SCP model similar to the one presented in Section 13.2 with multistage manufacturing, multiple constraining material (Fig 13.1). While our VCM method can be applied to the universal SCP paradigm, we will focus on a set of industry sector where we think the method is particularly effective:

- Consumer packaged goods—SC (for instance: CPG, foods, and beverages);
- Consumer durable goods—SC (for instance: CDG, electronic and component, white goods);
- Health and beauty—SC (for instance: cosmetics, personal care).

13.5 Variables clustering for the general SCP paradigm

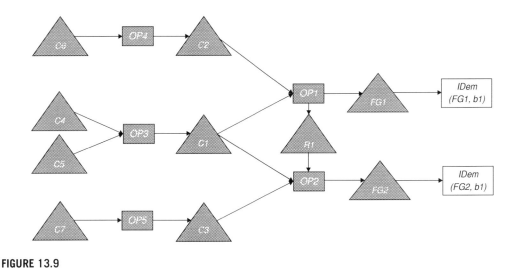

FIGURE 13.9

Universal SCP model for CPG.

The SCP model that can cover this broad category of industries can be characterized by the following aspects:

- Different finished goods of the same product family have similar supply-chain/product structure (BOM) [5];
- The majority of the corporate business is done with high volumes of finished goods;
- Different products of the same product family share the same set of manufacturing lines (packaging, filling, etc.)
- A large number of purchased materials must be coordinated (packaging materials, etc.)

Fig. 13.9 shows a typical, simple CPG (consumer packaged goods) model where two different finished goods are packaged with the same "production path"(same packaging line), use several common components (liquids or food ingredients), with the only difference of using two diverse packaging materials represented by the components C_2 and C_3 (for instance: FG_1 and FG_2 could be the same core product destined for two different countries and requiring country-specific labeling). These two components, C_2 and C_3, must be purchased from suppliers through purchasing operations OP_4, OP_5 (for instance, the supplier is transforming a "vanilla flavor"paper box or bottle by adding a country-specific label).

This example shows how the two finished goods (FG_1, FG_2) can share a significant part of the SC with just a small difference between the two separate packaging materials they use. This is likely to apply to most products that belong to the same product family, which is a widespread occurrence in the industries included in the category we are considering—a multitude of products within the same product family sharing the same SC model with just a few differences. Identifying an efficient method to systematically capture these "commonalities"could enable the "variables clustering" approach which will significantly reduce the SCP model complexity for these industry sectors. As a result of the

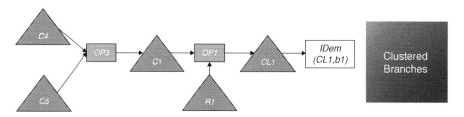

FIGURE 13.10

Clustered SCP model.

"clusterization," the underlying MIP model will work on "class variables" representing product families $cl_{c,j,k}$ rather than working on single products $x_{i,j,k}$

There will always be some abnormal cases characterized by special product structures or special SC structures that unexpectedly deviate from this standard scenario and threaten the effectiveness of our general "variables clustering" method. Nevertheless, it is always possible to measure and preliminary estimate the effectiveness of the "variables clustering" method for any given SCP instance as

$$\text{Clustering Factor (SCP)} = \left(1 - \frac{\#cl_{c,j,k}}{\#x_{i,j,k}}\right)$$

This clustering factor indicates how much the given "variables clusterization" reduces the complexity of an original SCP model, at least in terms of the reduction of the number of its decision variables.

Before describing our general VCM method, let us first see how the "clusterization" principle could work for the SCP example depicted in Fig. 13.9. As already noticed, the product structure (BOM) of FG_1 is similar to the structure of FG_2. The only difference is that FG_1 uses the component C_2, while FG_2 uses the component C_3. This indicates a clear opportunity for clustering the variables and splitting the SCP model into two parts: One "common" part for the SC shared by FG_1 and FG_2 (probably common to the entire products' family) and one product-specific part for the planning of either C_2 or C_3. Fig. 13.10 illustrates the resulting SCP clustered model.

It could be argued that this approach if extended to the whole SC, would just fragmentize the global SCP model. However, there is an important aspect to consider that makes this transformation effective and efficient. In the illustrated example, the low value of the C_2, C_3 components (printed labels) would make the calculation of their stock requirements based on their historical consumption data an acceptable simplification. In practice, C_2 and C_3 could be planned independently with a simpler planning algorithm (reorder point) without any significant impact on the quality of the overall plan but with a substantial reduction in the SCP complexity. Removing components such as C_2 and C_3 from the main planning model will enable the clustering of all FGs that belong to the same product family, which will reduce the overall SCP solution space.

To generalize this "variables clustering" principle for our general SCP paradigm, we can define the VCM conversion method as

$$\hat{C} : SCP \to CSCP$$

The method "$\hat{C}$" can transform an SCP instance into a clustered CSCP instance while identifying and evaluating all opportunities for "clustering." Let us consider in more detail how this conversion method works. It consists of two different steps (Step A and Step B):

Step A: Identify clustering opportunities

This step involves finding all candidate flows $x_{i,j,k}$ in *SCP* (BOM branches) that could generate clustering opportunities once removed. The procedure is simply based on the information that can be retrieved from the product structure (BOM) [5]. This information indicates for each "branch" of the product structure if the "branch" could be a good candidate for removal. This is a parameter that product engineers defining BOMs can easily set as they do with all other production parameters. Once potential candidates are identified, they must be tested/validated by analyzing their historical data to evaluate their predictability and suitability. This validation process can be fully automated and consists of checking average demand volumes and demand variances for each candidate. High demand volumes and low demand variances are good indicators of the candidate's suitability and promote the candidate to be "decoupled" from the global SCP model. On the contrary, if these indicators are below the minimum threshold, candidates are rejected for clustering as their decoupling might create SC synchronization issues. Lastly, all validated "planning BOMs" and "decoupled parts" are fed into the second step.

Step B: Define clustered BOMs

In the second step, all the individual "planning BOMs" defined in the previous step are examined and combined into clustered-BOMs. These constitute the baseline for the new CSCP model together with the decoupled components.

As summarized in Fig.13.11 the overall planning algorithm in the new CSCP scenario must solve two "easier" subproblems:

- The optimization of the clustered model with a reduced number of decision variables;
- The planning of the decoupled components (or semifinished goods).

As for the automotive special cases, the VCM procedure must execute a postprocess to allocate clustered quantities $\#cl_{c,j,k}$ to *FG* quantities $\#x_{i,j,k}$. In a few preliminary experiments conducted with a selection of CPG test models, the Clustered SC had a significantly lower number of decision variables compared to the corresponding original SCP model.

$$Clustering\ Factor\ (CPGx) \approx 0.6 \div 0.9$$

The complexity of the planning algorithms used to plan the decoupled branches is linear in terms of the number of branches involved ("reorder-point algorithm" based on historical consumption), and as such it does not represent a bottleneck. Moreover, for these CPG SCP examples, the clustering factor tends to increase with the size of the SCP model as clustering opportunities tend to grow with the number of products/locations modeled, as illustrated in Table 13.3 and Fig. 13.12.

These preliminary results provide some evidence of how our "VCM" enables the planning of larger SCs and allows us to enlarge the class of SCP instances that can be effectively resolved. Another important consideration is that our SCP "VCM" is also very well aligned with the more recent SCP methodologies such as "DDMRP"[14] and "Holistic Inventory Optimization"[15] which also aim to decouple and optimize stocking points that can be planned through a historical consumption baseline.

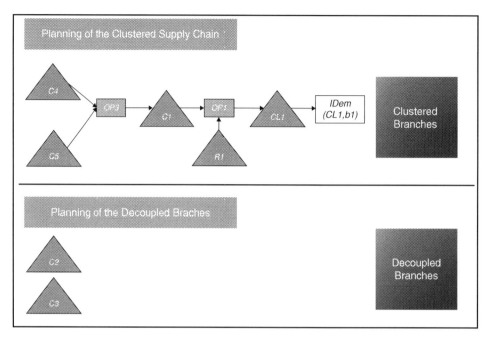

FIGURE 13.11

The two CSP planning algorithms.

Table 13.3 Clustering factor and SCP model size.			
SCP-CPG test case	SCP-CPG1	SCP-CPG2	SCP-CPG3
Number of products/locations	10,000	30,000	100,000
Clustering factor	0.57	0.73	0.9

13.6 Conclusions

This chapter shows how to map the general Bucketed SCP problem with its typical constraints derived from its SC and product structures into an MNFO model as a special type of a MOCO (Multi Objectives Combinatorial Optimization) problem. We have seen how the Variable Clustering principle already experimented for the automotive industry to solve specific planning problems can be generalized and applied to the general SCP/MOCO paradigm reducing the number of decision variables to be modeled without compromising on the quality of the optimization results. This new VCM (Variable Clustering Methodology) is extending the class of SCP problems that can be practically solved. Furthermore, VCM methodology is closely aligned with some of the latest innovative SCM concepts such as the "DDMRP" and "Holistic-Inventory-Optimization."

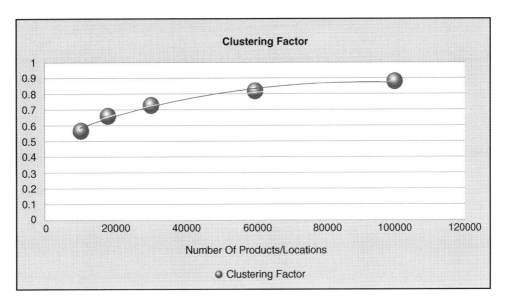

FIGURE 13.12

Clustering factor versus SCP model size.

References

[1] J.G.A.J. van der Vorst, Supply chain management: theory and practices, in: Theo Camps, Paul Diederen, Gert Jan Hofstede, Bart Vos (Eds.), The Emerging World of Chains & Networks, Elsevier, Hoofdstuk, 2004.

[2] M. Assey, J. Janvier, A New Introduction to Supply Chains and Supply Chain Management: Definitions and Theories Perspective, Glorious Sun School of Business and ManagementDonghua University Shanghai 200051, China, 2011.

[3] R. Ahuja, T. Magnanti, J. Orlin, Network Flows Theory, Algorithms, and Applications, Pearson Education (US), 1993.

[4] M. Ehrgott, Exact methods for multi-objective combinatorial optimisation, in: S. Greco, M. Ehrgott, J. Figueira (Eds.), Multiple Criteria Decision Analysis. International Series in Operations Research & Management Science, vol 233, SpringerSpringer, New York, NY., 2016, pp. 817–850.

[5] R. Sato, Y. Matsui, M. Furukado, Concept of Bills of Material for Supply Chain Planning and Its Prototyping with ERP, in: Proc. 2011 International Conference on Industrial Engineering and Operations Management Kuala Lumpur, Yokohama University, 2011.

[6] M. Ehrgott, A discussion of scalarization techniques for multiobjective integer programming, Ann. Oper. Res. 147 (2006) 343–360.

[7] E. Bertolotti, Al., Near-Optimal Object Packing through Dimensional Unfolding, in: Proc. the Eighth Annual Conference on Innovative Applications of Artificial Intelligence, AAAI, 1996, pp. 1433–1442.

[8] N. Gunantara, A review of multi-objective optimization: methods and its applications, Cogent Eng. J. 5 (Issue 1) (2018), ID 1502242.

[9] O. Turgut, Exact And Representative Algorithms For Multi-Objective Optimization, (Ph.D. Diss.) Wayne State University, 2013.

[10] IBM ILOG CPLEXOptimization Studio Documentation. 2019. URL: ILOG CPLEX Optimization Studio | IBM.
[11] C. Solnon, V. Cung, A. Nguyen, The car sequencing problem: overview of state-of-the-art methods and industrial case-study of the ROADEF'2005 challenge problem, Eur. J. Oper. Res. 191 (3) (2008) 912–927.
[12] V. Raymond, M. Brisson, Systema and Method of Automotive Production Planning, US Justia Patent #20190362286 (2019).
[13] G. Nemhauser, L. Wolsey, Integer and Combinatorial Optimization, John Wiley & Sons, Print ISBN:9780471828198, 1988.
[14] C. Ptak, C. Smith, Demand Driven Material Requirements Planning, ISBN: 978-0-8311-3598-0, Industrial Press, DDMRP Institute, 2017.
[15] A.S. Eruguz, Contributions to the Multi-echelon Inventory Optimization Problem using the Guaranteed-service Model Approach (Ph.D. Diss.), University of Eindhoven, 2014, https://pure.tue.nl/ws/portalfiles/portal/3939301/341232557774342.pdf.

Index

Page numbers followed by "*f*" and "*t*" indicate, figures and tables respectively.

A

Additive DEA model, 32
Alpha wolf, 138
Analytic hierarchy process (AHP), 124
Ant Colony Optimization (ACO), 19, 179
Artificial Bee Colony algorithm, 71

B

Beta wolf, 138
Binary gaining sharing knowledge (BGSK), 238, 254

C

Cairo, tourism in, 238
Cantilever beam design problem, 84*f*
Chaotic Bat algorithm, 71
Chaotic Crow-Search Algorithm, 72
Charged particles (CP), 122
Charged system search (CSS), 122
Coefficient of performance (COP), 219
Cognitive space, mapping, 3
Combinatorial optimization problems, 1, 9
Common set of weights (CSW), 30
Computational flowchart of NSGA-II., 166*f*
Computational optimization, 1
Confusion matrix, 86*t*
Constant returns to scale (CRS), 32
Constructive methods, 13
Coverage of two sets (CS), 121
Crowding distance (CD), 122
Crow optimization (CO) algorithm, 71
    materials and methods, 73
        arithmetic crossover based on genetic algorithm, 74
        crow search optimization, 73
        hybrid CO algorithm, 74
    outcomes, 75
Crow search algorithm (CSA), 73

D

Data envelopment analysis (DEA), 29, 32, 198, 214
    models, 30
    nonradial, 29
    radial, 29
    usage, 56
DC-OPF technique, 161
DDMRP (Demand Driven Material Requirements Planning), 177
Decision-making units (DMU), 29, 214
Decomposition methods, 13
Decision space, 266*f*
Delta wolf, 138
Differential evolution (DE), 21
Discriminant analysis (DA), 30
Dominance Front, 120
Downdraft gasifier, 134

E

Energy efficiency, 218
Energy systems, thermodynamic modeling of, 217
    energy balance, 217
    energy efficiency, 218
    entropy balance, 218
    exergy balance, 218
    exergy efficiency, 219
    mass balance, 217
Etropy balance, 218
Euclidean distances, 142
Evolutionary algorithms (EA), 16
Exergy
    efficiency, 219
    volume system, 218
Expansion planning framework, 163*f*
Expected energy not supplied (EENS), 159

F

Firefly algorithm, 72
First law of thermodynamics, 217

G

Gaining-sharing knowledge (GSK), 238
Gasification agents, advantages and weaknesses of, 136
Gasifier
    waste-to-energy plants, 134
    gasification system, 140
Genetic algorithm (GA), 17, 73, 119
    arithmetic crossover on, 74
    optimization, 134
Genetic Algorithm for Community Diagnosis (GACD), 179
Golden Ratio Optimization Method (GROM), 91, 94
Graph chart structure and community discovery, 184*f*

Grasshopper optimization approach (GOA), 72
Gray Wolf Optimization (GWO) method, 133
Greedy randomized adaptive search procedures (GRASP), 19

H
Harmonic Search (HS) algorithm, 72
Heuristic algorithms, 13
Heuristics methods, 12, 10
Higher heating value (HHV), 133
High-level planning framework, 275f
Hybrid optimization algorithm, 74
    flowchart, 75f
    metaheuristics, 20

I
"Interconnected decisions,", 267
Isentropic thermal efficiency, 219

J
Journal map (density), 6f

K
Knapsack Problems, 237

L
Lagrangian relaxation, 157
Linear programming technique for multidimensional analysis of preference (LINMAP), 142

M
"Master production scheduling" (MPS), 267
Mathematical model, 212
Metaheuristic algorithms, 71
Mine blast approach (MBA), 72
Mixed-integer linear programming (MILP), 203
Model formulation, 125
    manipulation methods, 14
    multisource problems, 125
Modularity-based discrete state transition algorithm (MDSTA) model, 179
Most preferred solution (MPS), 29
Multicriteria
    assignment problem, 178, 193
    combinatorial optimization, 186, 193
    multicriteria decision making, 124
Multicriteria gray wolf optimization (MCGWO), 138
    gasifier level, 143
    pseudo-code, 141, 142f
    WtEP level, 150
Multifacility weber problem (MFWP), 125

Multiobjective charged system search (MOCSS), 94, 95, 120
    algorithm, 122
    analytic hierarchy process, 124
    implementation, 127
    model formulation, 125
    utilized methods, 122
Multiobjective combinatorial optimization, 265
    data and basic statistics, 2
    future research, 7
    methodology, 1
    outcomes, 3
        mapping the cognitive space, 3
        mapping the social space, 4
Multiobjective Cuckoo Search Algorithm (MOCSA), 177
    model, 182
    model flowchart, 183f
Multiobjective discrete biogeography based optimization (MODBBO), 179
Multiobjective discrete flower pollination algorithm (MOFPA), 186
Multiobjective Genetic Algorithm (MOGA), 186
Multiobjective evolutionary algorithm (MOEA), 178
Multiobjective genetic algorithm, 178
Multiobjective harmony search (MOHS), 119
Multiobjective KP (MOKP), 242
"Multiobjective Network Flow Optimization" (MNFO), 181
Multiobjective optimization, 180
    complex energy systems, 217
    integrated biomass trigeneration system, 222
        exergoeconomic principles, 225
        exergy efficiency and CO_2 emission cost, 231
        multiobjective exergoeconomic optimization, 226
        sensitivity analysis, 232
        thermodynamic analysis, 223
    mathematical model, 212
    methods, 91
    optimization, thermoeconomics methodology, 220
        F and P rules, 222
        SPECO method, 221
    optimization criteria, 211
    optimization of energy systems, 215
        multiobjective thermoeconomic optimization, 217
        single-objective thermoeconomic optimization, 216
        thermodynamic optimization and economic optimization, 215
        thermoeconomic optimization, 215
    principle, 163
    problems, 91
        background and related work, 93
        definition of, 92
        literature review, 93
        simulation results, investigation, and analysis, 97
            fifth class, 113

            first class, 99
            fourth class, 111
            second class, 101
            third class, 103
        sensitivity analysis of energy systems, 222
        suboptimization, 212
        system boundaries, 211
        thermodynamic modeling of energy systems, 217
            energy balance, 217
            energy efficiency, 218
            entropy balance, 218
            exergy balance, 218
            exergy efficiency, 219
            mass balance, 217
        types of optimization problems, 213
            multiobjective data envelopment analysis, 214
            multiobjective evolutionary algorithms, 214
            multiobjective optimization, 213
            single-objective optimization, 213
        variables, 212
Multiobjective decision methods, 11
Multiobjective particle swarm optimization (MOPSO) algorithm, 119, 122
Multiobjectives combinatorial optimization model (MOCO), 196, 265
Multiple attribute decision making (MADM), 29
Multiple criteria decision making (MCDM), 29
Multiple objective decision making (MODM), 29, 31
    application of, 35
        classical DEA models, 35
        common set of weights, 46
        DEA-discriminant analysis, 51
        efficient units and efficient hyperplanes, 54
        secondary goal models, 43
        target setting, 37
        value efficiency, 41
    classification, 56
        efficient points, 56
    data envelopment analysis, 32
Multiple objective linear programming (MOLP), 29
Multiple objective programming (MOP), 30
Multisource Weber Problem (MSWP), 125

N

Neighborhood search
    methods, 14
    procedures, 16
Network flow integrity constraints, 270
Network visualization, 4f
Nondominated Sorting (NS), 123
Nondominated Sorting Genetic Algorithm II (NSGA-II), 119
    algorithm, 122

O

Optimal power flow (OPF), 161
Order-based MIX planning, 276f
Organic Rankine Cycle system, 135

P

Pairwise comparisons matrices, 126
Pareto frontier, 213
Pareto-optimal front (POF), 92
    dominance, 92
    nondominated, 92
    pareto optimal, 92
Pareto-optimal set (POS), 92, 213
Pareto-optimal solutions, 11, 213
Pareto set, 273f
Particle Swarm Optimization (PSO), 10, 18, 72, 122, 134
Performance metrics, 121
    coverage of two sets, 121
    spacing, 121
Physical supply chain., 266f
Planning of tour-visits problem (PTP), 238, 239
    future studies, 259
    mathematical model, 244
        Ramses Hilton Hotel, 248
    proposed methodology, 250
        binary gaining sharing knowledge, 254
        gaining sharing knowledge, 251
    real application case study, 247
    vs. TSP and KP, 240
        differences between, 243
        traveling salesman problem and variations, 240
Population-based methods, 17, 18
Probabilistic index, 162
Production possibility set (PPS), 29

R

Ramses Hilton Hotel, 248
    Baron palace, 249
    Cairo Tower, 248
    Egyptian museum, 249
    Giza Pyramids and Sphinx, 248
    Saladin citadel, 249
Rankine cycle, 135
Real-world optimization problem, 91
Relative closeness (RC), 143
Reversible work, 218

S

"Scalarization" methods, 272
Scatter search (SS), 19
Scientific community working

from different organizations, 7f
  on multiobjective combinatorial optimization problems, 6f
Scientometric analysis, 3
Scoups databases, 1, 2
Selective refining harmony search algorithm, 72
Simulate annealing (SA), 18, 119
Single-objective optimization, 213
  algorithms, 271
  thermoeconomic, 216
Six-bus test system, 170f, 170
Social space, mapping, 4
Social spider optimization method, 72
Spacing (S) metric, 121
Suboptimization, 212Multiobjective optimization
Supply chain planning (SCP), 265
  base model, 268f
  time dimension, 268f
Supply chains (SCs), 265
Support vector machine (SVM), 72

T

Teaching learning–based optimization (TLBO), 21
Technique for order of preference by similarity to ideal solution (TOPSIS) schemes, 142
Thermodynamic optimization and economic optimization, 215
Thermoeconomic optimization, 215

Three-bar truss design problem, 83f
Total relative satisfaction (TRS), 246
Total visited places (TVP), 246
Tourism
  Cairo, 238
  Egypt, 238
Transmission system, 163
Travelling salesman problem (TSP), 237
Two-phase algorithm, 203

V

Variables clustering method" (VCM), 265
Variable returns to scale (VRS), 32
Variables clustering method (VCM), 265, 275
"Vehicles orders variables" approach, 280
VIKOR decision maker, 169
Virtual database, 168f

W

Waste-to-energy plants (WtEP), 134
  modeling, 136
  systems description, 134
    downdraft gasifier, 134
    waste-to-energy plant, 135
  typical and advanced, 137f

Printed in the United States
by Baker & Taylor Publisher Services